Advanced Technologies in Power Quality and Power Disturbance Data Application

Advanced Technologies in Power Quality and Power Disturbance Data Application

Guest Editors

Yi Zhang
Ying Wang

MDPI

Basel • Beijing • Wuhan • Barcelona • Belgrade • Novi Sad • Cluj • Manchester

Guest Editors
Yi Zhang
College of Electrical
Engineering and Automation
Fuzhou University
Fuzhou
China

Ying Wang
College of Electrical
Engineering
Sichuan University
Chengdu
China

Editorial Office
MDPI AG
Grosspeteranlage 5
4052 Basel, Switzerland

This is a reprint of the Special Issue, published open access by the journal *Symmetry* (ISSN 2073-8994), freely accessible at: https://www.mdpi.com/journal/symmetry/special_issues/Advanced_Technologies_Power_Quality_Power_Disturbance_Data_Application.

For citation purposes, cite each article independently as indicated on the article page online and as indicated below:

Lastname, A.A.; Lastname, B.B. Article Title. *Journal Name* **Year**, *Volume Number*, Page Range.

ISBN 978-3-7258-5313-7 (Hbk)
ISBN 978-3-7258-5314-4 (PDF)
https://doi.org/10.3390/books978-3-7258-5314-4

Contents

Preface

The electrical power system can be regarded as a comprehensive symmetrical system of power supply and power consumption, with load microelectronics technology as the core of many new technologies and emerging industries. The most common PQ events are associated with sags (dips), harmonics/interharmonics, transients, and asymmetrical variations. PQ is related to the safe and stable operation of power systems and users' high-quality electricity consumption. Therefore, the monitoring, prevention, and mitigation of PQ disturbances are of great concern for both parties. To achieve these goals, the study of many PQ issues and their methods of analysis remains a challenging task. Moreover, with the widespread use of power quality monitoring tools, more and more users and developers are starting to realize that power disturbances can carry valuable information about the conditions of a system and its equipment. As a result, initiatives that explore the "useful" aspects of power disturbances using power-related big data have emerged.

Yi Zhang and Ying Wang
Guest Editors

symmetry

MDPI

Editorial

Special Issue: Advanced Technologies in Power Quality and Power Disturbance Data Application

Yi Zhang

College of Electrical Engineering and Automation, Fuzhou University, Fuzhou 350108, China; zhangyi@fzu.edu.cn

Power quality has been a rapidly growing area of research. In this Special Issue, we present twelve papers authored by a select group of experts in the area of power quality. These papers cover a wide spectrum of important problems and topics of current research interest. The topology of the static synchronous compensator of reactive power for a low-voltage three-phase utility grid is proposed in [1]. A voltage sag amplitude prediction method based on data fusion is proposed in [2]. A scheme that applies the cascaded H-bridge flexible fault current limiter to realize both voltage regulation and current limiting is proposed in [3]. A state machine-based droop control method (SMDCM) is proposed in [4]. A modified active voltage control algorithm is proposed in [5]. A flexible current-limiting device (FCLD) is proposed in [6]. A novel equivalent control method for voltage source inverters (VSIs) with disturbance observers (DOBs) is proposed in [7]. In [8], the authors use a particle swarm optimization algorithm (PSO) to determine the most suitable size and location of photovoltaic-based distributed generation (PVDG). A microgrid optimization scheduling strategy considering the integration of electric vehicles (EVs) is shown in [9]. An adaptive virtual inertia control strategy for a grid-connected converter of a DC microgrid based on an improved model prediction is proposed in [10]. An improved strategy based on variational mode decomposition (VMD) is proposed in [11]. A voltage sag loss assessment method based on a two-stage Taguchi quality perspective approach is proposed in [12].

In the following text, we comment on the main goals and results of these contributions.

In the first paper, "Cascaded Multilevel Inverter-Based Asymmetric Static Synchronous Compensator of Reactive Power", the authors propose the topology of a static synchronous compensator capable of asymmetric reactive power compensation in a low-voltage three-phase utility grid. It is implemented using separate, independent cascaded H-bridge multilevel inverters for each phase. Each inverter consists of two H-bridge cascades. The first cascade operating at grid frequency is implemented using thyristors, and the second one—operating at high frequency—is based on high-speed MOSFET transistors. The investigation shows that the proposed compensator is able to effectively compensate the reactive power in a low-voltage three-phase grid. Related results can be seen in [13–15].

In the second paper, "A Residual Voltage Data-Driven Prediction Method for Voltage Sag Based on Data Fusion", the authors propose a voltage sag amplitude prediction method based on data fusion. First, the multidimensional factors that influence voltage sag residual voltage are analyzed. Second, these factors are used as inputs, and a model for predicting voltage sag is constructed based on data fusion. Finally, the model is trained and debugged, which enables it to predict the voltage sag residual voltage accurately. The accuracy and feasibility of the method are verified using actual data from the power grid in East China. Related results can be found in [16–18].

In the third paper, "Research on a Non-PLL Control Strategy for a Flexible Fault Current Limiter and Its Application in Improving the FRT Capability of Microgrids", the

https://doi.org/10.3390/sym17020264

authors propose a scheme that applies a cascaded H-bridge flexible fault current limiter to achieve both voltage regulation and current limiting at the point of common coupling (PCC). The d-q axis components are extracted without a PLL in a double synchronous rotating coordinate system by setting the angular frequency of the coordinate axis. To address the issue of the set frequency deviating from the grid frequency, a calculation formula for frequency correction is derived. Through iterative correction, the set frequency is adjusted without feedback control. A sequence decomposition and compensation control strategy for the CHB-FFCL is presented. Finally, through simulation analysis, the effectiveness of this strategy is verified. Related results can be found in [19–22].

In the fourth paper, "A State Machine-Based Droop Control Method Aided with Droop Coefficients Tuning through Infeasible Range Detection for Improved Transient Performance of Microgrids", the focus is to address the issues and the role of a droop controller's dynamics on the stability of microgrids. A small-signal stability analysis is conducted, thereby identifying an infeasible range of droop values. Accordingly, safe values for droop coefficients are defined using the state machine concept. The proposed SMDCM is compared with the conventional constant droop control method and the fuzzy logic-based droop control method in terms of frequency, power, and voltage characteristics under different power factor loading conditions. Related results can be found in [23–26].

In the fifth paper, "Modern Active Voltage Control in Distribution Networks Including Distributed Generation, Using the Hardware-in-the-Loop Technique", the authors prove that active approaches can greatly lower connection costs while boosting the capacity of connectable distributed generation when used in place of the passive strategy. In this article, a modified active voltage control algorithm is applied to an IEEE 33-bus system to test the robustness and reliability of the control algorithm under severe conditions. The simulations are carried out using the hardware-in-the-loop method. Real-time simulations are used to test data transfer and the reliability of the control algorithm's implementation. This analysis is based on a three-phase symmetric power system. Related results can be found in [27–32].

In the sixth paper, "Research on the Fault-Transient Characteristics of a DC Power System Considering the Cooperative Action of a Flexible Current-Limiting Device and a Circuit Breaker", the authors propose a FCLD that improves the operational ability of the DC system under asymmetric conditions. First, a rectifier provides a set-slope current to each cascade inductor, which enables the voltage of the inductor to be clamped. Second, a controlled current source is applied to generate inverse flux to block the inductor from magnetic saturation. The protection action time of the DC circuit breaker is reformulated. Finally, by considering the synergistic action of the current-limiting device and the circuit breaker, as well as the transient characteristics of the DC grid fault, the protection scheme of the multi-terminal flexible DC system is proposed. Related results can be found in [33–36].

In the seventh paper, "Utilizing Full Degrees of Freedom of Control in Voltage Source Inverters to Support Micro-Grid with Symmetric and Asymmetric Voltage Requirements", the authors propose a novel equivalent control method for VSI that incorporates DOB. The method leverages degrees of freedom of the VSI under symmetric and asymmetric microgrid voltage conditions by utilizing the mean-point voltage of the MG. This method enables the three-phase inverter to generate voltages as needed by the MG in response to changing loads in the microgrid circuits or phases. The method is also insensitive to disturbances because of the DOB, which is part of the controller. The proposed method is validated under both the balanced and unbalanced voltage demands of the microgrid. Related results can be found in [37–39].

In the eighth paper, "Optimal Location and Sizing of Photovoltaic-Based Distributed Generations to Improve the Efficiency and Symmetry of a Distribution Network by Handling Random Constraints of Particle Swarm Optimization Algorithm", the authors discuss the use of a PSO algorithm. The goal is to determine the appropriate sizes of PVDG and find the best locations for PVDG. It is thus expected that this algorithm will provide an efficient and consistent solution to improve the overall performance of the power system. Placement and sizing of the distributed generation aim to minimize power losses, enhance the voltage profile, which brings symmetry to the voltage profile of the system, and provide maximum cost savings. The simulation results are successful, indicating its viability. Related results can be found in [40–42].

In the ninth paper, "Research on Optimal Scheduling Strategy of Microgrid Considering Electric Vehicle Access", the authors propose a microgrid optimization scheduling strategy considering the integration of EVs. Firstly, to reduce the impact of random integration of EVs on power system operation, a schedulable model of an EV cluster is constructed based on the Minkowski sum. Then, based on the wavelet neural network (WNN), the renewable energy output is forecasted to reduce the influence of its output fluctuation on the operation of the power system. Considering the operational constraints of each unit in the microgrid, the network active power loss and node voltage deviation are considered as the optimization objectives, and the established microgrid model is transformed equivalently using second-order cone relaxation to improve solution efficiency. Through network reconfiguration and flexible load participation in demand response, the economic efficiency and reliability of system operation are improved. Finally, the feasibility and effectiveness of the proposed method are verified based on the simulation examples. Related results can be found in [43–46].

In the tenth paper, "Adaptive Virtual Inertia Control Strategy for a Grid-Connected Converter of DC Microgrid Based on an Improved Model Prediction", the authors study an adaptive virtual inertia control strategy for a grid-connected converter of a DC microgrid based on an improved model prediction. Firstly, an adaptive analog virtual synchronous generator is introduced into the voltage outer loop by combining the inertial parameters with the rate of change of the voltage, realizing flexible adjustment of the inertial parameters. Secondly, the improved model predictive control is introduced into the current inner loop to achieve fast tracking of the reference current value and improve the dynamic performance of the control system. Finally, a system model is established based on Matlab/Simulink for simulation. The results show that the proposed control strategy can effectively improve the stability of DC bus voltage and the operational capability of the system under asymmetric conditions. Related results can be found in [47,48].

In the eleventh paper, "A New Method for the Analysis of the Broadband Oscillation Mode in New Energy Stations", the authors propose a method for analyzing the broadband oscillation mode based on VMD and Prony. This method can achieve improved decomposition performance through VMD, eliminating the dimensionality issues that are prone to the traditional Prony algorithm, which improves the identification accuracy. The amplitude, frequency, initial phase, and damping factor of the signal can be obtained, which helps better identify and analyze the broadband oscillation mode. Finally, the effectiveness of the proposed method is verified by analyzing examples and simulation data. Related results can be found in [49,50].

In the twelfth paper, "Research on Voltage Sag Loss Assessment Based on a Two-Stage Taguchi Quality Perspective Method", a voltage sag loss assessment method based on a two-stage Taguchi quality perspective method is proposed to perform a quantitative analysis of voltage sag economic losses. Initially, using the Taguchi quality perspective approach, single-index quality loss functions are separately established for voltage sag magnitude and fault duration. Subsequently, by introducing a comprehensive load tolerance curve, sensitivity parameters in the quality loss function are accurately calculated. This yields a deterministic model for voltage sag assessment. Based on this, the relative impact of the two indices on voltage sag loss is evaluated using the quality loss function. Consequently, a comprehensive loss model under the influence of multiple indices is formed by integrating the two single-index evaluation models. The simulation results indicate that this method reduces the computational complexity of loss assessment through the consolidation process of intervals with similar sensitivity parameters. Related results can be found in [51–53].

Conflicts of Interest: The author declares no conflicts of interest.

References

1. Šapurov, M.; Baskys, A.; Pomarnacki, R.; Serackis, A.; Jankauskas, M.; Huynh, V.K.; Bleizgys, V.; Dervinis, A.; Bielskis, E.; Paulikas, S.; et al. Cascaded Multilevel Inverter-Based Asymmetric Static Synchronous Compensator of Reactive Power. *Symmetry* **2022**, *14*, 483. [CrossRef]
2. Zheng, C.; Dai, S.; Zhang, B.; Li, Q.; Liu, S.; Tang, Y.; Wang, Y.; Wu, Y.; Zhang, Y. A Residual Voltage Data-Driven Prediction Method for Voltage Sag Based on Data Fusion. *Symmetry* **2022**, *14*, 1272. [CrossRef]
3. Zheng, F.; Wu, G.; Qu, B.; Xiang, X.; Zheng, S. Research on a Non-PLL Control Strategy for a Flexible Fault Current Limiter and Its Application in Improving the FRT Capability of Microgrids. *Symmetry* **2022**, *14*, 2652. [CrossRef]
4. Mandarapu, S.; Yellapragada, K.P.V. A State Machine-Based Droop Control Method Aided with Droop Coefficients Tuning through In-Feasible Range Detection for Improved Transient Performance of Microgrids. *Symmetry* **2022**, *15*, 1. [CrossRef]
5. Fanos, B.N.F.; Soliman, M.H.; Talaat, H.E.A.; Attia, M.A. Modern Active Voltage Control in Distribution Networks, including Distributed Generation, Using the Hardware-in-the-Loop Technique. *Symmetry* **2022**, *15*, 90. [CrossRef]
6. Zheng, F.; Peng, Y.; Wang, W.; Zheng, S. Research on the Fault-Transient Characteristics of a DC Power System Considering the Cooperative Action of a Flexible Current-Limiting Device and a Circuit Breaker. *Symmetry* **2023**, *15*, 134. [CrossRef]
7. Rasool, A.; Ahmad, F.; Fakhar, M.S.; Kashif, S.A.R.; Matlotse, E. Utilizing Full Degrees of Freedom of Control in Voltage Source Inverters to Support Micro-Grid with Symmetric and Asymmetric Voltage Requirements. *Symmetry* **2023**, *15*, 865. [CrossRef]
8. Ali, M.A.; Bhatti, A.R.; Rasool, A.; Farhan, M.; Esenogho, E. Optimal Location and Sizing of Photovoltaic-Based Distributed Generations to Improve the Efficiency and Symmetry of a Distribution Network by Handling Random Constraints of Particle Swarm Optimization Algorithm. *Symmetry* **2023**, *15*, 1752. [CrossRef]
9. Wu, Z.; Zou, Y.; Zheng, F.; Liang, N. Research on Optimal Scheduling Strategy of Microgrid Considering Electric Vehicle Access. *Symmetry* **2023**, *15*, 1993. [CrossRef]
10. Zheng, F.; Su, M.; Liu, B.; Liu, W. Adaptive Virtual Inertia Control Strategy for a Grid-Connected Converter of DC Microgrid Based on an Improved Model Prediction. *Symmetry* **2023**, *15*, 2072. [CrossRef]
11. Guo, C.; Yang, L.; Dai, J.; Chen, B.; Yin, K.; Dai, J. A New Method for the Analysis of Broadband Oscillation Mode of New Energy Stations. *Symmetry* **2024**, *16*, 278. [CrossRef]
12. Guo, C.; Zhang, X.; He, M.; Wang, L.; Yang, X. Research on Voltage Sag Loss Assessment Based on a Two-Stage Taguchi Quality Perspective Method. *Symmetry* **2024**, *16*, 328. [CrossRef]
13. Chang, W.-N.; Liao, C.-H. Design and Implementation of a STATCOM Based on a Multilevel FHB Converter with Delta-Connected Configuration for Unbalanced Load Compensation. *Energies* **2017**, *10*, 921. [CrossRef]
14. Chaudhari, P.; Rane, P.; Bawankar, A.; Shete, P.; Kalange, K.; Moghe, A.; Panda, J.; Kadrolkar, A.; Gaikwad, K.; Bhor, N.; et al. Design and implementation of STATCOM for reactive power compensation and voltage fluctuation mitigation in microgrid. In Proceedings of the 2015 IEEE International Conference on Signal Processing, Informatics, Communication and Energy Systems (SPICES), Kozhikode, India, 19–21 February 2015; pp. 1–5.
15. Wang, Q.; Wang, B.; Xu, W.; Xu, J. Research on STATCOM for reactive power flow control and voltage stability in microgrid. In Proceedings of the 2018 13th IEEE Conference on Industrial Electronics and Applications (ICIEA), Wuhan, China, 31 May–2 June 2018; pp. 2474–2479.
16. dos Santos, A.; Rosa, T.; de Barros, M.T.C. Stochastic Characterization of Voltage Sag Occurrence Based on Field Data. *IEEE Trans. Power Deliv.* **2019**, *34*, 496–504. [CrossRef]
17. Zambrano, X.; Hernandez, A.; Izzeddine, M.; de Castro, R.M. Estimation of Voltage Sags from a Limited Set of Monitors in Power Systems. *IEEE Trans. Power Deliv.* **2017**, *32*, 656–665. [CrossRef]
18. Wang, Y.; Yang, M.-H.; Zhang, H.-Y.; Wu, X.; Hu, W.-X. Data-driven prediction method for characteristics of voltage sag based on fuzzy time series. *Int. J. Electr. Power Energy Syst.* **2022**, *134*, 107394. [CrossRef]
19. Yang, H.; Nian, H. Stability analysis of grid-connected converter based on interconnected system impedance modeling under unbalanced grid conditions. In Proceedings of the 2014 17th International Conference on Electrical Machines and Systems (ICEMS), Hangzhou, China, 22–25 October 2014; pp. 2631–2636.

20. Cespedes, M.; Jian, s. Modeling and mitigation of harmonic resonance between wind turbines and the grid. In Proceedings of the Energy Conversion Congress & Exposition, Phoenix, AZ, USA, 17–22 September 2011.
21. Nian, H.; Cheng, P.; Zhu, Z.Q. Coordinated Direct Power Control of DFIG System Without Phase-Locked Loop Under Unbalanced Grid Voltage Conditions. *IEEE Trans. Power Electron.* **2016**, *31*, 2905–2918. [CrossRef]
22. Cheng, P.; Nian, H. Direct power control of voltage source inverter in a virtual synchronous reference frame during frequency variation and network unbalance. *IET Power Electron.* **2016**, *9*, 502–511. [CrossRef]
23. Guan, M.; Pan, W.; Zhang, J.; Hao, Q.; Cheng, J.; Zheng, X. Synchronous Generator Emulation Control Strategy for Voltage Source Converter (VSC) Stations. *IEEE Trans. Power Syst.* **2015**, *30*, 3093–3101. [CrossRef]
24. Kumar, Y.V.P.; Bhimasingu, R. Fuzzy logic based adaptive virtual inertia in droop control operation of the microgrid for improved transient response. In Proceedings of the 2017 IEEE PES Asia-Pacific Power and Energy Engineering Conference (APPEEC), Bangalore, India, 8–10 November 2017; pp. 1–6. [CrossRef]
25. Coelho, E.A.A.; Cortizo, P.C.; Garcia, P.F.D. Small Signal Stability for Single Phase inverter Connected to Stiff AC System. In Proceedings of the Conference Record of the 1999 IEEE Industry Applications Conference, Thirty-Forth IAS Annual Meeting (CatNo.99CH36370), Phoenix, AZ, USA, 3–7 October 1999; Volume 4, pp. 2180–2187.
26. Rui, W.; Oiuve, S.; Pinia, Z.; Yonghao, G.; Dehao, O.; Peng, W. Reduced-0rder Transfer Function Model of the Droop-Controlled Inverter via Jordan Continued-Fraction Expansion. *IEEE Trans. Energy Convers.* **2020**, *35*, 1585–1595. [CrossRef]
27. Maharjan, S.; Khambadkone, A.M.; Peng, J.C.-H. Robust Constrained Model Predictive Voltage Control in Active Distribution Networks. *IEEE Trans. Sustain. Energy* **2021**, *12*, 400–411. [CrossRef]
28. Hou, Z.; Liu, Y.; Yan, J.; Liao, Y.; Miao, C.; Zhao, R. Voltage control strategy of AVC system based on Data Mining. In Proceedings of the 2022 14th International Conference on Measuring Technology and Mechatronics Automation (ICMTMA), Changsha, China, 15–16 January 2022.
29. Ping, B.; Zhang, X.; Song, Q.; Yu, Y.; Wu, N.; Ji, X. Voltage control strategy for integrated medium and low voltage distribution network based on active-reactive power coordination optimization. In Proceedings of the 2020 Chinese Automation Congress (CAC), Shanghai, China, 6–8 November 2020; pp. 1187–1192.
30. Sun, X.; Qiu, J.; Tao, Y.; Ma, Y.; Zhao, J. Coordinated Real-Time Voltage Control in Active Distribution Networks: An Incentive-Based Fairness Approach. *IEEE Trans. Smart Grid* **2022**, *13*, 2650–2663. [CrossRef]
31. Gerdroodbari, Y.Z.; Razzaghi, R.; Shahnia, F. Decentralized Control Strategy to Improve Fairness in Active Power Curtailment of PV Inverters in Low-Voltage Distribution Networks. *IEEE Trans. Sustain. Energy* **2021**, *12*, 2282–2292. [CrossRef]
32. Yu, P.; Wan, C.; Sun, M.; Zhou, Y.; Song, Y. Distributed Voltage Control of Active Distribution Networks with Global Sensitivity. *IEEE Trans. Power Syst.* **2022**, *37*, 4214–4228. [CrossRef]
33. Yang, S.; Xiang, W.; Lu, X.; Zuo, W.; Wen, J. An Adaptive Reclosing Strategy for MMC-HVDC Systems with Hybrid DC Circuit Breakers. *IEEE Trans. Power Deliv.* **2020**, *35*, 1111–1123. [CrossRef]
34. Wei, X.; Zou, G.; Zhang, S.; Xu, C.; Sun, W. Adaptive Reclosing, Scheme for flexible Multi-Terminal Dc Distribution Grid 2021. In Proceedings of the 2021 6th Asia Conference on Power and Electrical Engineering (ACPEE), Chongqing, China, 8–11 April 2021; pp. 746–750.
35. Mei, J.; Ge, R.; Zhu, P.; Fan, G.; Wang, B.; Yan, L. An Adaptive Reclosing Scheme for MMC-HVDC Systems Based on Pulse Injection from Parallel Energy Absorption Module. *IEEE Trans. Power Deliv.* **2021**, *36*, 1809–1818. [CrossRef]
36. Zheng, T.; Lv, W.; Li, R. An Adaptive Reclosing Scheme Based on Phase Characteristics for MMC-HVDC Systems. *IEEE Trans. Power Deliv.* **2022**, *37*, 2986–2996. [CrossRef]
37. Vijay, A.; Doolla, S.; Chandorkar, M.C. Unbalance mitigation strategies in microgrids. *IET Power Electron.* **2020**, *13*, 1687–1710. [CrossRef]
38. Wodyk, S.; Iwanski, G. Three-phase converter power control under grid imbalance with consideration of instantaneous power components limitation. *Int. Trans. Electr. Energy Syst.* **2020**, *30*, e12389. [CrossRef]
39. Rasool, A. Control of Three Phase Converters as Source for Microgrid. Ph.D. Thesis, Sabanci University, Tuzla, Istanbul, Turkey, 2017.
40. Atwa, Y.M.; El-Saadany, E.F.; Salama, M.M.A.; Seethapathy, R. Optimal Renewable Resources Mix for Distribution System Energy Loss Minimization. *IEEE Trans. Power Syst.* **2009**, *25*, 360–370. [CrossRef]
41. Kennedy, J.; Eberhart, R. Particle Swarm Optimization. In Proceedings of the lCNN'95-International Conference on Neural Networks, Perth, Australia, 27 November–1 December 1995; Volume 4, pp. 1942–1948.
42. Eberhart, R.; Kennedy, J. A new optimizer using particle swarm theory. In Proceedings of the Sixth international Symposium on Micro Machine and Human Science, Nagoya, Japan, 4–6 October 1995; pp. 39–43.
43. Xiao, Z.; Zhang, K.; Feng, J. Hierarchical optimal dispatching of wind /PV/diesel islanded microgrid with EVs charging station. *Tianjing Univ.* **2022**, *41*, 61–74.
44. Shao, S.; Ma, X.; Yuan, W.; Zhang, K.; Fu, X.; Huang, C. Robust optimal dispatching method for uncertain microgrid including electric vehicles. *Electr. Eng.* **2023**, *18*, 201–209.

45. Li, C.; Xu, M.; Cai, S. Two-stage optimal scheduling strategy for micro-grid considering EV default uncertainty. *Trans. China Electrotech. Soc.* **2023**, *38*, 1838–1851.
46. Zhou, M.; Wu, Z.; Wang, J.; Li, G. Forming Dispatchable Region of Electric Vehicle Aggregation in Microgrid Bidding. *IEEE Trans. Ind. Informatics* **2021**, *17*, 4755–4765. [CrossRef]
47. Liu, J.; Vazquez, S.; Wu, L.; Marquez, A.; Gao, H.; Franquelo, L.G. Extended State Observer-Based Sliding-Mode Control for Three-Phase Power Converters. *IEEE Trans. Ind. Electron.* **2017**, *64*, 22–31. [CrossRef]
48. Zheng, X.; Shi, X.; Li, H. VSG Control Strategy Based on Model Predictive for Islanded Microgrid Inverter. In Proceedings of the IECON 2021-47th Annual Conference of the IEEE Industrial Electronics Society, Toronto, ON, Canada, 13–16 October 2021; pp. 1–7.
49. Li, Y.; Liang, X.; Yang, Y.; Xu, M.; Huang, W. Early Fault Diagnosis of Rotating Machinery by Combining Differential Rational Spline-Based LMD and K–L Divergence. *IEEE Trans. Instrum. Meas.* **2017**, *66*, 3077–3090. [CrossRef]
50. Wang, R.; Huang, W.; Hu, B.; Du, Q.; Guo, X. Harmonic Detection for Active Power Filter Based on Two-Step Improved EEMD. *IEEE Trans. Instrum. Meas.* **2022**, *71*, 9001510. [CrossRef]
51. Xu, M.Z.; Tang, Z.; Wang, J. Research on the Applicability of Taguchi Method. *Sci. Technol. Manag. Res.* **2021**, *41*, 216–223.
52. de Almeida, F.A.; Santos, A.C.O.; de Paiva, A.P.; Gomes, G.F.; Gomes, J.H.d.F. Multivariate Taguchi loss function optimization based on principal components analysis and normal boundary intersection. *Eng. Comput.* **2022**, *38*, 1627–1643. [CrossRef]
53. Sun, J.H.; Kainz, J. Optimization of hybrid pulse power characterization profile for equivalent circuit model parameter identification of Li-ion battery based on Taguchi method. *J. Energy Storage* **2023**, *70*, 108034. [CrossRef]

symmetry

Article

Research on Voltage Sag Loss Assessment Based on a Two-Stage Taguchi Quality Perspective Method

Cheng Guo [1], Xinyuan Zhang [1,*], Mi He [2], Linling Wang [1] and Xuanming Yang [1]

1 School of Electric Power Engineering, Kunming University of Science and Technology, Kunming 650500, China; 20210090@kust.edu.cn (C.G.)
2 Kunming Power Supply Bureau of Yunnan Power Grid Co., Ltd., Kunming 650000, China
* Correspondence: 20212202176@stu.kust.edu.cn

Abstract: Voltage sags resulting from symmetrical or asymmetrical faults pose a significant threat to power quality. In response to this challenge, a voltage sag loss assessment method based on a two-stage Taguchi quality perspective approach is proposed to address the quantitative analysis of voltage sag economic losses. Initially, using the Taguchi quality perspective method, single-index quality loss functions are separately established for voltage sag magnitude and fault duration. Subsequently, by introducing a comprehensive load tolerance curve, sensitivity parameters within the quality loss function are accurately calculated. This yields a deterministic model for voltage sag assessment. Building upon this, the relative impact of the two indices on voltage sag loss is evaluated using the quality loss function. Consequently, a comprehensive loss model under the influence of multiple indices is formed by integrating two single-index evaluation models. The simulation results indicate that this method can effectively assess the economic losses of voltage sags under the combined influence of multiple factors. Compared to the original economic loss assessment method, it improves quantitative accuracy by approximately 3.72%. Moreover, the method reduces the computational complexity of loss assessment through the consolidation of intervals with similar sensitivity parameters.

Keywords: economic assessment; quality loss function; sensitivity parameters; Taguchi quality perspective; voltage sag

1. Introduction

With continuous socio-economic development, the usage of power-sensitive equipment is increasing, and the voltage sag problem is receiving more and more widespread attention [1,2]. Due to lightning, wind, external damage, power equipment insulation level reduction, and other factors, asymmetric faults such as single-phase grounding, two-phase grounding, and two-phase short-circuit will inevitably occur in the power system. It is also possible to have a symmetrical fault of three-phase short-circuit. Furthermore, the resulting voltage drop will affect the normal operation of power-sensitive equipment [3] and may cause serious economic losses [4,5]. Assessing the losses caused by voltage sag incidents will aid in formulating effective response measures, enhancing the reliability and robustness of the power system, reducing economic losses for users, and ensuring the quality and availability of the power supply [6].

The economic fallout from voltage dips encompasses a spectrum of losses, spanning production line shutdowns, data losses, compromised product quality, and damages incurred due to equipment malfunctions [7]. In the context of assessments based on actual measurement data, [8] proposed a user-centric approach for analyzing the economic losses of power quality. In [9], process-immune time is incorporated into the consideration of economic losses due to distribution network faults, representing a further refinement of the loss model. From an economic indicator perspective, [10] established a consequence model for voltage sag fault levels in sensitive equipment. As a practical matter, [11] developed a model assessing the impact of seasonal weather on predicting disturbances in power quality in distribution networks, representing a practical advancement in loss prediction methods under specific environmental conditions. The study by [12] proposed a low data-dependent assessment method for interruption probability in industrial processes, overcoming challenges related to logical relationships among various equipment in industrial processes through a heuristic search algorithm-based interruption probability solving method. To move forward a single step, [13] suggested optimizing and adjusting simulated loss values using BETA parameter correction based on empirical data. Combining prediction results with sag tolerance characteristics, [14] classified voltage sag risk levels, providing an evaluation that benefits from both ample simulation data and a method driven by actual measurements, making it more closely aligned with objective reality. The authors of [15] introduced an approach using fuzzy probability and possibility distribution to handle uncertainty, evaluating equipment tripping by transforming rigorously enforced statistical data into a fuzzy possibility distribution function. The above-mentioned studies can be categorized into two main types: combined assessment methods and fuzzy comprehensive assessment methods. While the combination assessment method is noted for its subjectivity and limited logical relationships [16], the fuzzy comprehensive assessment method faces challenges stemming from an unclear mapping between risk indicators and economic losses, thereby compromising assessment accuracy [15]. These inherent difficulties significantly impact the precision of assessment outcomes. However, a paradigm shift occurs when the improved fuzzy comprehensive assessment method is integrated with the change curve of process parameters. This integration not only enhances the adaptability of assessment results but also provides novel insights for evaluating transient losses in voltage. This innovative approach augments the discourse on the critical facet of power system analysis, offering valuable perspectives for the assessment of voltage transients.

The choice of assessment model building method—the Taguchi method [17]—as an experiment design and quality optimization method [18], provides new ideas and solutions for voltage-transient loss assessment [19]. The advantage of using Taguchi's method to assess the economic loss of a voltage-dropout event is that, compared to traditional methods, Taguchi's method does not need to subdivide and calculate the losses one by one according to the classifications of direct loss, restoration loss, and other losses that are difficult to quantify, but only needs to focus on the maximum loss at the time of the interruption, and then it can convert the loss toward the system interruption loss according to the degree of deviation of the status of the dropout and the system interruption [20]. The degree of deviation is expressed as a sensitivity parameter. Existing studies have mainly used the method of discounting the transient loss to the interruption loss by the transient drop amplitude as a constant or univariate functional relationship [21]. Obviously, the voltage-dropout loss model under the influence of multiple factors is more in line with the reality. Problems in the existing research on voltage sag loss under the influence of multiple factors focus on simplifying assumptions and lack of data. For example, in the study, it is assumed that the relationship between dropout amplitude and duration is linear and the effects on losses are independent of each other, i.e., they are regarded as separate variables [22] and their effects on losses were considered separately in the assessment process. This approach of looking at the amplitude and duration of the temporary drop independently ignores the possible nonlinear relationship and interaction effects between the two, as well as the combined effect of the two on the losses in real situations. In addition, existing studies usually assume the sensitivity parameter of the mass loss function under different states in single-indicator assessment models as a certain value empirically, and for multiple-indicator models, the expansion of dimensionality makes that the single certain value determined by the original empirical value can no longer approximately satisfy the actual conditions under more constraints. Therefore, the value of the sensitivity parameter in the quality loss function becomes the main factor limiting the establishment of the multi-indicator model, and how to refine the model assumptions on the basis of determining the value of the parameter is the key to establishing an optimal assessment model and improving the accuracy of the assessment of voltage drop loss.

Based on this, this paper proposes an economic assessment method for voltage transients based on a two-stage Taguchi quality view approach, aiming to address the threat to power quality brought about by the inputs of sensitive equipment in industrial processes and the economic loss brought about by voltage transients to the normal operation of industrial processes. In this paper, the methodology introduces the Taguchi quality loss function in two stages: in the first stage, the Taguchi quality loss function is introduced, and the effects of the transient drop time and fault duration on the losses are modeled, respectively. By introducing the signal-to-noise ratio, the interrelationships and interactions between different influencing factors are considered, but due to the nonlinear relationship between the influencing factors, it still may not be possible to completely eliminate the effects brought about by the factors being independent of each other. Therefore, the Taguchi quality loss function is introduced again in the second stage to assess the magnitude of the influence of the two parameters on the loss, respectively, and a comprehensive quality loss model is established in the case of inconsistent magnitude of the influence of multiple indicators on the loss. Through the assignment of the multivariate quality loss function, the combined effect of multiple influencing factors on the loss is more accurately reflected. The experimental results show that the method can effectively assess the economic loss level of voltage dips in the target system under the joint influence of multiple indicators.

2. Taguchi's Method

2.1. Taguchi's View of Quality

Taguchi's view of quality is an important branch of Taguchi's method. The Taguchi view of quality recognizes that the quality of a product or process begins at the design stage and continues to improve during manufacturing and service. Taguchi's view of quality emphasizes the economic benefits of quality, which can be achieved by improving quality, which can reduce product or process variability, increase efficiency and productivity, and reduce costs [23].

Product quality characteristics refer to the inherent features of a product, process, or system related to requirements. From the perspective of measurable characteristics, products can be categorized into three types based on different target values: "larger is better" characteristics, "smaller is better" characteristics, and "nominal best" characteristics. In other words, the ideal states for these characteristics are +∞, 0, and a certain specified value k, respectively. The number of quality characteristic dimensions for measurable characteristics can only be a non-negative number.

2.2. Quality Loss Function

In order to achieve a quantitative description of the mapping relationship between the magnitude of the deviation of a quality characteristic from the target value and the resulting economic loss, Taguchi's quality loss function for the quality view, in general form, is

$$L_i(x) = K_i \cdot F_i(x - A_i) \tag{1}$$

where $i \in \{1, 2, 3\}$, representing the look-ahead, small, and visual characteristics, respectively; x is the quality characteristic; A_i is the target value; x-A_i is the deviation of the quality characteristic from the target value; F_i (-) is a functional expression describing this deviation; and K is the maximum value of loss caused by the deviation of the quality characteristic x from the target value.

$$A_i = \begin{cases} +\infty, i = 1 \\ 0, i = 2 \\ N^+, i = 3 \end{cases} \tag{2}$$

Common quality loss functions are:

(1) Mean square error loss function

The mean square error loss function calculates the squared error between the predicted value and the true value and averages it out, also known as the quadratic mass loss function.

$$L_i(x) = \begin{cases} K_1 \cdot (1/x^2), i = 1 \\ K_2 \cdot x^2, i = 2 \\ K_3 \cdot (x - A)^2, i = 3 \end{cases} \tag{3}$$

where $(x\text{-}A)^2$, x^2, and $1/x^2$ reflect the degree of proximity of the quality characteristics to the target value, i.e., the degree of fluctuation of the quality characteristics of the lookout, lookout small, and lookout large, respectively.

(2) Inverse normal mass loss function

The inverse probability mass loss function is a loss function used to evaluate classification models. The basic idea is to view the classification problem as a probability distribution function, where each sample has a probability value indicating the probability that it belongs to each category. For a classifier, its goal is to make the predicted probability distribution function as close as possible to the true probability distribution function. Its general form is:

$$L(x) = K\left[1 - \frac{\pi(x)}{m}\right] \tag{4}$$

where: $\pi(x)$ is the probability density of the random distribution of the mass characteristic x, and m is the upper definite bound of $\pi(x)$.

For the normal distribution there is

$$\pi(x) = \frac{1}{\sqrt{2\pi\sigma}} e^{-\frac{(x-A)^2}{2\sigma^2}} \tag{5}$$

Thus an upper definite bound for Π(x) can be computed:

$$m = \frac{1}{\sqrt{2\pi\sigma}} \tag{6}$$

Therefore, the inverse normal mass loss function is:

$$L(x) = K \cdot [1 - e^{-\frac{(x-A)^2}{2\sigma^2}}] \tag{7}$$

Unlike traditional loss functions, the inverse probability mass loss function focuses on assessing the difference between the predicted distribution and the true distribution, rather than the specific values of the predicted distribution. Specifically, it calculates the ratio between the predicted distribution and the true distribution for each category, which can be thought of as a weight for each category that is used to adjust for the effects of different categories in an unbalanced dataset.

In addition to the above two methods, there are other quality loss functions such as the cross-entropy loss method and the logarithmic loss method. In contrast to the above methods, the inverse normal quality loss function is advantageous in that it can handle unbalanced datasets and can focus more on the predictive performance of a few categories. In addition, it does not need to explicitly define the category weights because these weights can be automatically inferred from the data set.

2.3. Integrated Approach

Signal-to-noise ratio η is an important indicator to reflect the stability of product quality. η represents dimensionless data; the larger the value, the more stable the product quality, and the smaller the loss caused.

Let us assume that the value of the quality characteristic of the product is a random variable, its mathematical expectation is μ, its variance is σ^2, and there exists a target value

A. For the value of the product characteristic x, it is desirable that μ = A and that σ^2 is as small as possible.

$$\eta_i = \begin{cases} 10\log(\sigma^2 + \mu^2), i = 1 \\ 10\log(\frac{1}{\sigma^2+\mu^2}), i = 2 \\ 10\log(\mu^2/\sigma^2), i = 3 \end{cases} \tag{8}$$

The assignment is accomplished based on the signal-to-noise ratio values. The mathematical expectation of the η value in the same row is taken as the reciprocal, the values obtained at different amplitudes are summed, and the values in each row are compared with the sum to obtain the magnitude of the contribution of the transient amplitude to the fluctuations.

$$\omega_i = \frac{1}{\eta_i} \Big/ \sum_{j=1}^{n} \frac{1}{\eta_j} \tag{9}$$

2.4. Quality Loss Function Model with Multiple Indicators

From the definition of the function and the objective reality, it may be assumed that the overall quality obeys an n-dimensional normal distribution, i.e., Y~N_n (μ, Σ), where μ is the overall mean vector and the positive definite matrix Σ is the covariance matrix

$$\mu = \begin{pmatrix} \mu_1 & \mu_2 & \cdots & \mu_3 \end{pmatrix} \tag{10}$$

$$\Sigma = \begin{pmatrix} \sigma_{11} & \sigma_{12} & \cdots & \sigma_{1n} \\ \sigma_{21} & \sigma_{22} & \cdots & \sigma_{2n} \\ \cdots & \cdots & \cdots & \cdots \\ \sigma_{n1} & \sigma_{n2} & \cdots & \sigma_{nn} \end{pmatrix} \tag{11}$$

Thus the ith quality feature y_i also obeys normal distribution, i.e., y_i ~N(μ_i, σ_{ii}). Therefore, according to the size of the signal-to-noise ratio, based on the size of the "contribution" of different quality indicators to the fluctuations, different weights λ can be assigned, and the assignment method is the same as the ω.

The loss function for the ith quality indicator is determined based on the economic significance of the indicator and its quality characteristics. Thus, the multivariate quality loss function is:

$$L(y_1, y_1 \cdots y_n) = \sum_{i=1}^{n} \lambda_i L(y_i) \tag{12}$$

3. Tolerance Curves for Sensitive Equipment

The occurrence of voltage sag events with a specific duration and magnitude may or may not lead to a system interruption. The numerical probability of this occurrence is inherently linked to the proportion γ of economic losses caused by incidents characterized by these features. Therefore, the economic loss situation can be characterized by solving for the probability of voltage sag events leading to a system interruption under different conditions.

Different sensitive equipment have different electrical characteristics, and accordingly the degree of tolerance to voltage dips is also different, so the sensitive load tolerance curve there is uncertainty region. According to the sensitive load tolerance curve, the sensitive equipment working area is divided into normal operation area, uncertainty area, and fault area shown in Figure 1. When the voltage sag eigenvalue index is above curve 1, the equipment is not affected by the transient drop, and the failure rate is 0. When the transient drop eigenvalue index is below curve 2, the equipment cannot operate, and the failure rate is 1. The intermediate region between curve 1 and curve 2 cannot determine the operation of the equipment, and it is necessary to assess its probability [24].

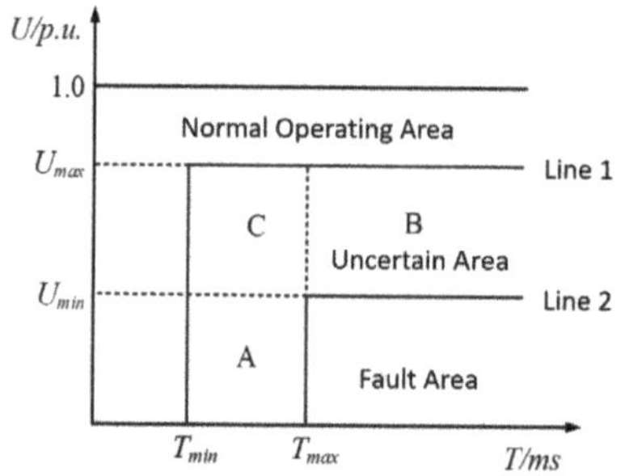

Figure 1. Uncertain region of sensitive load tolerance curve.

According to the different influencing factors, the uncertainty region can be further divided into three parts, A, B, and C, where the equipment failure rate in region A is a one-dimensional function about the transient time T, and the failure rate is proportional to T. Thus the failure probability model for region A can be obtained:

$$P_A = \frac{T - T_{\min}}{T_{\max} - T_{\min}}, (U,T) \in A \tag{13}$$

The equipment failure rate in region B is a one-dimensional function about U, and the failure rate is inversely proportional to the square of U. Thus the failure probability model for region B can be obtained:

$$P_B = \frac{U_{\max}^2 - U^2}{U_{\max}^2 - U_{\min}^2}, (U,T) \in B \tag{14}$$

The failure rate of region C is a two-dimensional function with respect to U and T. Thus the failure probability model for region C is:

$$P_C = P_A P_B = \frac{T - T_{\min}}{T_{\max} - T_{\min}} \times \frac{U_{\max}^2 - U^2}{U_{\max}^2 - U_{\min}^2}, (U,T) \in C \tag{15}$$

Based on the above analysis, the failure rate of the sensitive equipment on the load side can be obtained as the severity of voltage dips on the load side, i.e.,

$$E = \sum P_i \alpha_i \tag{16}$$

where α_i is the percentage of sensitive device i in the load side.

Based on the sensitive load withstand curve and the voltage withstand capability of the sensitive equipment, it is possible to calculate the probability that a transient event with a fault duration of any value and a transient drop amplitude of any value may lead to a system interruption, i.e., the value γ of the ratio of the economic loss corresponding to the maximum loss when an interruption occurs within any given time and amplitude range.

There is a relationship between this ratio value and the sensitivity parameter [25]:

$$\sigma^2 = -\frac{(x - A)^2}{2\ln(1 - \gamma)} \tag{17}$$

In summary, the solution of the key parameter σ^2 for calculating the voltage-transient loss in Taguchi's quality view can be realized by using the sensitive load tolerance curve and the voltage tolerance capability of the sensitive equipment.

4. Economic Evaluation Process of Voltage Dips

The process of economic evaluation of voltage transients is shown in Figure 2, as described below.

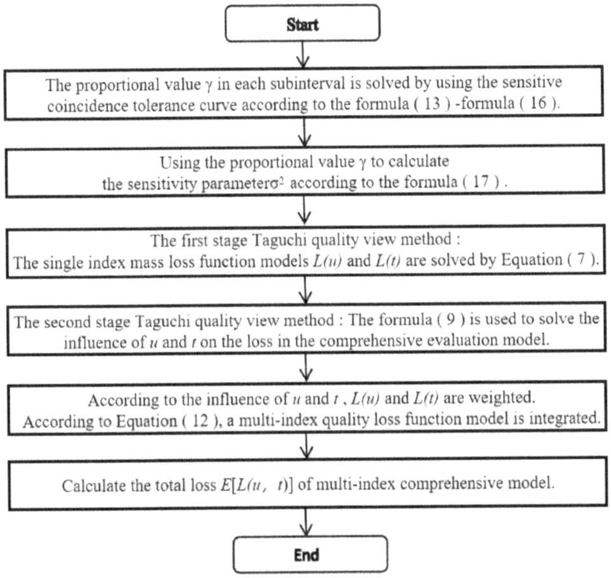

Figure 2. Flowchart for economic evaluation of voltage dips.

(1) In the first stage of Taguchi's quality view, based on Equation (7), the quality loss functions $L(u)$ and $L(t)$, are solved for the voltage-transient economic loss with respect to the transient amplitude and fault duration, respectively.

The solution steps include:

a. Establishment of single-index models: Utilizing the Taguchi quality perspective, single-index quality loss functions $L(u)/L(t)$ are developed separately for sag magnitude and fault duration. These functions are employed to depict the losses incurred by deviations from the target values at varying degrees.
b. Continuous processing: According to the voltage-transient density index of the target power system, obtain the continuous type probability distribution function $F(u)/F(t)$ and the probability density function $f(u)/f(t)$.
c. Discrete calculation: Firstly, select the unit length of transient drop amplitude and fault duration, and divide the evaluation range into several subintervals of equal length according to this setting; use the probability density function obtained from the continuous processing to calculate the mean and variance in each unit amplitude/time.
d. Weight calculation: Using Equations (8) and (9), substitute the mean and variance within each amplitude/time subinterval to calculate the weight ω of each amplitude/time subinterval.
e. Proportional value γ solution: Using Formulas (13) to (15), calculate the probability of failure in each region of each sensitive load tolerance curve combined with the system of each sensitive load share, and use Formula (16) to further calculate the probability of failure in the system within the amplitude/time subinterval; that is, the range of interruptions occurring when the corresponding economic loss accounted for the proportion of the maximum loss of the value of the proportion of γ.

f. Sensitivity parameter σ^2 solution: Introduce the continuous function again and use Equation (17) to integrate and solve the sensitivity parameter of the mass loss function $L(u)/L(t)$ in each time t/amplitude u subinterval.
g. Establishment of deterministic model: Incorporate the critical parameters obtained from the aforementioned steps into the quality loss functions $L(u)/L(t)$ to derive the single-index deterministic model.

(2) In the second stage of Taguchi's quality view, again utilizing Taguchi's quality view, the magnitude of the impact of the two parameters of transient drop amplitude and duration on the loss is quantified according to Equation (9), respectively. The combination of the two single-indicator assessment models is realized by assigning weights to each of them, so as to establish a comprehensive quality loss model under the influence of multiple indicators as shown in Equation (12).

The specific realization is as follows: the third use of the probability density functions $f(u)$ and $f(t)$, respectively, to calculate the transient drop amplitude and fault duration under the entire definition domain of the expectation, the variance, and thus the use of the signal-to-noise ratio and the calculation of weights. A quality loss function model with multiple indicators is established.

(3) The expected value of the voltage-dip economic loss function is then used to characterize the level of voltage-dip economic loss for that time period. Based on Equation (25), the economic loss under the joint influence of the amplitude of the sag and the duration of the fault is calculated.

5. Example Analysis

Tests were conducted using the IEEE 39-node test system, the structure of which is shown in Figure 3.

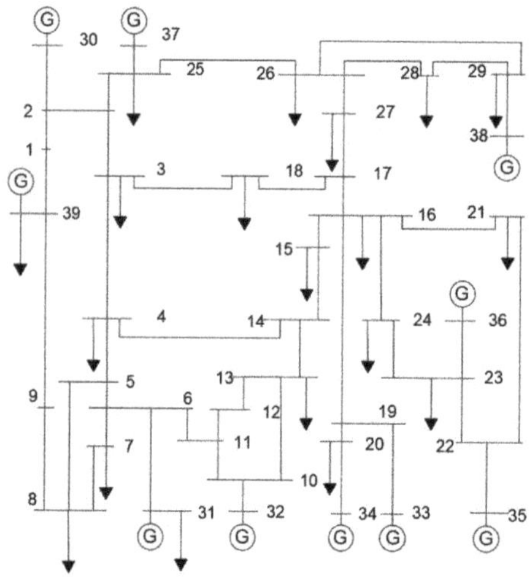

Figure 3. IEEE 39-node system architecture.

The load distribution parameter is set to be 0.5 and the number of Monte Carlo simulations is 5000 to obtain the voltage sag density metrics as shown in Table 1.

The total load of load point 15 is 140.5 MW, where the ratio of PC, ASD, PLC, and insensitive equipment is 1/15, 1/5, 1/5, 8/15, respectively. Knowing that the cost of interruption of sensitive loads counted under this ratio is USD 0.5363/kW, the maximum value of economic loss for load point 15 is USD 75,350.15.

Table 1. Voltage sag density index of load point 15.

Voltage Sag Magnitude	Fault Duration/s				
	0.0~0.2	0.2~0.4	0.4~0.6	0.6~0.8	>0.8
0~10%	0	0	0.023	0.031	0.012
10~20%	0	0	0.041	0.078	0.031
20~30%	0	0.003	0.073	0.121	0.061
30~40%	0	0.005	0.073	0.102	0.069
40~50%	0.003	0.007	0.064	0.138	0.081
50~60%	0.017	0.027	0.217	0.462	0.254
60~70%	0.074	0.105	0.409	0.750	0.384
70~80%	0.125	0.241	0.700	1.106	0.649
80~90%	0.161	0.331	0.576	0.698	0.352

5.1. *Solving the Mass Loss Function*

The first stage of the Taguchi mass-viewing method is used to obtain the mass loss functions $L(u)$, $L(t)$ under the influence of transient drop amplitude and fault duration, respectively.

An example of this is the quality loss function $L(t)$ of the voltage-transient economic loss with respect to fault duration.

5.1.1. Continuous Type Conversion of Discrete Interval Data

Taking load point 15 as an example, its voltage sag density data are studied, and the continuum type probability distribution function under different transient drop amplitudes are fitted separately using numerical analysis, and from this, the probability density function of voltage sag is calculated.

$$f(t) = -22.4336t^4 + 20.6672t^3 + 2.9340t^2 -2.0109t + 0.3474 \tag{18}$$

Further, a continuous two-dimensional probability density function may be calculated based on a quality loss function $f(t)$ for voltage-transient economic losses with respect to fault duration and a loss function $f(u)$ with respect to transient amplitude:

$$f(u,t) = f(u)\cdot f(t) \tag{19}$$

The probability density metrics of voltage transients after the continuumization process are shown in Figure 4.

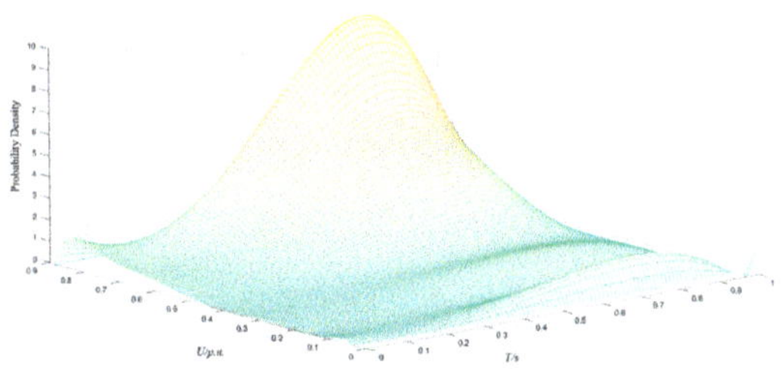

Figure 4. Continuous two-dimensional probability density function image.

Accordingly, we can visually analyze the trend of the probability of the occurrence of a transient-dropout accident with the transient-dropout amplitude and duration. As can be seen in Figure 4, the probability of voltage dips with low amplitude (0.6~0.9) is much larger than that with large amplitude (<0.6); the probability of a voltage dip being instantaneous (0~0.5 s) is much smaller than that of being temporary (0.5~1 s).

The distribution of voltage sag amplitude in Figure 4 is close to the probability density shown in Table 1, and the distribution trend is that most of the voltage sag amplitudes are above 0.6 and the duration is above 0.4 s, which shows that the fitting results are accurate.

5.1.2. Calculation of Weights

It is worthwhile to select 0.1 p.u. and 0.2 s as the unit lengths of the transient drop amplitude and fault duration, respectively, and to divide the analyzed range of 0–0.9 p.u. and 0–1 s into 9/5 subintervals according to the transient drop amplitude/fault duration. The mean and variance of each subinterval are calculated by using the probability density function obtained from the continuum processing, and the weight ω of each unit amplitude is calculated accordingly; the related parameters in each subinterval of the transient amplitude calculated are shown in Table 2.

Table 2. Taguchi quality view parameters under different sag amplitude.

Voltage Sag Magnitude	EX_j	DX_j	η_j	ω_j
0~10%	0.045210	0.001603	9.055668	0.239014
10~20%	0.133704	0.000837	13.293675	0.162817
20~30%	0.257539	0.000728	19.592860	0.110470
30~40%	0.343885	0.000798	21.707538	0.099709
40~50%	0.461008	0.000794	24.277935	0.089152
50~60%	0.557246	0.000779	26.007863	0.083222
60~70%	0.653269	0.000826	27.132266	0.079773
70~80%	0.754876	0.000818	28.428599	0.076136
80~90%	0.824429	0.000161	36.251479	0.059706

5.1.3. Calculation of Sensitivity Parameters

The regions of uncertainty in the operational status of each sensitive load are shown in Figure 5:

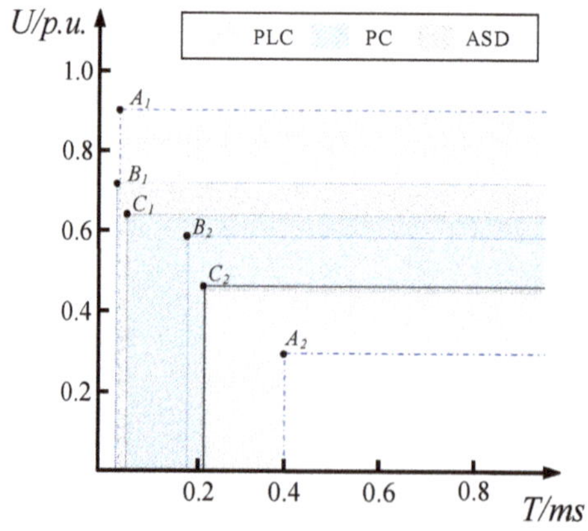

Figure 5. Uncertainty region of each sensitive load's operating status.

Of these, A_1 (0.02,0.90), A_2 (0.40,0.30), B_1 (0.015,0.71), B_2 (0.175,0.59), C_1 (0.04,0.63), and C_2 (0.205,0.46).

In Figure 5, the region enclosed by the two lines with turning points A_1 and A_2 (shaded square area) represents the uncertainty region of the sensitive device PLC. Similarly, the uncertainty regions for PC and ASD are indicated by two other shaded areas.

From the minimum and maximum values of the voltage tolerance amplitude of each sensitive load shown in Figure 5 and the minimum and maximum values of the equipment transient drop tolerance duration, the fault rate of each sensitive load in the region of uncertainty of the operating state calculated according to Equations (13) to (15) is shown in Table 3.

Table 3. Failure rates of sensitive loads in the region of uncertain operating conditions.

Equipment Type	Failure Rate		
	P_A	P_B	P_C
PC	0.1005	0.0894	0.0087
ASD	0.0800	0.0618	0.0049
PLC	0.1900	0.3500	0.0665

The probability that a device fails in each transient subinterval is:

$$G_i = S_i(A) \cdot P_A + S_i(B) \cdot P_B + S_i(C) \cdot P_C + S_i(e) \cdot 1 \tag{20}$$

where $i \in \{1, 2, 3\}$, representing PC, ASD, and PLC, respectively; G_i denotes the probability of failure of the ith device in this staging range; P_A, P_B, and P_C are the probabilities of failure of the device in the area of A, B, and C; S_i (A), S_i (B), S_i (C), and S_i (e) denote the magnitude of the proportion of the areas of A, B, C, and the area of the completely failed area to the total area in this staging range.

Since there are multiple loads in the system and the percentage of each load is different, the value of the discounted percentage of interruption loss γ for the combined consideration of multiple loads can be calculated by the following equation:

$$\gamma = \sum_{i=1}^{4} G_i \cdot M(i) \tag{21}$$

where $M(i)$ represents the proportion of the i middle load in the system.

The important parameters of the loss function under each amplitude subinterval calculated from Equation (21) and the sensitivity parameters obtained from Equation (17) are shown in Table 4.

Table 4. The parameters of mass loss function under different sag amplitudes.

Voltage Sag Magnitude	γ_j	${\sigma_j}^2$
0~10%	0.356347	0.378276
10~20%	0.356347	0.378276
20~30%	0.356347	0.378276
30~40%	0.268961	0.531992
40~50%	0.249164	0.581596
50~60%	0.204019	0.730417
60~70%	0.057408	2.819052
70~80%	0.048089	3.381751
80~90%	0.047054	3.458027

In summary, the loss function of voltage sag with respect to fault duration can be obtained as

$$L_1(t) = 75350.15 \times \sum_{j=1}^{9} \varpi_j \cdot (1 - e^{-\frac{t^2}{2\sigma_j^2}}) \tag{22}$$

From Table 4, it is noted that when the amplitude is in the range of 0 to 0.3 p.u., the sensitivity parameters of the three subintervals are the same, so they can be combined.

Similarly, the loss function of the voltage transient with respect to the transient amplitude is given as

$$L_2(u) = 75350.15 \times \sum_{k=1}^{5} \varpi_k \cdot (1 - e^{-\frac{(u-1)^2}{2\sigma_k^2}}) \tag{23}$$

where ω_k = [0.035515, 0.071567, 0.120867, 0.228174, 0.543877] and ${\sigma_k}^2$ = [5.855566, 1.046046, 0.555858, 0.555858, 0.555858]. Similarly, due to the same sensitivity parameters, the three subintervals within the fault duration of 0.4 to 1 s can be combined and processed.

5.2. Multi-Indicator Quality Loss Function Modeling

The second stage of the Taguchi mass view is used to quantify the magnitude of the effect of the two parameters—transient drop amplitude and duration, respectively—on losses.

There are two important factors affecting the voltage-dropout loss in the model: dropout magnitude x, and fault duration y. Among them, x has a look-ahead property with a target value of 1, and y has a look-small property.

The magnitude of the effect of the variables u, t on the transient loss calculated by using the Taguchi mass view method in combination with the expectation and variance of the transient loss amplitude calculated by the probability density function $f(u), f(t)$ using Equations (8) and (9) is shown in Table 5.

Table 5. Algorithm parameters under multi-index.

Index	EX	DX	η	ω
t	0.631392	0.040943	3.569439	0.779779
u	0.678700	0.025087	12.639008	0.220221

According to the influence of transient drop amplitude and fault duration on the transient loss, the transient drop amplitude and duration are given a weight of 0.779779 and 0.220221, respectively, and the weight is used to realize the combination of the two single-indicator models, which establishes a multi-indicator quality loss function model:

$$L(u,t) = 0.779779L_1(t) + 0.220221L_2(u) \tag{24}$$

Then the total loss of the node is

$$E[L(u,\mathrm{t})] = \iint_D L(u,t)\cdot f(u,t)dtdu = 50521 \tag{25}$$

where D denotes a region with a transient drop amplitude of 0 to 0.9 p.u. and a fault duration of 0 to 1 s.

5.3. Analysis of Results

Taking into account both the voltage sag amplitude and duration's impact on sag losses, under the assumptions of the case study in this paper, the economic loss due to voltage sags is approximately USD 50,521. According to the reasoning process of the traditional method, if the characteristic matrix is transposed, meaning if duration is considered as the primary evaluation criterion, this method should be equivalent to the original method [25]. However, as is shown in Table 6, the results show that when using the traditional quality engineering theory to analyze this case study, the losses calculated based on sag amplitude (traditional method) or duration (variation of the traditional method) as the primary evaluation criteria are USD 48,643 and USD 79,343, respectively. It is evident that traditional methods yield significantly different loss calculation results for the same case study.

Table 6. Evaluation results of voltage sag losses under different methods.

<table>
<tr><th colspan="2">Quality Engineering Theory and Methods</th><th>Methodology of This Paper</th></tr>
<tr><td>Voltage Sag Magnitude</td><td>Fault duration</td><td rowspan="2">50,521</td></tr>
<tr><td>48,643</td><td>79,343</td></tr>
</table>

This is because, in traditional assessment methods, simplifying assumptions are often made to streamline calculations. Firstly, the impact of sag duration and amplitude on losses is often independently considered. Secondly, the assessment model may only account for the influence of a single factor on voltage sag losses [20], or it might categorize sag duration into large intervals (instantaneous, temporary, short-term) and treat the influence factor of sag duration within each interval as a constant. In practical production, the majority of sag durations are less than 1 s. Therefore, when assessing losses for sag events within this time range, the influence factor of variable time is considered constant, essentially ignoring the impact of time changes on losses within this range.

The method proposed in this paper, from the perspective of model construction, comprehensively considers the impact of both sag duration and amplitude on economic losses, making the assessment more comprehensive.

From the computational results, the losses obtained using the method proposed in this paper fall between the values obtained using the traditional method (considering only the influence of sag amplitude) and its variation (considering only the influence of duration). Compared to the results of the traditional one-dimensional quality loss function, the relative error is 3.72%, and the deviation is less than 10%, within a reasonable range. This is because the method in this paper, based on the two-stage Taguchi quality perspective approach, accurately calculates the quality loss functions for each influence, considering the joint impact and relative importance of these two critical factors on losses. The essence of the method is to improve the accuracy of traditional methods and their variations, then obtain a weighted average of sag losses influenced by different factors. In practical applications, the correctness of the loss values obtained by this method can be verified by comparing them with the values obtained using traditional methods and their variations.

6. Conclusions

In this paper, for the problem of quantitative analysis of voltage-transient economic loss, a voltage-transient assessment method based on two-stage Taguchi's quality view is proposed on the basis of the simplified model assumed by the traditional one-dimensional mass loss function. The method not only overcomes the problem that the single multivariate mass loss function weakens the interrelationships among multiple influencing factors, but also gives an optimization scheme for the weakening assumption of the time factor in the traditional quality engineering theoretical approach. At the same time, the sensitivity conformity tolerance curve is used to realize the problem of taking values of sensitivity parameters in the construction of a multivariate loss function model.

The case analysis indicates that, through the aforementioned improvements, the accuracy of the assessment results has been enhanced by approximately 3.72%. In practical applications, the correctness of the loss values obtained through this method can be verified by comparing them with the losses calculated using traditional methods and their variations.

Simultaneously, in the assessment calculation process, by merging intervals with similar attributes and combining the use of discrete probability values and continuous probability density functions, the calculation precision was improved without excessively complicating the assessment process.

Author Contributions: Conceptualization, supervision, writing—review and editing, C.G.; methodology, software, validation, writing—original draft, X.Z.; project administration, funding acquisition, M.H.; formal analysis, data curation, visualization, L.W.; investigation, data curation, X.Y. All authors have read and agreed to the published version of the manuscript.

Funding: This research was funded by National Natural Science Foundation of China, grant number 52367002 and Yunnan Provincial Department of Science and Technology Joint Special Fund, grant number 202201BE070001-15.

Data Availability Statement: Data are contained within the article.

Conflicts of Interest: Mi He was employed by the company Kunming Power Supply Bureau of Yunnan Power Grid Corporation. The remaining authors declare that the research was conducted in the absence of any commercial or financial relationships that could be construed as a potential conflict of interest.

Nomenclature

Symbols	Definitions
A_i	target value
F_i (-)	a functional expression describing this deviation
K	the maximum value of loss caused by the deviation of the quality characteristic x from the target value
L_i (-)	loss function
$\pi(x)$	the probability density of the random distribution of the mass characteristic x
m	the upper definite bound of $\pi(x)$.
η	an important indicator to reflect the stability of product quality
ω_i	the contribution of the temporary decline value to the fluctuation
μ	population mean vector
Σ	the covariance matrix
λ	
P_A	the fault probability of area A
T	the transient time
U	voltage
E	the failure rate of the sensitive equipment on the whole load side
α_i	the percentage of sensitive device i in the load side.
σ^2	the sensitivity parameter
γ	the value of the corresponding economic loss as a proportion of the maximum loss when an interruption occurs
f(-)	probability density function
EX	mathematical expectation
DX	variance

References

1. Dash, S.K.; Ray, P.K. A New PV-Open-UPQC Configuration for Voltage Sensitive Loads Utilizing Novel Adaptive Controllers. *IEEE Trans. Ind. Inform.* **2021**, *17*, 421–429. [CrossRef]
2. Zheng, C.; Dai, S.; Zhang, B.; Li, Q.; Liu, S.; Tang, Y.; Wang, Y.; Wu, Y.; Zhang, Y. A Residual Voltage Data-Driven Prediction Method for Voltage Sag Based on Data Fusion. *Symmetry* **2022**, *14*, 1272. [CrossRef]
3. Shen, Y.P.; Liu, D.; Liang, W.H.; Zhang, X.Z. Current Reconstruction of Three-Phase Voltage Source Inverters Considering Current Ripple. *IEEE Trans. Transp. Electrif.* **2023**, *9*, 1416–1427. [CrossRef]
4. Guddanti, B.; Orrego, J.R.; Roychowdhury, R.; Illindala, M.S. Sensitivity Analysis Based Identification of Key Parameters in the Dynamic Model of a Utility-Scale Solar PV Plant. *IEEE Trans. Power Syst.* **2022**, *37*, 1340–1350. [CrossRef]
5. Xiao, S.; Wang, Z.; Wu, G.; Guo, Y.; Gao, G.; Zhang, X.; Cao, Y.; Zhang, Y.; Yu, J.; Liu, P.; et al. The impact analysis of operational overvoltage on traction transformers for high-speed trains based on the improved capacitor network methodology. *IEEE Trans. Transp. Electrif.* **2023**, 1. [CrossRef]
6. Zhang, X.; Gong, L.J.; Zhao, X.Y.; Li, R.R.; Yang, L.; Wang, B. Voltage and frequency stabilization control strategy of virtual synchronous generator based on small signal model. *Energy Rep.* **2023**, *9*, 583–590. [CrossRef]
7. Park, C.; Jang, G. Systematic method to identify an area of vulnerability to voltage sags. *IEEE Trans. Power Del.* **2017**, *32*, 1583–1591. [CrossRef]
8. Zhen, X.C.; Tao, S.; Xiao, X.N. An evaluation model of plant-level economic loss due to voltage dips. *Power Syst. Prot. Control* **2013**, *41*, 104–111.
9. Cebrian, J.C.; Milanovic, J.V.; Kagan, N. Probabilistic assessment of financial losses in distribution network due to fault-induced process interruptions considering process immunity time. *IEEE Trans. Power Del.* **2015**, *30*, 1478–1486. [CrossRef]
10. Liu, B.J.; Li, H.Q.; Xiao, X.Y. Risk assessment for failure level of sensitive equipment caused by voltage sag. *Proc. CSU-EPSA* **2016**, *28*, 87–92.
11. Michałowska, K.; Hoffmann, V.; Andresen, C. Impact of seasonal weather on forecasting of power quality disturbances in distribution grids. In Proceedings of the 2020 International Conference on Smart Energy Systems and Technologies (SEST), Istanbul, Turkey, 7–9 September 2020.

12. Wang, Y.; Yang, H.; Xiao, X.; Yang, Y. A Low-Data-Dependence Assessment Method of Industrial Process Interruption Probability Due to Voltage Sag. *IEEE Trans. Power Del.* **2022**, *37*, 5267–5277. [CrossRef]
13. Guo, C.; Zhang, X.Y.; Wang, L.L. Economic Evaluation of Voltage Sag under Uncertain Conditions Based on Beta Distribution Parameter Correction. In Proceedings of the 2023 IEEE International Conference on Power Science and Technology (ICPST), Kunming, China, 5–7 May 2023.
14. Zhang, Y.; Wu, Y.F.; Li, C.D. Voltage Sag Risk Assessment Based on Fusion of Simulated and Measured Data. *Autom. Electr. Power Syst.* **2023**, *47*, 174–185.
15. Behera, C.; Reddy, G.H.; Chakrapani, P.; Goswami, A.K.; Gupta, C.P.; Singh, G.K. Assessment of Equipment Trip Probability Due to Voltage Sags Based on Fuzzy Possibility Distribution Function. *IEEE Access* **2018**, *6*, 76889–76899. [CrossRef]
16. Cebrian, J.C.; Kagan, N.; Milanovic, J.V. Probabilistic estimation of distribution network performance with respect to voltage sags and interruptions considering network protection setting—Part I: The methodology. *IEEE Trans. Power Del.* **2018**, *33*, 42–51. [CrossRef]
17. Xu, M.Z.; Tang, Z.; Wang, J. Research on the Applicability of Taguchi Method. *Sci. Technol. Manag. Res.* **2021**, *41*, 216–223.
18. de Almeida, F.A.; Santos, A.C.O.; de Paiva, A.P.; Gomes, G.F.; Gomes, J.H.D.F. Multivariate Taguchi loss function optimization based on principal components analysis and normal boundary intersection. *Eng. Comput.* **2022**, *38*, 1627–1643. [CrossRef]
19. Sun, J.H.; Kainz, J. Optimization of hybrid pulse power characterization profile for equivalent circuit model parameter identification of Li-ion battery based on Taguchi method. *J. Energy Storage* **2023**, *70*, 108034. [CrossRef]
20. Mcgranaghan, M.; Rorttger, B. Economic evaluation of power quality. *IEEE Power Eng. Rev.* **2002**, *22*, 8–12. [CrossRef]
21. Xiao, X.Y.; Tan, X.M.; Wang, Y. Voltage sag economic losses assessment based on electrical characteristic-physical attribute-perceived loss. *Proc. CSEE* **2018**, *38*, 105–110.
22. Wang, L.; Xiao, X.Y.; Wang, Y. DNN-based estimation model of economic loss caused by voltage sag. *Electr. Power Autom. Equip.* **2020**, *40*, 156–165.
23. Parisa, M.; Shahram, K.; Ata, C.; Mohsen, A. Multi-objective optimization of a power generation system based SOFC using Taguchi/AHP/TOPSIS triple method. *Sustain. Energy Technol. Assess.* **2020**, *38*, 10067.
24. Zhao, F.; Du, Z.B. Impact Analysis of Voltage Sensitive Load on Planning of Distribution Network with New Energy. *Electr. Power Eng. Technol.* **2018**, *37*, 56–63.
25. Zhou, M.; Zhang, B.; Li, G.Y. Economic assessment of voltage sags based on quality engineering theory. *Trans. China Electrotech. Soc.* **2007**, *22*, 152–158.

symmetry

MDPI

Article

A New Method for the Analysis of Broadband Oscillation Mode of New Energy Stations

Cheng Guo [1], Lingrui Yang [1,*], Jianbo Dai [1], Bo Chen [1], Ke Yin [2] and Jing Dai [1]

1 Faculty of Power Engineering, Kunming University of Science and Technology, Kunming 650500, China; 20210090@kust.edu.cn (C.G.); daijianbo@stu.kust.edu.cn (J.D.); chenbo626@stu.kust.edu.cn (B.C.); 20212105016@stu.kust.edu.cn (J.D.)

2 Beijing Xinleineng Technology Co., Ltd., Chengdu Branch, Chengdu 610213, China; ink_monica@163.com

* Correspondence: yanglingrui@stu.kust.edu.cn

Abstract: The accurate identification of the broadband oscillation mode is the premise of solving the resonance risk of new energy stations. Reviewing the traditional Prony algorithm, the problems of the high model order and poor noise immunity in broadband oscillation mode are identified. The accuracy and running time of the Variational mode decomposition (VMD) is a symmetric trade-off problem. An improved strategy based on VMD is proposed. Firstly, the optimal value of the number of modes and penalty factors obtained by a particle swarm optimization algorithm is input into VMD to decompose the signal into multiple modes. Then, combined with the energy threshold method, the denoising and signal reconstruction of each mode component after decomposition are carried out. Finally, the Prony algorithm is used to identify the oscillation mode of the original signal and the reconstructed signal, respectively. The Signal-to-noise ratio (SNR) and model order are compared and analyzed. Through the analysis of the example and simulation data, it is shown that the proposed method effectively solves the problem of poor engineering adaptability of the traditional Prony algorithm. It also can accurately obtain the time-domain characteristics of broadband oscillation, which has a promising future in the engineering application.

Keywords: broadband oscillation; damping factor; model order; Prony; VMD algorithm

1. Introduction

Due to the global shortage of fossil energy, climate warming and environmental deterioration, vigorously developing new energy is one of the important measures to help prevent these problems [1]. The large-scale grid-connection of wind power and photovoltaic will bring a new problem of broadband oscillation ranging from several Hz to thousands of Hz [2]. For example, on 1 July 2015, a 20~40 Hz oscillation occurred in the northwest of China, resulting in the shutdown of thermal power units more than 300 km away [3]. Broadband oscillation has the characteristics of broadband time-varying and temporal and spatial distribution, which will cause great harm to the stable operation of power systems [4]. Therefore, accurate pattern recognition of broadband oscillation is significant for further research on the principle of broadband oscillation.

At present, the analysis methods of the influence factors of broadband oscillation are mainly analytical calculation [5] and numerical simulation [6]. Commonly used analytical calculation methods are the complex torque coefficient method, the state space method and the impedance analysis method, etc. [7]. These methods are only suitable for the oscillation mechanism analysis of small systems [8]. Due to the time-variability and complexity of broadband oscillation, the accuracy and stability of the numerical simulation method need to be improved [9]. With the development of broadband phasor measurement and synchronous phasor measurement, the analysis of broadband oscillation is changing from model-driven to data-driven [10–12]. Fourier transform is the most commonly used method to analyze broadband oscillation mode in engineering. Fourier transform shows the merits of fast calculation speed and simple principle, yet it is easy to produce spectrum leakage and a fence effect [13]. Moreover, the damping factor of oscillation, the initial phase angle and other time-domain characteristic quantities cannot be obtained. The wavelet transform is actually a set of adjustable window Fourier transforms; there will still be energy leakage. In addition, it needs to determine a set of basic functions in advance, and the selection of basic functions has a great influence on the subsequent processing results [14]. Its relative reconstruction error is quadratical growth with the upper bound of the frequency [15]. Empirical mode decomposition (EMD) is based on its own time scale, which means there is no need to choose a basis function in advance. However, it is sensitive to noise and prone to spectrum aliasing [16]. By introducing the Hilbert spectrum analysis method into EMD, the Hilbert–Huang method with strong adaptability but obvious end effect is formed [17]. To improve the accuracy of recognition, ref. [18] adopts Savitzky–Golay (S-G) filter denoising and total least squares estimation in two stages, the optimum running time of the S-G filter is obtained after several previous runs. In addition, it studies the identification problem of broadband oscillation based on an artificial intelligence algorithm. A large amount of historical data is needed for training to obtain more accurate analysis results [19–23].

Prony algorithm is a widely used method to analyze broadband oscillation in recent years. Due to the sensitivity to noise, it is necessary to increase the order of the model to improve the analysis accuracy. It often leads to the problem of "dimensional disaster" in engineering applications, which extremely limit the engineering applicability of the algorithm [24,25]. The common solution is to denoise the signal, such as Kalma filter [26], Wavelet denoising [27], Singular Value Decomposition (SVD) [28], a denoising method combining local mean decomposition and robust independent component analysis (LMD-RobustICA) [29] and Ensemble Empirical Mode Decomposition (EEMD) [30], etc.

To improve the identification accuracy, this paper proposes a method for analyzing broadband oscillation mode based on VMD and Prony. This method can achieve a better decomposition effect through VMD, eliminating the dimension disaster prone to the traditional Prony algorithm, improving the identification accuracy. The amplitude, frequency, initial phase and damping factor of the signal can be obtained, so as to better identify and analyze the broadband oscillation mode. Finally, the effectiveness of the proposed method is verified by analyzing the example and simulation data.

This paper is organized as follows: Section 2 introduces the variational mode decomposition, its parameter optimization and signal noise elimination; Section 3 introduces the Prony algorithm. Section 4 introduces the process of this method; Sections 5 and 6 analyze the example and simulation data; Section 7 concludes the methods in this paper.

2. VMD and Parameter Optimization Algorithm

2.1. Variational Mode Decomposition

Variational mode decomposition is a method that can decompose multi-component signals into multiple single-component amplitude-frequency signals with different center frequencies; what matters most is to establish variational constraints based on Hilbert transform, center frequency correction and computational bandwidth. By introducing a penalty factor α and the Lagrange factor λ, the variational constraint problem is transformed into variational unconstrained problem and solved.

First, the original signal $f(t)$ is decomposed to obtain multiple modal components $u(t)$, then the modal component is $u(t)$ transformed by Hilbert, and the corresponding unilateral spectrum is obtained.

$$[\delta(\mathrm{t}) + \mathrm{j}/(\pi\mathrm{t})]u(t) \tag{1}$$

where $\delta(t)$ is pulse function, satisfy $\delta(t) = 0$ when $\mathrm{t} = 0$ and $\delta(t) = \infty$ for others.

$$\int_{-\infty}^{+\infty} \delta(t)dt = 1$$

The spectrum is modulated by adding exponential terms, adjusting to the corresponding fundamental frequency band.

$$[(\delta(t)+j/(\pi t))u(t)]e^{-j\omega_k t} \tag{2}$$

where ω_k is the frequency of center. The bandwidth of each mode component is calculated according to the Gaussian smoothing of the demodulation signal. The constrained variational problem of VMD can be expressed as,

$$\begin{cases} \min\limits_{\{u_k\},\{\omega_k\}} \left\{ \sum\limits_{k=1}^{K} \left\| \partial_t[\delta(t)+\frac{j}{\pi t}u_k(t)]e^{-j\omega_k t} \right\|_2^2 \right\} \\ s.t. \sum\limits_{k=1}^{K} u_k = f(t) \end{cases} \tag{3}$$

In order to transform the above constrained variational problem into an unconstrained variational problem, the damping factor α and the Lagrange factor λ are introduced to ensure the accuracy of the reconstruction signal and the strictness of the constraints, resulting in the Lagrange multiplier.

$$L(u_k,\omega_k,\lambda)=\alpha\sum_{k=1}^{K}\left\|\partial_t[\delta(t)+\frac{j}{\pi t}u_k(t)]e^{-j\omega_k t}\right\|_2^2+\left\|f(t)-\sum_{k=1}^{K}u_k(t)\right\|_2^2+\left\langle\lambda(t),f(t)-\sum_{k=1}^{K}u_k(t)\right\rangle \tag{4}$$

Then, the alternate direction multiplier method was introduced to obtain the optimal solution of Equation (4), and the modal component and center frequency update expressions were obtained.

$$\hat{u}_k^{n+1}(\omega)=\frac{\hat{f}(\omega)-\sum\limits_{i\neq k}\hat{u}_i(\omega)+(\hat{\lambda}(\omega)/2)}{1+2\alpha(\omega-\omega_k)^2} \tag{5}$$

$$\omega_k^{n+1}=\frac{\int_0^\infty \omega|\hat{u}_k(\omega)|^2 d\omega}{\int_0^\infty |\hat{u}_k(\omega)|^2 d\omega} \tag{6}$$

Finally, the VMD algorithm is calculated and solved. The specific steps are as follows:

Step 1: Initialization u_k^1, ω_k^1, λ^1, $n = 0$.

Step 2: Let $n = n + 1$, according to Equations (5) and (6), the amplitude and center frequency of each modal component are iterative updated from 1 to K.

Step 3: Update the Lagrange factor λ,

$$\lambda^{n+1}=\lambda^n+\tau(f(t)-\sum_{k=1}^{K}\hat{u}_k^{n+1}) \tag{7}$$

Step 4: Set the termination condition of iteration. If Equation (8) is satisfied, the loop will be terminated; otherwise, repeat steps 2~4.

$$\sum_{k=1}^{K}\left\|\hat{u}_k^{n+1}-\hat{u}_k^n\right\|_2^2/\left\|\hat{u}_k^n\right\|_2^2<\xi \tag{8}$$

2.2. Optimize VMD Parameters

This paper adopts the particle swarm optimization algorithm to optimize some parameters in the VMD algorithm such as number of modes K and damping factor α, in order to improve the accuracy of decomposition results. Its fitness function is envelope entropy. The envelope entropy function of signal $f(t)$ can be expressed as

$$H_P=-\sum_{i=1}^{n}p_i \lg p_i \tag{9}$$

$$p_i = a_i / \sum_{i=1}^{n} a_i \tag{10}$$

where a_i is the envelope signal of the decomposed component of signal $f(t)$ after Hilbert transformation. The envelope entropy value can reflect the signal decomposition effect of the VMD algorithm. The smaller the value, the better the decomposition effect. The algorithm flow of parameter optimization is shown in Figure 1 below.

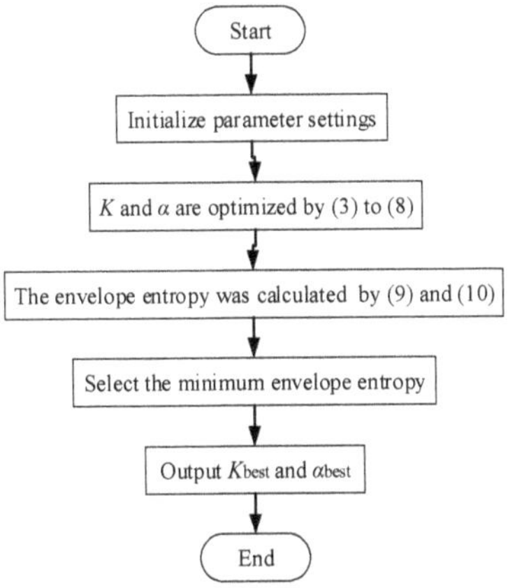

Figure 1. Flow chart of the genetic particle swarm optimization algorithm.

2.3. Signal Denoising Based on VMD

White noise is the main noise source in the power system, due to the frequency distribution characteristics of power signals. Improving VMD decomposition, the low-frequency modal component is mainly composed of the dominant harmonic component. The low-frequency harmonic component has a high signal-to-noise ratio, but there are still a few noise components distributed in the whole frequency band. In the middle and high frequency modal component, the high-order harmonic content is low, the amplitude is small, and the signal is straightforward to be submerged by noise. As the signal-to-noise ratio is very low, noise has a great impact on the detection accuracy. Based on the above analysis, it can be seen that the effective signals in the middle and high frequency modes have limited influence on the accuracy of the analysis of broadband oscillation signals. The high noise content in the high frequency modes is prone to excessive order determination in the solution of Prony. Therefore, the middle and high frequency mode components can be regarded as noise components. Based on the above discussion, this paper proposes a denoising method based on energy threshold. According to the distribution of signal and noise in different frequency bands, the energy threshold is set, and the modal components with different energy levels are retained and screened to achieve a better denoising effect [31].

Considering the EMD, it is easy to lose part of the effective harmonic component in the signal, and all the noise in the high frequency band cannot be completely screened out. Therefore, the energy threshold method is adopted for denoising. That is, the decomposed modal components have different levels of energy, the IMF with larger energy a is harmonic component, and the IMF with smaller energy is a noise component and false component. By setting the threshold, the modal components that only contain noise in the signal are screened out.

$$c_i = \max(u_i) \quad i = 1, 2, \cdots, K \tag{11}$$

$$m_i = c_i / \max(c) \quad i = 1, 2, \cdots, K \tag{12}$$

where K is number of modal components, u_i, c_i are the modal component and its maximum value, respectively, c is the set of maximum values of each modal component, m_i is the coefficient of energy. Threshold g needs to be set according to the noise content of the signal. If m_i is less than g, the corresponding modal component is retained. Otherwise, the mode is regarded as noise and screened out.

3. Prony Algorithm and Evaluation Parameters

3.1. Prony Algorithm

Prony algorithm is a mathematical model based on a linear combination of exponential functions, which can achieve the fitting of signal with the same sampling interval and obtain the signal amplitude, frequency, phase and damping factor. According to Euler's formula, the signal with sinusoidal component $x(t)$ is converted into N equally spaced sampling points, written as:

$$x(t) = \sum_{k=1}^{P} A_k e^{\alpha_k t} \cos(2\pi f_k t + \theta_k) \tag{13}$$

$$\hat{x}(n) = \sum_{k=1}^{P} b_k z_k^n \qquad n = 0, 1, \cdots, N-1 \tag{14}$$

$$b_k = A_k e^{j\varphi_k} \tag{15}$$

$$z_k = e^{(\alpha_k + j2\pi f_k)\Delta t} \tag{16}$$

where A_k is amplitude, f_k is frequency, θ_k is phase, α_k is damping factor, P is order of model, N is sampling number, Δt is sample interval, $\hat{x}(n)$ is signal of fit.

Equation (17) is the objective function of the minimum fitting error.

$$\text{minz} = \sum_{n=0}^{N-1} |x(n) - \hat{x}(n)|^2 \tag{17}$$

We constructed a linear difference equation with a constant coefficient, in which Equation (14) is the homogeneous solution and Equation (17) is the fitting error objective function.

$$\hat{x}(n) = -\sum_{k=1}^{P} a_k \hat{x}(n-k) \qquad P \le n \le N-1 \tag{18}$$

$$x(n) = -\sum_{k=1}^{P} a_k x(n-k) + \sum_{k=0}^{P} a_k e(n-k) \tag{19}$$

where $e(n)$ is error of fitting.

Taking error $e(n)$ as the excitation of P-order autoregressive model $\hat{x}(n)$, and achieving the actual signal $x(n)$ to solve the regular equation. Hence, getting the parameter a_k. Substitute a_k into (20) and source the root of it to find the parameter z_k.

$$\sum_{k=0}^{P} a_k z^{P-k} = 0 \tag{20}$$

According to Equation (14), the matrix equation is

$$\begin{bmatrix} 1 & 1 & \cdots & 1 \\ z_1 & z_2 & \cdots & z_P \\ \vdots & \vdots & \cdots & \vdots \\ z_1^{N-1} & z_2^{N-1} & \cdots & z_P^{N-1} \end{bmatrix} \begin{bmatrix} b_1 \\ b_2 \\ \vdots \\ b_P \end{bmatrix} = \begin{bmatrix} \hat{x}(0) \\ \hat{x}(1) \\ \vdots \\ \hat{x}(N-1) \end{bmatrix} \tag{21}$$

In combination with the matrix equation of Equation (21), the parameter b_k is obtained based on the idea of the least square method,

$$b = (V^H V)^{-1} V^H \hat{x} \tag{22}$$

Finally, the amplitude, frequency, initial phase and damping factor of each component are obtained by z_k, b_k.

3.2. Order of Model

The model order of the Prony algorithm can reflect the number of harmonic components in the system. The closer the order is to the number of harmonic components, the higher the performance of the Prony algorithm and the better the fitting effect; SNR can reflect the noise content of the signal and the fitting degree of the Prony algorithm. The noise content of a signal is inversely proportional to the SNR. Therefore, the model order and signal-to-noise ratio are selected as the evaluation parameters of the algorithm performance.

The traditional model order selection methods generally include a singular value method of autoregressive model and an empirical method based on the number of sampling points, etc. In this paper, the model order selection method with SNR and mean square error as indicators is adopted. SNR can represent the fitting effect of the signal. When the order of the selected model approaches the real value of the system, SNR will increase. According to the experimental results, when SNR < 20 dB, the signal fitting effect is poor. When SNR > 50 dB, the signal can be accurately fitted [31]. In addition, the mean square deviation can reflect the change degree of SNR, when the SNR finally tends to be stable, the signal fitting is accurate.

4. Flow of Broadband Oscillation Mode Identification Algorithm

The identification accuracy and algorithm performance of the traditional Prony algorithm are greatly affected by noise, and the fitting effect of signals with low SNR is poor, resulting in the failure to accurately identify the characteristic parameters of signals. In addition, it is difficult to select the optimal decomposition parameters of the VMD decomposition algorithm, which can lead to the unsatisfactory signal decomposition effect and, ultimately, affect the signal denoising effect. Therefore, this paper adopts a new method combining particle swarm optimization (PSO), VMD and the Prony algorithm to analyze harmonic signals. The basic steps are as follows:

Step 1: The number of modes K and damping factor α, which have great influence on VMD decomposition, are optimized by the optimization algorithm to achieve the best effect of VMD decomposition.

Step 2: The optimal number of modes and damping factors obtained in step 1 are substituted into VMD decomposition, and the original signal is decomposed to obtain a series of modal components with different frequency bands.

Step 3: Combined with the idea of energy threshold method, the energy coefficient of each modal component is calculated, and the threshold value g is set for comparative analysis. The modal component containing only medium and high frequency noise is screened out.

Step 4: The real harmonic components are reconstructed, and the Prony algorithm is used to identify the reconstructed signal, and the amplitude, frequency, phase and damping factor of the signal are obtained.

Finally, the broadband oscillation mode is analyzed according to the obtained parameters. The flow chart is shown in Figure 2.

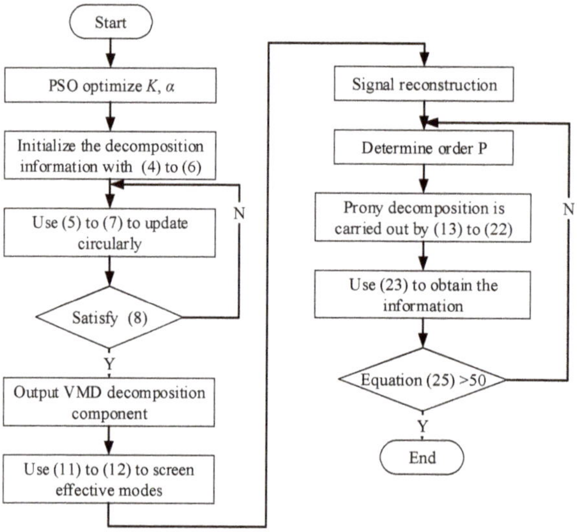

Figure 2. Flow chart of the harmonic detection method.

5. Examples Analysis

In order to test the effectiveness of the proposed method for broadband oscillation mode identification, an electric power signal containing fundamental waves, harmonics and inter-harmonics is constructed in this paper.

$$\begin{aligned} y(t) &= 45e^{-0.08t}cos(50 \times 2\pi t + 0.5\pi) + 25e^{-0.32t}cos(28 \times 2\pi t - 0.6\pi) + \\ &18.5e^{0.14t}cos(72 \times 2\pi t + 0.8\pi) + 8.5e^{-1.04t}cos(150 \times 2\pi t - 0.3\pi) \end{aligned} \tag{23}$$

The sampling frequency is set as 1000 Hz, and the number of sampling points is 500. The wave forms of the four groups of signals with noise and the noiseless signals are shown in Figure 3.

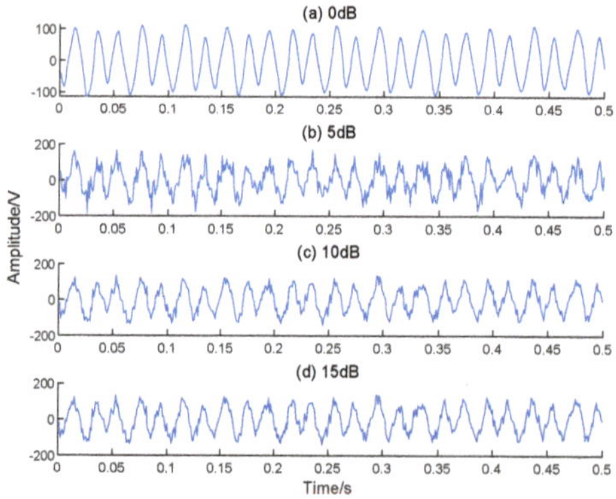

Figure 3. Waveform of the noisy signal and the noiseless signal.

5.1. Analysis of Denoising Performance

In order to improve the decomposition effect of the VMD algorithm, distinguish effective signals from noise and false components, an optimization algorithm (PSO) is used to optimize the VMD decomposition parameters. It improves the performance of the Prony algorithm. First, set the initial parameters of the particle swarm optimization algorithm, including acceleration factor c_1 and c_2 are 1.8, the position interval of damping factor α is (1000, 8000), number of modes K position interval is (2, 8).

The particles were iterative updated according to the above parameters, and the envelope entropy was used as the fitness function. The optimal positions of particles under different noise levels are shown in Table 1.

Table 1. Parameter values for different noise levels.

	Noiseless	5 dB	10 dB	15 dB
(K, α)	(45, 686.2)	(55, 593.8)	(56, 877.6)	(55, 101.8)

From Table 1, the number of modes K remains at 5 under different noise levels, but the value of the damping factor is inconsistent. Thus, the noise affects the value of the penalty factor while the number of modes K is not sensitive to noise.

Then, we chose the data with 15 dB noise added for subsequent analysis. The optimal number of modes and damping factors are substituted into the VMD algorithm to decompose the waveform and spectrum of each mode, as shown in Figures 4 and 5.

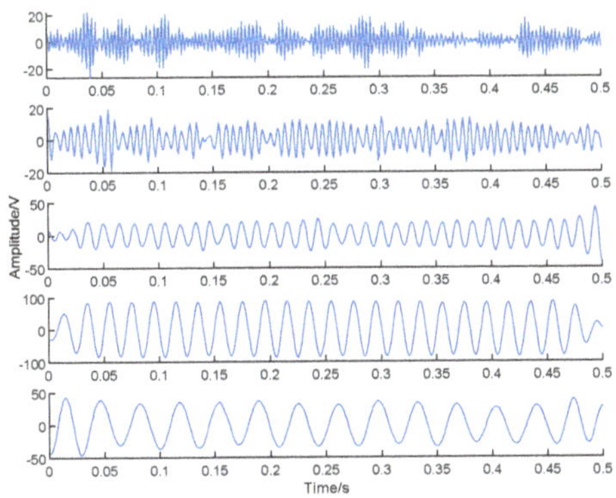

Figure 4. Waveform of the signal with 15 dB noise.

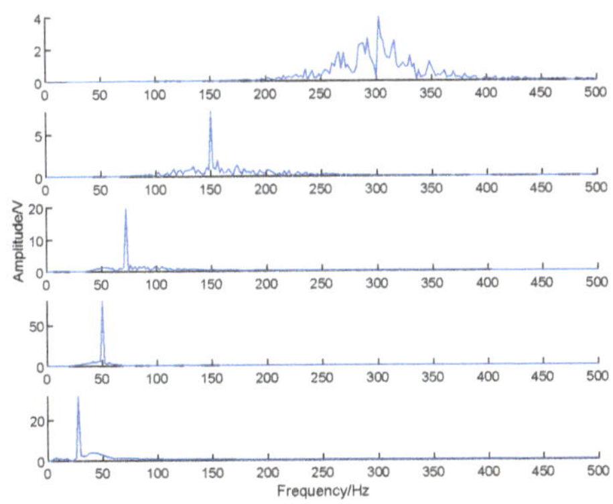

Figure 5. FFT analysis of the signal with 15 dB noise.

According to Figures 4 and 5, VMD decomposition can accurately obtain the amplitude and frequency of the modal component. Equations (11) and (12) calculate the energy level of each modal component and the results are shown in Table 2. Setting the threshold g is 5×10^{-2} according to Table 2; IMF_2~IMF_5 are true harmonic components and they are retained for Prony analysis. It is concluded that the VMD decomposition method based on parameter optimization can accurately separate the effective component and noise component of the signal.

Table 2. Energy coefficients for each modal component of the example signal.

Modal Component	IMF_1	IMF_2	IMF_3	IMF_4	IMF_5
Energy coefficient	0.0374	0.0863	0.2399	1	0.3948

5.2. Test Performance Analysis

In order to further verify the effectiveness and superiority of the proposed method in parameters identification, Prony identification is carried out on the signal reconstructed by VMD and the noiseless signal, respectively. Under different noise contents, when the fitted SNR is stable, the results of each fitted SNR are shown in Table 3. As can be seen from Table 3, the proposed method in this paper always has a stable SNR of de-noising fitting under different noise content, and the de-noising fitting effect shows a steady growth trend, indicating its robustness.

Table 3. Fitting SNR under different noise content.

Noise Content	5 dB	10 dB	15 dB	20 dB
Fitted SNR/dB	12.3814	18.2488	25.5612	32.3470

The result of different fitting orders for the signal with 15 dB noise added is shown in Figure 4. From Figure 6, the fitting effect of the proposed method is obviously better than that of Prony. The fitting SNR value of the proposed method is stable near 25.5 dB with Prony stable near 20 dB. It is proved that the proposed VMD based on particle swarm denoising is effective and can significantly improve the fitting effect of Prony.

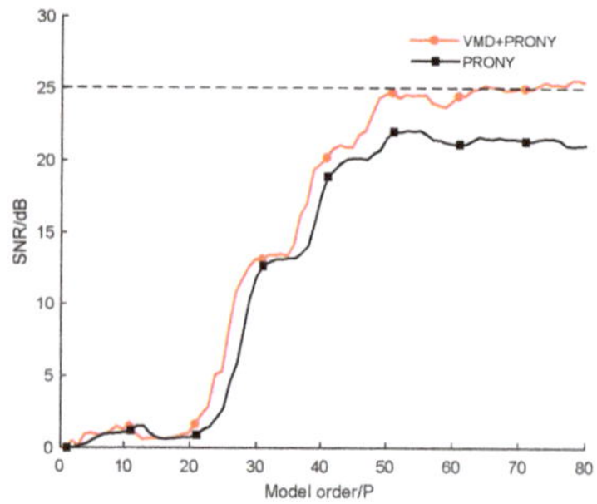

Figure 6. Comparison of fitting effect of example signal.

When the model order is 75, the fitting SNR of VMD + Prony is 25.5475 dB, and the fitting SNR of Prony is 20.9977 dB. The identification results of specific parameters are shown in Table 4.

From Table 4, it can be seen that the average identification errors of amplitude and frequency of the proposed method are 4.4034% and 0.0291%, that of Prony is 7.1808% and 0.0827% and that of EMD+FFT is 5.9997% and 0. The identification results of the proposed method are better than Prony in amplitude and frequency, except the frequency of EMD+FFT. The identification errors in phasor and damping factor are smaller than Prony. Although EMD+FFT has a better frequency identification effect than the proposed method and Prony, the amplitude identification effect is worse than the proposed method due to the existence of the damping factor, and the damping factor cannot be obtained.

Table 4. Identification result based on three methods from example data.

	Proposed Method				Prony				EMD+FFT	
	A/V	**f/Hz**	**φ/rad**	**α**	**A/V**	**f/Hz**	**φ/rad**	**α**	**A/V**	**f/Hz**
Value	85.2287	50.0033	1.2605	−0.1055	82.5226	50.0048	1.2792	−0.0361	83.8355	50
error (%)	0.2691%	0.0066%	19.75%	31.85%	0.6148%	0.0096%	18.56%	54.88%	1.37%	0
Value	34.0082	28.0172	−1.7570	−0.3299	38.0395	28.0218	−1.7446	−0.0361	33.3253	28
error (%)	2.8337%	0.0614%	6.7883%	3.094%	8.6848%	0.0779%	7.4461%	88.72%	4.7849%	0
Value	19.2786	71.9910	2.9456	0.1835	16.3884	71.9708	3.0091	0.5002	19.3774	72
error (%)	4.2087%	0.0125%	17.202%	31.07%	11.414%	0.0406%	19.73%	289.5%	4.7427%	0
Value	9.3757	149.946	1.4044	−0.9773	7.8192	149.696	0.3270	−0.4769	7.3864	150
error (%)	10.302%	0.036%	49.01%	80.98%	8.0094%	0.2027%	65.3%	11.69%	13.101%	0

6. Simulation Analysis

In order to further verify the superiority of the algorithm, a system model with fan was built on the simulation platform in this paper. The simplified model is shown in Figure 7 below:

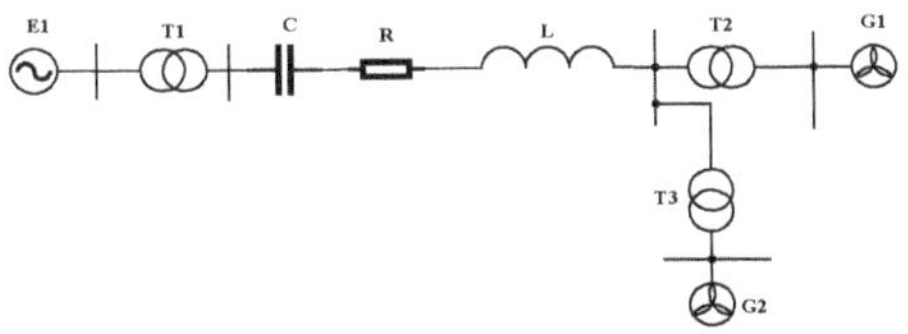

Figure 7. This is a figure. Simplified model diagram with wind power generation. E1 represents equivalent system, G1 and G2 represent wind turbines, and L, R and C represent transmission line parameters.

The simulation data of voltage, current and function of broadband oscillation are obtained by running the simulation model. The voltage waveform obtained by simulation is shown in Figure 8, where the sampling frequency is 500 Hz and the sampling time is 0.4 s.

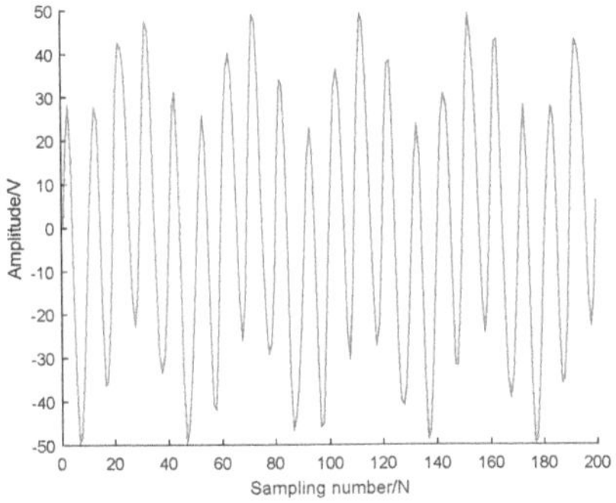

Figure 8. Waveform of the signal.

Fourier transform is applied to this set of data to obtain the amplitude and frequency of this set of signals. The signal waveform and FFT analysis results are shown in Figure 9, and the identification results are shown in Table 5.

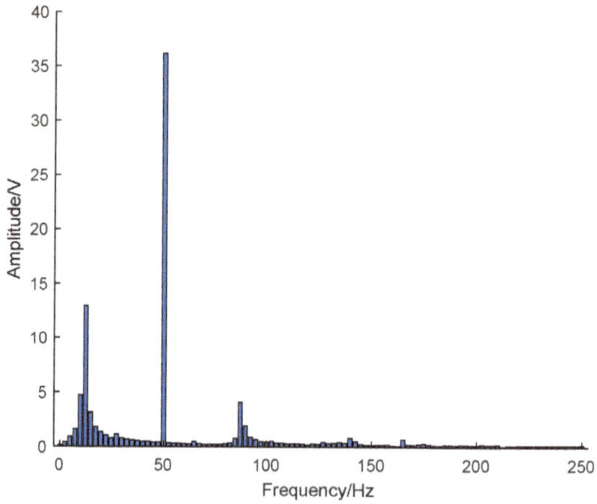

Figure 9. FFT analysis spectrum of the signal.

Table 5. FFT Analysis of the Signal Example.

IMF_i	f/Hz	A/V
IMF_1	11.5789	14.3803
IMF_2	50	36.6267
IMF_3	88.4211	4.8791

6.1. Analysis of Denoising Performance

For the simulation signals, this paper changed the position interval of the number of modes to [2,5], which can ensure the accuracy of VMD decomposition results. The position interval of punishment factor is (1000, 8000), with other parameters unchanged. The optimal position of the particle is (5, 6819.1); it was substituted into the VMD algorithm. The decomposition results are shown in Figures 10 and 11.

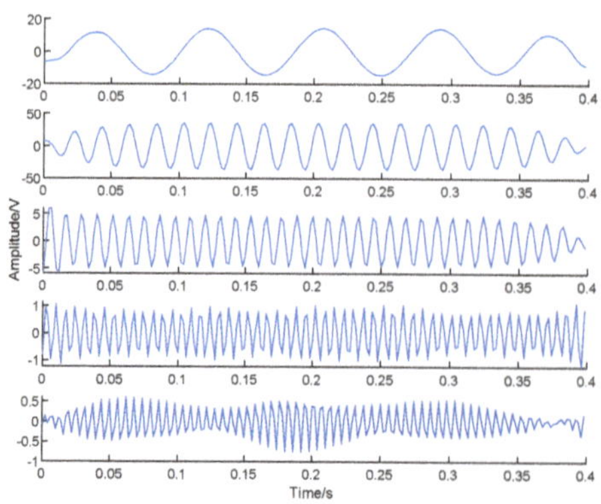

Figure 10. Intrinsic Mode Functions waveform.

According to Figures 10 and 11, IMF_1~IMF_3 are true harmonic components and are retained for Prony analysis. It is concluded that the VMD decomposition method based on parameter optimization can accurately separate the effective component and noise component of the signal. The energy coefficients of each component are shown in Table 6.

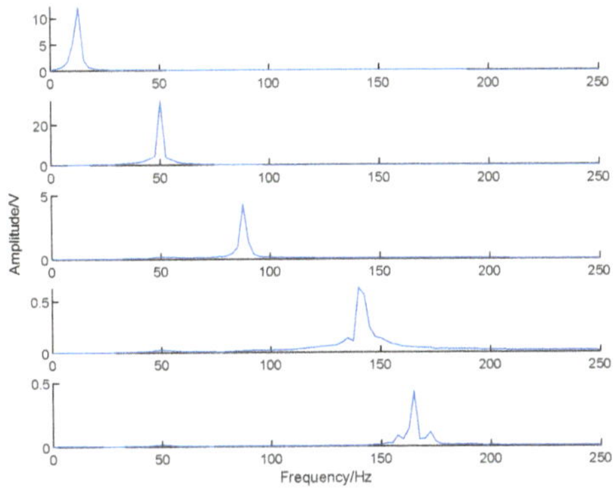

Figure 11. Intrinsic Mode Functions spectrum.

Table 6. Energy coefficients for each modal component of the simulation signal.

Modal Component	IMF_1	IMF_2	IMF_3	IMF_4	IMF_5
Energy coefficient	0.4016	1	0.1654	0.0319	0.0169

In order to test the denoising effect of the method presented in this paper, SNR and root mean square error (RMSE) are used as evaluation indexes in Table 7. SNR reflects the noise content in the signal. The denoising effect is proportional to the SNR. RMSE reflects the restoration degree of the denoised signal. RMSE decreases as the deviation between the denoised signal and the noiseless original signal decreases.

Table 7. Denoising effect of different algorithms.

	db2 Wavelet Hard Threshold	Proposed Method
SNR/dB	28.6656	37.6178
RMSE	1.0277	0.3719

According to the evaluation index results in Table 8, the denoising ability of db2 wavelet hard threshold method and the method in this paper is compared and analyzed. It can be seen that the denoising effect of the proposed method is better than that of the db2 wavelet hard threshold method in terms of SNR and RMSE.

Table 8. The fitting effect of different algorithms.

Evaluation Index	Order of Model	SNR/dB
Original signal	47	37.7888
Reconstructed signal	47	38.4783
Wavelet signal	47	31.0155

6.2. Test Performance Analysis

After signal reconstruction, in order to verify the effectiveness of the proposed method, the Prony algorithm is used to carry out signal fitting and harmonic identification for the original signal and reconstructed signal, respectively. The fitting effect is shown in Figure 12.

As can be seen from Figure 12, the trend of the two curves is almost the same, indicating that the proposed method and the original Prony method have almost the same data processing capability. This is because Figure 12 analyzes the simulated signal, and no artificial noise is added, so the noise content of the final signal is very low, which can also be seen from Figure 9.

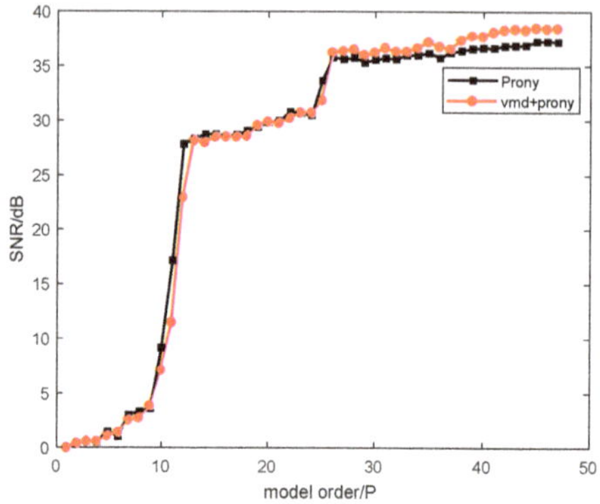

Figure 12. Comparison of fitting effect of simulated signal.

Under the condition of the same model order, the SNR results obtained by Prony analysis of different signals are shown in Table 8. The fitting SNR of reconstructed signal is 38.4783 dB, which is optimal. Under the condition that the model order is 47, the reconstructed signal and the original signal and the results denoised by the db2 wavelet hard threshold are identified. Results of identification are shown in Table 9.

Table 9. Identification results based on three methods for simulation data.

	Proposed Method				Prony				EMD+FFT	
	A/V	**f/Hz**	**φ/rad**	**α**	**A/V**	**f/Hz**	**φ/rad**	**α**	**A/V**	**f/Hz**
Value error (%)	36.5722 0.1488%	50 0	−1.0829 /	0.0003 /	36.5478 0.2154%	49.9985 0.003%	−1.0803 /	−0.0036 /	36.5154 0.3039%	49.9997 0.0006%
Value error (%)	14.5146 0.9339%	11.7642 1.6003%	0.4327 /	0.0033 /	14.7154 2.3303%	11.7692 1.73%	0.4301 /	−0.0467 /	14.6212 1.6752%	11.765 1.6072%
Value error (%)	4.9136 0.7071%	88.2317 0.2142%	−0.6053 /	−0.0251 /	4.939 1.2277%	88.2113 0.2373%	−0.5886 /	−0.0495 /	4.7846 1.9368%	88.227 0.2195%

It can be seen that the frequency and amplitude identification errors of the EMD+FFT are almost the same as those of the traditional Prony algorithm. However, the frequency and amplitude identification errors of the proposed method are smaller than those of the EMD+FFT and the traditional Prony algorithm. This method has the highest accuracy in harmonic identification.

Through the above simulation analysis of broadband oscillation, it can be seen that:

(1) Based on time-domain analysis, the Prony algorithm cannot only obtain the frequency and amplitude obtained by FFT transformation, but can also obtain time-domain information such as the damping factor;

(2) The VMD-Prony algorithm, proposed in this paper, is more accurate than EMD+FFT and the traditional Prony method. The effectiveness of the proposed algorithm is verified by comparing it with the traditional Prony algorithm.

In summary, the proposed method can obtain more time domain analysis results than FFT, such as the damping factor. Compared with the traditional Prony algorithm, the proposed method uses VMD denoising to solve the problem that the model order is too large, so that the identification accuracy is less affected by the model order. VMD denoising is better than the Kalman filter to solve the problem of insufficient precision of a nonlinear model. The denoising effect based on VMD is better than that based on EMD+FFT. The proposed method significantly reduces the influence of noise on signal identification. Therefore, the method in this paper is superior in engineering practicability.

7. Conclusions

It is prone to induce widebroad oscillation when new energy with high permeability is connected to a weak regional power grid. The accurate identification of the wide-frequency oscillation mode is the prerequisite to solve the resonant operation risk of new energy station. This paper proposes a new method for wind farm wide-band oscillation mode identification based on the combination of VMD and Prony, and focuses on the parameter optimization, noise removal and performance detection of VMD in the algorithm. The research conclusions are as follows:

(1) The number of modes K and penalty factor α in the VMD algorithm have great influence on the algorithm, which can be augmented by particle swarm optimization.
(2) Noise signals in measured data are likely to lead to poor analysis effect of the Prony algorithm. In this paper, a method combining the VMD algorithm and the energy threshold idea is adopted to effectively eliminate noise interference in data and retain effective components in signals.
(3) In this paper, SNR and RMSE are used as evaluation indexes. The SNR reflects the size of noise content in a signal. The denoising effect is better with the increase in SNR. The RMSE reflects the restoration degree of the denoised signal. The smaller the RMSE, the smaller the deviation between the denoised signal and the noiseless original signal. In order to obtain accurate identification results, SNR is usually required to be greater than 20 dB.

The composition and distribution of noise are more complex in the actual environment, so further research on noise identification and removal for measured analysis can be carried out in the future. Furthermore, in Prony detection, when the fitting order is half of the number of sampled data points, the fitted SNR sudden drop phenomenon is still unclear, and the specific mechanism can be further studied in the next research.

Author Contributions: Conceptualization, C.G. and L.Y.; methodology, C.G., L.Y., B.C. and J.D. (Jianbo Dai); writing—original draft preparation, C.G. and L.Y.; writing—review and editing, J.D. (Jing Dai) and K.Y. All authors have read and agreed to the published version of the manuscript.

Funding: This research was funded by the National Natural Science Foundation of China (52367002) and Yunnan Province joint fund key project (202201BE070001-15).

Data Availability Statement: The raw/processed data required to reproduce these findings are contained within the article.

Acknowledgments: We would like to express our sincere gratitude to Kunming University of Science and Technology for their support.

Conflicts of Interest: Author Ke Yin was employed by the company Beijing Xinleineng Technology Co., Ltd., Chengdu Branch. The remaining authors declare that the research was conducted in the absence of any commercial or financial relationships that could be construed as a potential conflict of interest.

References

1. Bai, J.; Xin, S.; Liu, J.; Zheng, K. Roadmap of Realizing the High Penetration Renewable Energy in China. *Proc. CSEE* **2015**, *35*, 14.
2. Xie, X.; Liu, H.; He, J.; Liu, H.; Liu, W. On New Oscillation Issues of Power Systems. *Proc. CSEE* **2018**, *38*, 10.
3. Chen, G.; Li, M.; Xu, T.; Liu, M. Study on Technical Bottleneck of New Energy Development. *Proc. CSEE* **2017**, *37*, 1.
4. Ma, N.; Xie, X.; Tang, J.; Chen, L. Wide-area measurement and early warning system for wide-band oscillation in "double-high" power systems. *J. Tsinghua Univ. (Sci. Technol.)* **2021**, *61*, 5.
5. Zhang, D.; Wang, Y.; Hu, J.; Ma, S.; He, Q.; Guo, Q. Impacts of PLL on the DFIG-based WTG's electromechanical response under transient conditions: Analysis and modeling. *CSEE J. Power Energy Syst.* **2016**, *2*, 30–39. [CrossRef]
6. Tong, Y.; Yin, Y. *Multi-Scale Simulation and Test Technology of Power System*; China Electric Power Press: Beijing, China, 2013; pp. 184–188.
7. Ma, N.; Xie, X.; Kang, P.; Zhang, F. Wide-area Monitoring and Analysis of Subsynchronous Oscillation in Power Systems With High-penetration of Wind Power. *Proc. CSEE* **2021**, *41*, 1.
8. Ma, N.; Xie, X.; He, J.; Wang, H. Review of broadband Oscillation in Renewable and Power Electronics Highly Integrated Power Systems. *Proc. CSEE* **2020**, *40*, 15.

9. Wang, W.; Zhu, Y.; Liu, C.; Dong, P.; Hu, T.; Li, B.; Li, Y.; He, F.; Zhang, Y. Realization of Electromagnetic Real-time Simulation of Large-scale Grid Based on HYPERSIM. *Power Syst. Technol.* **2019**, *43*, 4.
10. Zhang, F.; Cheng, L.; Gao, W.; Huang, R. Synchrophasors-based identification for subsynchronous oscillations in power systems. *IEEE Trans. Smart Grid* **2018**, *10*, 2224–2233. [CrossRef]
11. Yang, X.; Zhang, J.; Xie, X.; Xiao, X.; Gao, B.; Wang, Y. Interpolated DFT-based identification of sub-synchronous oscillation parameters using synchrophasor data. *IEEE Trans. Smart Grid* **2019**, *11*, 2662–2675. [CrossRef]
12. Xu, Y.; Liu, H.; Cheng, Y. Key Influencing Factors on Propagation of Sub-Synchronous Oscillations in AC and DC Grids. *Mod. Electr. Power* **2022**, 1–11. [CrossRef]
13. Mortensen, A.N.; Johnson, G.L. A power system digital harmonic analyzer. *IEEE Trans. Instrum. Meas.* **1988**, *37*, 537–540. [CrossRef]
14. Yu, Y.; Zhao, W.; Li, S.; Huang, S. A Two-Stage Wavelet Decomposition Method for Instantaneous Power Quality Indices Estimation Considering Interharmonics and Transient Disturbances. *IEEE Trans. Instrum. Meas.* **2021**, *70*, 9001813. [CrossRef]
15. Dragomiretskiy, K.; Zosso, D. Variational Mode Decomposition. *IEEE Trans. Signal Process.* **2014**, *62*, 531–544. [CrossRef]
16. Riaz, F.; Hassan, A.; Rehman, S.; Niazi, I.K.; Dremstrup, K. EMD-Based Temporal and Spectral Features for the Classification of EEG Signals Using Supervised Learning. *IEEE Trans. Neural Syst. Rehabil. Eng.* **2016**, *24*, 28–35. [CrossRef] [PubMed]
17. Laila, D.S.; Messina, A.R.; Pal, B.C. A Refined Hilbert–Huang Transform with Applications to Interarea Oscillation Monitoring. *IEEE Trans. Power Syst.* **2009**, *24*, 610–620. [CrossRef]
18. Chen, K.; Jin, T.; Mohamed, M.A.; Wang, M. An Adaptive TLS-ESPRIT Algorithm Based on an S-G Filter for Analysis of Low Frequency Oscillation in Wide Area Measurement Systems. *IEEE Access* **2019**, *7*, 47644–47654. [CrossRef]
19. Feng, S.; Cui, H.; Chen, J.; Tang, Y.; Lei, J. Applications and Challenges of Artificial Intelligence in Power System broadband Oscillation. *Proc. CSEE* **2021**, *41*, 23.
20. Mo, W.; Lv, J.; Pawlak, M.; Annakkage, U.D.; Chen, H. Power System Oscillation Mode Prediction Based on the Lasso Method. *IEEE Access* **2020**, *8*, 101068–101078. [CrossRef]
21. He, J.; Luo, G.; Cheng, M.; Liu, Y.; Tan, Y.; Li, M. A Research Review on Application of Artificial Intelligence in Power System Fault Analysis and Location. *Proc. CSEE* **2020**, *40*, 17.
22. Ding, R.; Shen, Z. Modal Identification of Low-frequency Oscillation in Power System based on EMO-EDSNN. *Autom. Electr. Power Syst.* **2020**, *44*, 122–131.
23. Gupta, A.K.; Verma, K. PMU-ANN based real timemonitoring of power system electromechanical oscillations. In Proceedings of the 2016 IEEE 1st International Conference on Power Electronics, Intelligent Control and Energy Systems (ICPEICES), Delhi, India, 4–6 July 2016; pp. 1–6.
24. Chen, C.I.; Chang, G.W. Virtual instrumentation and educational platform for time-varying harmonics and interharmonics detection. *IEEE Trans. Ind. Electron.* **2010**, *57*, 3334–3342. [CrossRef]
25. Hauer, F. Application of Prony analysis to the determination of modal content and equivalent models for measured power system response. *IEEE Trans. Power Syst.* **1991**, *6*, 1062–1068. [CrossRef]
26. Pantaleon, C.; Souto, A. Comments on "An aperiodic phenomenon of the extended Kalman filter in filtering noisy chaotic signals". *IEEE Trans. Signal Process.* **2005**, *53*, 383–384. [CrossRef]
27. Smith, C.B.; Agaian, S.; Akopian, D. A Wavelet-Denoising Approach Using Polynomial Threshold Operators. *IEEE Signal Process. Lett.* **2008**, *15*, 906–909. [CrossRef]
28. Janik, P.; Rezmer, J.; Ruczewski, P.; Waclawek, Z.; Lobos, T. Adaptation of SVD and Prony method for precise computation of current components in networks with wind generation. In Proceedings of the 2009 International Conference on Clean Electrical Power, Capri, Italy, 9–11 June 2009; pp. 624–629. [CrossRef]
29. Li, Y.; Liang, X.; Yang, Y.; Xu, M.; Huang, W. Early Fault Diagnosis of Rotating Machinery by Combining Differential Rational Spline-Based LMD and K–L Divergence. *IEEE Trans. Instrum. Meas.* **2017**, *66*, 3077–3090. [CrossRef]
30. Wang, R.; Huang, W.; Hu, B.; Du, Q.; Guo, X. Harmonic Detection for Active Power Filter Based on Two-Step Improved EEMD. *IEEE Trans. Instrum. Meas.* **2022**, *71*, 9001510. [CrossRef]
31. Zhang, Z.; Tan, Z.; Zhang, C.; Wang, X.; Liu, X.; Yu, Y. Speech Endpoint Detection Based on Bayseian Decision of Logarithmic Power Spectrum Ratio in High and Low Frequency Band. *Comput. Sci.* **2021**, *48*, 6A.

symmetry

MDPI

Article

Adaptive Virtual Inertia Control Strategy for a Grid-Connected Converter of DC Microgrid Based on an Improved Model Prediction

Feng Zheng, Minghong Su, Baojin Liu * and Wanling Liu

College of Electrical Engineering and Automation, Fuzhou University, Fuzhou 350100, China
* Correspondence: lbj@fzu.edu.cn

Abstract: Aiming at the problem that the bus voltage in a low-inertia DC microgrid is prone to be affected by internal power fluctuations, an adaptive virtual inertia control strategy for a grid-connected converter of a DC microgrid based on an improved model prediction is proposed. Firstly, the adaptive analog virtual synchronous generator (AVSG) is introduced into the voltage outer loop by combining the inertial parameters with the voltage change rate, and the flexible adjustment of the inertial parameters is realized. Secondly, the improved model predictive control is introduced into the current inner loop to realize the fast-tracking of the given current value and improve the dynamic characteristics of the control system. Finally, a system model is established based on Matlab/Simulink for simulation. The results show that compared with the traditional virtual inertia control strategy, the proposed control strategy has smaller bus voltage fluctuation amplitude and better dynamic performance; when a 10 kW load mutation occurs, the magnitude of bus voltage drop is reduced by 60%, and the voltage recovery time is shortened by 30%. The proposed control strategy can effectively improve the stability of DC bus voltage and the operation ability of the system under asymmetric conditions.

Keywords: DC microgrid; bidirectional grid-connected converter; adaptive virtual inertia control; model predictive control; dynamic performance

1. Introduction

In order to cope with the increasing depletion of traditional fossil energy sources and the resulting global environmental pollution, new energy sources represented by solar and wind power have been utilized on a large scale, and the proportion of renewable energy sources in the distribution network has been increasing [1–3]. With the development of power electronics technology, the converter-based DC microgrid has received widespread attention due to the high efficiency of system operation and the absence of frequency and rotor angle stability problems compared to the AC microgrid [4,5]. The DC microgrid is a converter-driven low-inertia system; DC bus voltage is very sensitive to power fluctuations of an intermittent power supply and load; when a disturbance caused by power fluctuation of the power supply and load occurs in the network, it will cause the fluctuation of the DC bus voltage, which is harmful to the stable operation of the DC microgrid [6,7]. The DC microgrid is connected to the grid through a bidirectional grid-connected converter (BGC). The BGC controls the energy exchange between the DC microgrid and the AC grid and plays a key role in stabilizing the DC bus voltage [8,9].

To enhance the inertia of the DC microgrid, some scholars have proposed virtual inertia control strategies applicable to the DC microgrid. At present, the virtual inertia control strategies for the DC microgrid are mainly divided into the following three main categories: variable droop coefficient control, additional differential inertia control, and AVSG control [10–15]. In literature [16], a variable droop coefficient control method is proposed. The droop coefficient is changed according to the voltage change rate of the DC bus, which changes the inertia margin of the microgrid. However, the introduction of the voltage differential term may lead to a decrease in system stability. In literature [17], by adding an additional virtual inertia control loop in the control loop, the released power of the controllable power supply is adjusted to provide inertia support for the microgrid. However, the introduction of a high-pass filter in the control loop will bring high-frequency interference. Literature [18] proposed AVSG control for a DC microgrid by analogy with a virtual synchronous generator (VSG) in an AC microgrid, and this control strategy can make the BGC provide additional power quickly when a power difference occurs in the DC microgrid, effectively suppressing the fluctuation of the DC bus voltage and enhancing the inertia of the DC microgrid. However, the inertia parameters in the control strategy are constant and cannot be adaptively adjusted according to the system dynamics.

In the virtual inertial control strategies described above, the current inner loop still uses the traditional PI control, but for nonlinear systems, it is difficult to obtain good dynamic performance by using PI control. In order to obtain better dynamic performance, nonlinear algorithms can be considered in the virtual inertial control strategy. Literature [19] introduced passivity-based control in the virtual inertial control strategy, verified the passivity of the BGC, and designed the current inner loop passive controller from the perspective of system energy to achieve good tracking of the expected current value, but the tracking effect of passive control is not good when the system model and parameters change due to interference. Literature [20] proposed a second-order sliding mode control of a grid-connected power converter. The controller adopts a cascade structure composed of two controllers. The current inner loop uses a sliding mode algorithm to track the actual current value to the expected value. Sliding mode control is not easily affected by system model and parameter changes and, at the same time, can effectively improve the dynamic performance of the DC microgrid system. However, the introduction of the sliding mode algorithm may lead to the problem of sliding mode chattering.

With the development of computer control technology, model predictive control (MPC) was proposed and applied in industrial practice. MPC can predict and calculate the output of the system at the next moment according to the state variables of the current system. The actual output of the system will track the given reference value with good transient steady-state characteristics. Literature [21] proposed a virtual inertial control strategy based on model prediction for DC microgrid battery systems, which can provide inertial support during the transient period and enhance the dynamic characteristics of DC bus voltage. Literature [22] proposed a VSG control strategy based on MPC for isolated island microgrids. The MPC control strategy replaces the traditional voltage-current double loop control, which eliminates the parameter setting and improves the dynamic response of the system. However, the proposed control strategy is aimed at the AC microgrid and cannot be applied to the DC microgrid.

Based on the shortcomings of the above studies, in order to improve the dynamic performance of the system and enhance the inertia of the DC microgrid, this paper proposes an adaptive virtual inertia control strategy for the DC microgrid grid-connected converter based on an improved model prediction. The main contributions are as follows:

(1) The traditional droop control is a non-inertial control method, which cannot provide inertial support for the DC microgrid. In order to solve this problem, an adaptive AVSG control is introduced into the voltage outer loop of the control strategy. The inertia parameter can be adjusted according to the voltage change rate, which improves the stability of DC bus voltage and the operation ability of the system under asymmetric conditions.

(2) In order to improve the dynamic response of the control system, an improved model predictive control is introduced into the current inner loop of the control strategy, which eliminates the parameter setting, eliminates the traditional PI controller and PWM regulator, realizes the fast-tracking of the given current value, and improves the dynamic characteristics of the system.

The rest of this paper is organized as follows. The topology of the DC microgrid and its traditional control strategy are presented in Section 2. The adaptive virtual inertia control strategy based on model prediction is introduced in Section 3. Section 4 analyzes the stability of the proposed control strategy. The proposed strategy is validated in Section 5, and the conclusion is given in Section 6.

2. DC Microgrid Topology and Control Strategy

2.1. DC Microgrid Topology

The topology of the DC microgrid studied in this paper is shown in Figure 1, including the PV unit, energy storage unit, AC main network, DC constant power load, and corresponding power electronic converter. BGC is responsible for balancing the power in the grid when the DC microgrid is in the grid-connected mode. The energy storage unit is connected to the DC bus through a bidirectional DC–DC converter (BDC) to realize bidirectional energy flow with the DC microgrid. The PV is connected to the DC bus through a boost converter, using maximum power point tracking control. The DC constant power load is connected to the DC bus through a buck converter, using maximum power point tracking control.

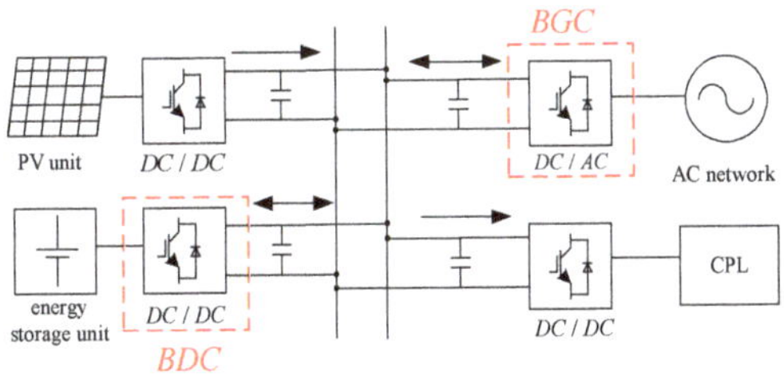

Figure 1. DC microgrid topology.

The focus of this paper is to analyze the operation control strategy of BGC in the grid-connected mode of the DC microgrid, so the DC microgrid is simplified into an equivalent topology containing BGC, as shown in Figure 2. In the equivalent topology, the PV unit, the energy storage unit, and their corresponding power electronic converters can be equivalent to current sources [18]. In Figure 2, $V_1 \sim V_6$ are the six IGBTs, $u_k(k = a, b, c)$ is the phase voltage of the three-phase AC supply, $e_k(k = a, b, c)$ is the AC side voltage of the grid-connected converter, $i_k(k = a, b, c)$ is the three-phase line current, u_{bus} is the DC bus voltage, i_{dc} is the DC side current, i_0 is the DC side output current, i_R is the current flowing through the load, L is the inductance of the filter reactor, R_L is the resistance of the filter reactor, C is the DC side capacitor, and R is the constant power load.

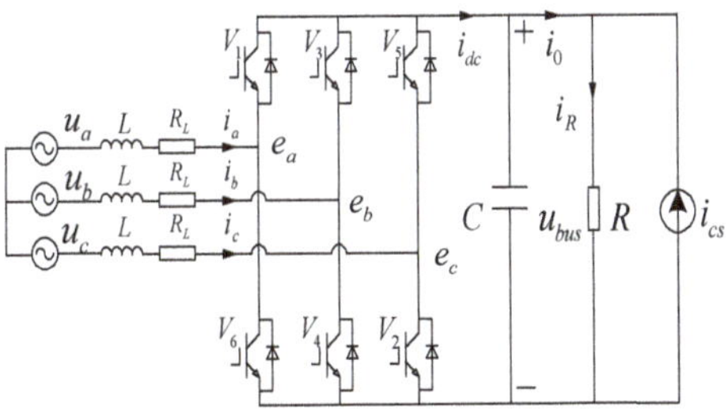

Figure 2. Equivalent topology of DC microgrid with BGC.

2.2. Traditional Control Strategy of DC Microgrid

Among many coordinated control methods of the DC microgrid system, one of the more commonly used traditional control strategies is droop control, and the typical voltage-current droop control equation is as follows:

$$u_{dc}^{*} = U_0 - ki_0 \tag{1}$$

where u_{dc}^{*} is the reference value of the DC bus voltage. U_0 is the reference value of the DC side voltage when BGC is not loaded; k is the droop coefficient, which measures the relationship between voltage and current.

The specific control block diagram is shown in Figure 3. The voltage reference value u_{dc}^{*} is calculated by U_0 and i_0 using the droop curve. Then, the voltage outer loop control is applied to obtain the current reference value, which is further utilized in the current inner loop control to generate the control signals for the converter. In practical applications, the traditional droop control has defects such as long transient response time, large response overshoot, and large output voltage deviation. In addition, the traditional droop control is an inertia-free control method, which cannot provide inertia and damping for the DC microgrid to improve the system stability.

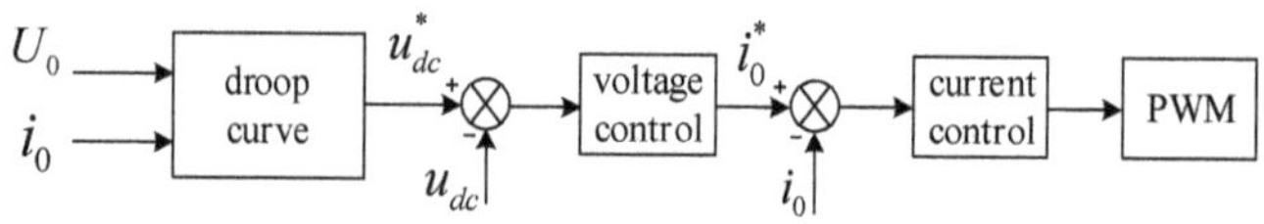

Figure 3. Diagram of droop control.

3. Adaptive Virtual Inertia Control Strategy

3.1. Voltage Outer Loop

In an AC microgrid, the VSG control technique enables the inverter to simulate inertia, droop, and damping characteristics similar to the synchronous generator by introducing virtual inertia and damping. Assuming that the number of poles of the synchronous generator is 1, the mechanical equations of VSG can be expressed as:

$$P_{set} - P_e - D_p(\omega - \omega_n) = J\omega\frac{d\omega}{dt} \approx J\omega_n\frac{d\omega}{dt} \tag{2}$$

where: P_{set} and P_e are the active power given and electromagnetic power respectively; D_p is the frequency damping factor, which describes the change in active power output of a VSG in response to a unit change in frequency; ω and ω_n are the angular frequency of VSG and the rated angular frequency of the grid respectively; J is the virtual rotational inertia.

For an AC power grid, its inertia can be reflected in both voltage stability and frequency stability. Reactive power control is a commonly used method for voltage control, while VSG control simulates the behavior of a traditional synchronous generator to achieve frequency regulation; when the frequency of the AC grid changes abruptly, the grid exhibits a large inertia due to the existence of J, and the VSG is able to quickly adjust the active output to achieve grid frequency support. As for the DC microgrid, the inertia is expressed as the ability of the system to hinder the sudden change of DC bus voltage. In the DC microgrid, the electric energy stored by the DC bus capacitor can be used to hinder the sudden change of the DC bus voltage. The stored electric energy can be expressed as:

$$W_c = \frac{1}{2}Cu_{bus}^2 \tag{3}$$

However, the DC microgrid is a small inertia system connected by a power electronic converter. When disturbances caused by intermittent power supply and load power fluctuations occur in the network, the DC bus voltage will fluctuate, which will adversely affect the safe and reliable operation. The power stored in the DC bus capacitor alone cannot suppress this fluctuation.

The AC microgrid and DC microgrid have many control strategy variables that can correspond to each other, as shown in Table A1. Therefore, according to the correspondence of variables, the AVSG control equation suitable for the DC microgrid can be obtained by analogy reasoning:

$$i_{set} - i_0 - D_v(u^*_{bus} - u_n) = C_v u^*_{bus}\frac{du^*_{bus}}{dt} \approx C_v u_n \frac{du^*_{bus}}{dt} \tag{4}$$

where: i_{set} is the given value of BGC output current; D_v is the voltage damping coefficient, which characterizes the ability of BGC to dampen the oscillations in the DC bus voltage; u_n is the DC side voltage rating value; u^*_{bus} is the DC voltage reference value of the virtual inertial outer loop output; C_v is the virtual capacitance value.

When the DC bus voltage changes abruptly, the DC microgrid exhibits great inertia due to the existence of C_v, and the AVSG can quickly regulate the DC side output current, thus suppressing the DC bus voltage fluctuations. After adopting the AVSG control strategy, BGC virtualizes a virtual capacitor on the DC side that is larger than the actual capacitance value, thereby providing inertial support for the DC microgrid.

In the traditional AVSG control, the virtual capacitance and virtual voltage damping parameters are fixed values. The system can obtain a large inertia, but its inertia parameters are constant and cannot be adaptively adjusted according to the system dynamics. For a DC microgrid with AVSG control, when a large power disturbance of intermittent power or load occurs, it will cause a large bus voltage change rate, and then the system is expected to generate a large virtual inertia to reduce the voltage fluctuation. When the large disturbance is over, the system power reaches balance, and the voltage change rate is small, then the system is expected to generate a small virtual inertia to achieve a fast recovery of voltage.

To meet the above requirements, the adaptive virtual inertia control strategy is considered to combine the inertia coefficient in AVSG with the voltage variation rate to achieve flexible and adjustable inertia parameters. The virtual inertia expression designed in this paper is:

$$C_V = \begin{cases} C_{vo}, & \left|\frac{dU_{dc}}{dt}\right| < M_0 \\ C_{vo} + k_1\frac{dU_{dc}}{dt}, & M_0 \leqslant \left|\frac{dU_{dc}}{dt}\right| < M_1 \\ C_{vo} + k_2\left|\frac{dU_{dc}}{dt}\right|k_3, & \left|\frac{dU_{dc}}{dt}\right| \geqslant M_1 \end{cases} \tag{5}$$

where: C_{vo} is the virtual capacitance value under the stable state of system voltage; dU_{dc}/dt is the voltage change rate ; M_0, M_1 is the voltage threshold when the virtual capacitance value changes ; k_1, k_2, k_3 are related control parameters for flexible adjustment of virtual inertia.

When $|dU_{dc}/dt| < M_0$, C_v is equal to the fixed value C_{vo}, which avoids frequent switching of virtual capacitance and maintains the normal operation of the system. When $|dU_{dc}/dt| > M_0$, C_v is an expression containing dU_{dc}/dt. In this case, C_v is changed to adjust the BGC output current, and then the inertia of the DC microgrid is adjusted, which prevents the sudden rise or fall of the DC bus voltage when the DC microgrid is affected by impulse disturbance.

In the selection of parameter k_1, k_1 is the adjustment coefficient when the capacitance value changes linearly. In this case, the voltage change rate is not large, and the virtual inertia is expected to be small, so the smaller k_1 can be selected. In the selection of parameters k_2 and k_3, when the system is subjected to large disturbance, the voltage change rate is large, and the virtual inertia is expected to be large. Therefore, the virtual capacitance value should be increased by increasing the value of k_3 on the premise of ensuring the stability of the system. Otherwise, the value of k_3 can be appropriately reduced. After selecting the appropriate value of k_3, k_2 are selected according to the stability requirements of the system. Figure 4 shows the influence of the change of k_2 andk_3 on the value of virtual capacitance. As can be seen from Figure 5, an increase of k_2 increases the value of C_v during the transient process. For k_3, when the DC voltage change rate $|dU_{dc}/dt| < 1$, the value of C_v in the transient process will gradually decrease with the increase of k_3. When $|dU_{dc}/dt| > 1$, the value of C_v gradually increases with the increase of k_3. With the increase of exponential coefficient k_3, the change rate and range of C_v along with $|dU_{dc}/dt|$ also increase, indicating that k_3 mainly affects the change rate of C_v.

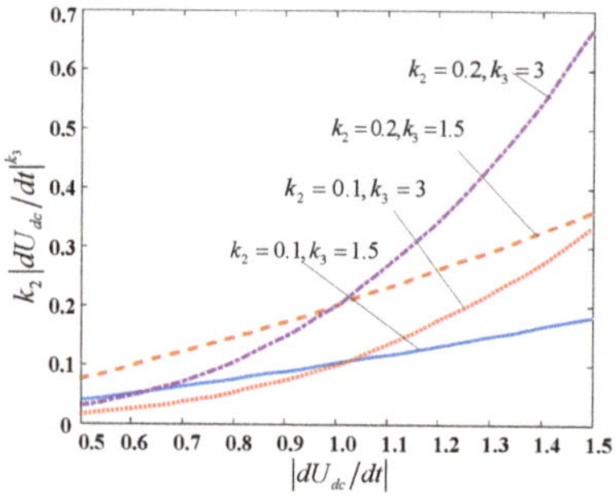

Figure 4. The influence of the change of k_2 and k_3 on the value of virtual capacitance.

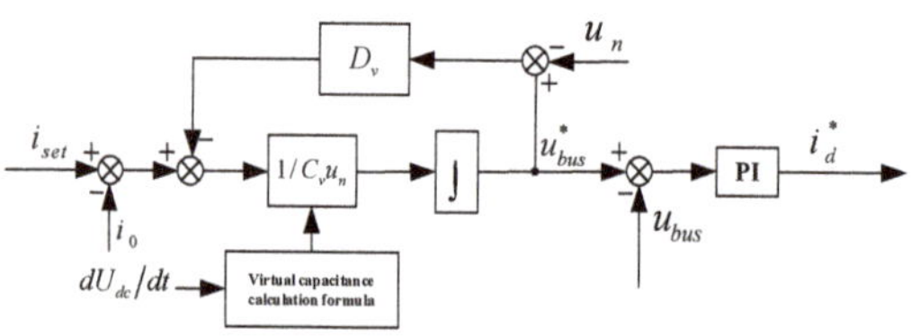

Figure 5. The outer loop control block diagram.

After introducing adaptive virtual inertia, the outer loop control block diagram shown in Figure 5 can be obtained.

3.2. Current Inner Loop

(1) Principle of current model predictive control

BGC often adopts the voltage and current double closed-loop control structure based on PI control. However, the traditional PI-based current inner loop control uses the measured current value for hysteresis regulation, which makes it difficult to obtain better dynamic performance. Therefore, this paper designs an adaptive virtual inertia strategy for a grid-connected converter based on improved model prediction.

For the topology of a two-level, three-phase BGC, the value of each switch is either 0 or 1. In order to avoid direct conduction of each bridge arm switch, it is common to design each bridge arm with two complementary switches, resulting in a total of $2^3 = 8$ effective switch combinations for the inverter. The system structure diagram of the Finite Control Set Model Predictive Control (FCS-MPC) is shown in Figure 6.

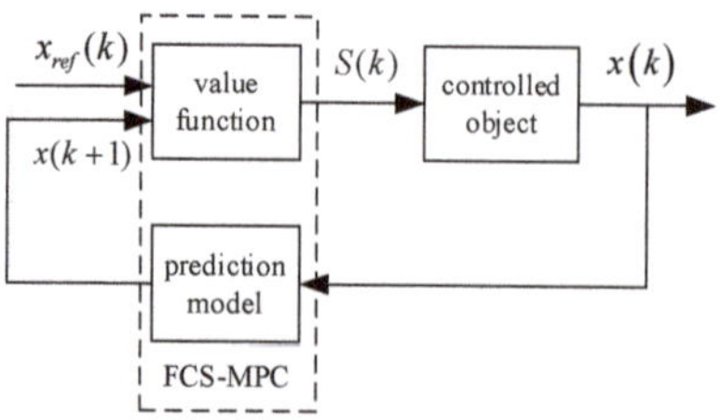

Figure 6. Structure diagram of FCS-MPC.

In Figure 6, $x_{ref}(k)$ represents the reference output signal, $S(k)$ denotes the switching sequence triggered by the inverter during the interval from k to $k+1$, and $x(k+1)$ corresponds to the output signal of the predictive model, which represents the alternative predicted output signal. The specific algorithmic process of FCS-MPC can be described as follows: at time k, the sampling module measures the output state of the inverter and feeds the measured electrical quantity information into the predictive model. Based on the acquired electrical quantities, the predictive model calculates all the alternative predicted output signals $x_i(k+1)$ at time $k+1$ $(i = 0,1,2,3,4,5,6,7)$. The value function compares each alternative predicted output signal with the reference signal and selects the switching sequence corresponding to the alternative predicted output signal that minimizes the value function for triggering.

(2) Prediction model of BGC

Assume that the three-phase power supply on the grid side is symmetrical, the switching devices are ideal devices, and define the switching function as:

$$S_k = \begin{cases} 1, the\ upper\ bridge\ arm\ is\ open \\ 0, the\ upper\ bridge\ arm\ is\ cloesd \end{cases} \tag{6}$$

where $k = a, b, c$. According to the topology diagram of BGC in Figure 2, the mathematical model of BGC under three-phase can be obtained as:

$$\begin{cases} L\frac{di_a}{dt} = e_a - R_L i_a - u_a \\ L\frac{di_b}{dt} = e_b - R_L i_b - u_b \\ L\frac{di_c}{dt} = e_c - R_L i_c - u_c \\ C\frac{du_{bus}}{dt} = S_a i_a + S_b i_b + S_c i_c - i_0 \end{cases} \tag{7}$$

After the $abc/\alpha\beta$ coordinate transformation, the two-phase decoupling expression of BGC is:

$$\begin{cases} L\frac{di_\alpha}{dt} = e_\alpha - u_\alpha - R_L i_\alpha \\ L\frac{di_\beta}{dt} = e_\beta - u_\beta - R_L i_\beta \\ C\frac{du_{bus}}{dt} = S_\alpha i_\alpha + S_\beta i_\beta - i_0 \end{cases} \tag{8}$$

where, e_α and e_β are the voltage components of three-phase grid voltages e_a, e_b and e_c in axis $\alpha\beta$, i_α and i_β are the current components of three-phase grid-connected currents i_a, i_b and i_c in axis $\alpha\beta$, u_α and u_β are the voltage components of AC side voltages u_a, u_b and $u_c$in axis $\alpha\beta$, S_α and S_β are the voltage components of the switching function in the axis $\alpha\beta$. In order to establish the prediction model of BGC, the continuous model shown in Equation (7) is discretized at time k and $k+1$, and collated as follows:

$$\begin{cases} i_\alpha(k+1) = \frac{T_s(e_\alpha(k) - u_\alpha(k) - R_L i_\alpha(k))}{L} + i_\alpha(k) \\ i_\beta(k+1) = \frac{T_s(e_\beta(k) - u_\beta(k) - R_L i_\beta(k))}{L} + i_\beta(k) \end{cases} \tag{9}$$

The voltage relationship between the AC and DC sides of BGC is:

$$\begin{cases} u_\alpha = \sqrt{\frac{2}{3}}(S_a - \frac{1}{2}S_b - \frac{1}{2}S_c)u_{dc} \\ u_\beta = \frac{\sqrt{2}}{2}(S_b - S_c)u_{dc} \end{cases} \tag{10}$$

where T_s is the sampling period. Since there are eight switching states of BGC, the values of u_α and u_β for each state can be obtained according to Equation (10), as shown in Table A2. By substituting the values of u_α and u_β from Table A2 into Equation (9), the values of $i_\alpha(k+1)$ and $i_\beta(k+1)$ at moment k+1 can be obtained. Taking the input current of the converter as the control object, the expression of the value function is:

$$\begin{cases} f_{i1} = (i_\alpha{}^* - i_{\alpha 1}(k+1))2 + (i_\beta{}^* - i_{\beta 1}(k+1))2 \\ f_{i2} = (i_\alpha{}^* - i_{\alpha 2}(k+1))2 + (i_\beta{}^* - i_{\beta 2}(k+1))2 \\ \vdots \\ f_{i8} = (i_\alpha{}^* - i_{\alpha 8}(k+1))2 + (i_\beta{}^* - i_{\beta 8}(k+1))2 \end{cases} \tag{11}$$

where $i_\alpha{}^*$ and $i_\beta{}^*$ are the reference currents. By comparing the 8 groups of function values obtained by Equation (11) and taking the minimum value $f_{\min}$, a group of switch sequences S_a, S_b and S_c satisfying the minimum value of the value function are applied to the converter. The current inner loop control block diagram can be obtained as shown in Figure 7.

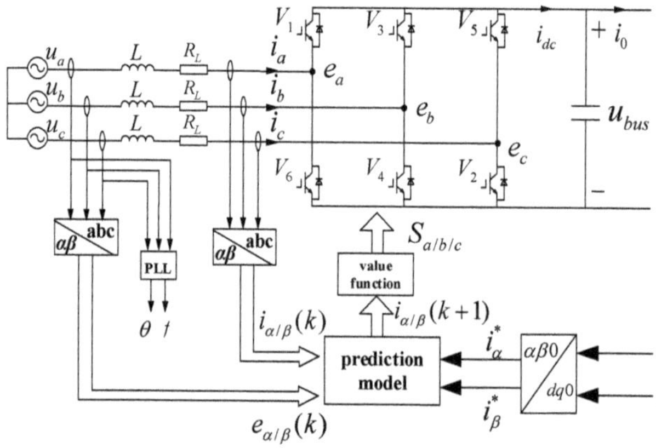

Figure 7. The current inner loop control block diagram.

(3) Improved model predictive control

Control delay is a major issue faced by predictive control methods. In practical applications, predictive control algorithms require extensive computations, which inevitably introduce time delays between inputs and optimization drives. The control delay can result in errors in the prediction of the next time step. When considering the delay issue without compensation, the operational flowchart of the FCS-MPC algorithm is depicted in Figure 8a. At time k, the system samples and performs calculations to predict the output value at time $k+1$. Based on the value function, a switching sequence is selected, and the system outputs switch signals to drive the inverter at time k'. Similarly, at time $k+1$, the system samples and performs calculations to predict the output value at time $k+2$. The switching sequence is selected based on the value function, and the system outputs switch signals to drive the inverter at time $k+1'$. After the above analysis, the FCS-MPC algorithm continues to employ the switch sequence implemented in the previous control cycle before the output switch sequence. For instance, during the period from $k+1$ to $k+1'$, the switch sequence S_2 used is the optimal switch sequence employed during the period from k to $k+1$, and so forth. Consequently, this leads to deviations of the inverter's output voltage from the reference voltage, resulting in a deterioration of the quality of the inverter's output power.

In order to address the aforementioned issues, this study improves the traditional algorithm based on the characteristic of multi-step prediction in MPC. The improved algorithm modifies the logical sequence of algorithm implementation by triggering the optimal switching sequence at the beginning of each control cycle. The specific algorithm process is illustrated in Figure 8b.

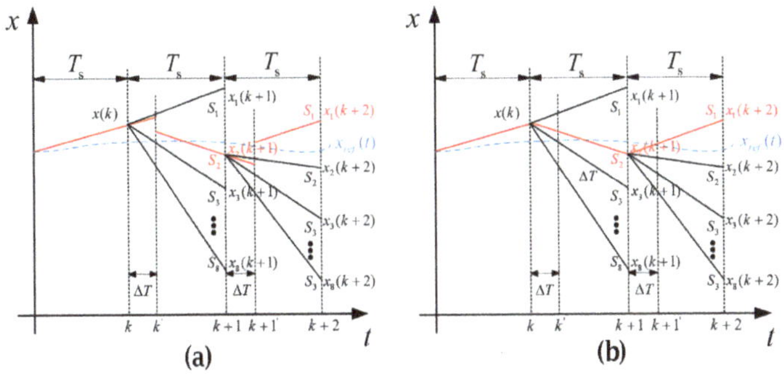

Figure 8. The running process of FCS-MPC algorithm. (**a**) shows the running process of FCS-MPC algorithm considering delay but without compensating; (**b**) shows the running process of FCS-MPC algorithm considering delay and compensating.

The specific algorithm process is as follows:

Step 1: Implement the optimal switching sequence S_2 obtained from the previous control cycle calculation at time k.

Step 2: Measure the state vector and disturbance vector of the inverter at time k and predict the optimal output vector $x_2(k+1)$ at time $k+1$ based on the input vector corresponding to the switching sequence S_2.

Step 3: Starting from $x_2(k+1)$, predict the eight sets of candidate output vectors for time $k+2$ based on the predictive model. Evaluate the candidate output vectors using the cost function and select the switching sequence S_1 that minimizes the cost function.

Step 4: Implement the optimal switching sequence S_1 at time $k+1$.

From the above algorithmic process, it can be observed that by changing the triggering logic of the optimal switching sequence, the switching sequence implemented from $k+1$ to $k+2$ is computed at k to $k+1$. This alignment of the switching sequence's action period with the prediction period of the predictive model reduces the issue of poor-quality inverter output power caused by delay problems. In order to visually illustrate the operation process of the delay compensation algorithm, this paper presents a flowchart of the FCS-MPC algorithm considering delay compensation, as shown in Figure 9.

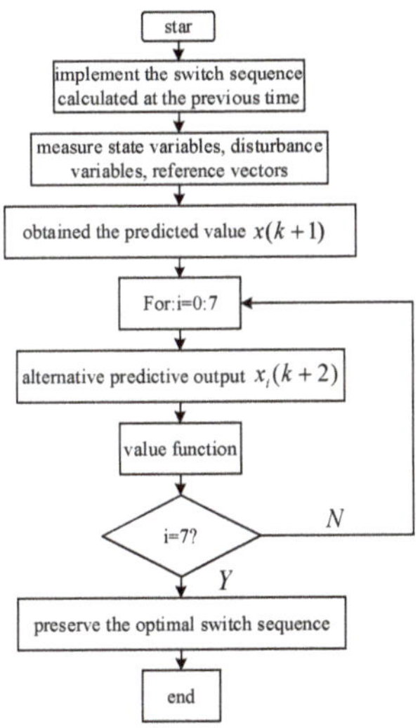

Figure 9. Flowchart of the FCS-MPC algorithm considering delay compensation.

3.3. Inertial Control Strategy under DC Inter-Pole Fault

One of the most severe faults in a DC system is the inter-pole short circuit fault in the DC line, as shown in Figure 10. When an inter-pole short circuit fault occurs in the DC line, the DC-side capacitors discharge rapidly, and the fault current reaches its peak within a few milliseconds. When the voltage across the fault-side capacitor drops to zero and starts to reverse charge, it can easily lead to the damage of reverse-parallel freewheeling diodes and capacitors. If virtual inertia control is considered and the inertia of the DC system is increased, the time for the capacitor voltage to drop to zero during a fault is delayed, and the peak fault current is reduced. This provides more response time for fault detection, protection actions, and reduces the impact of faults on grid operation.

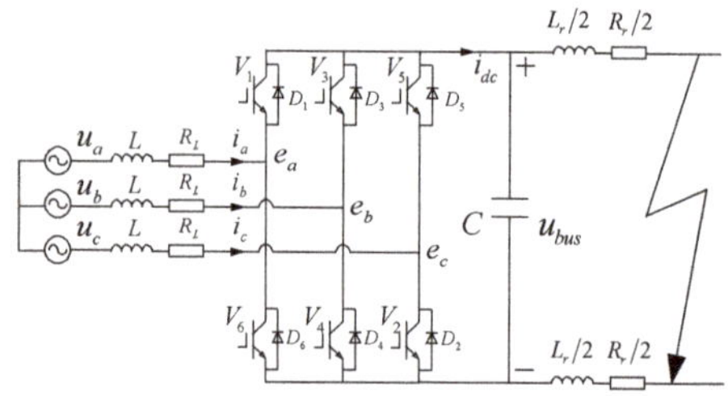

Figure 10. Equivalent circuit of inter-pole short circuit.

According to the circuit response characteristics of the inter-pole short circuit fault in the DC line, the fault process can be divided into three stages: the DC-side capacitor discharge stage, the uncontrolled rectifier initial stage, and the uncontrolled rectifier steady-state stage. Figure 11 shows the equivalent circuit diagrams for each stage.

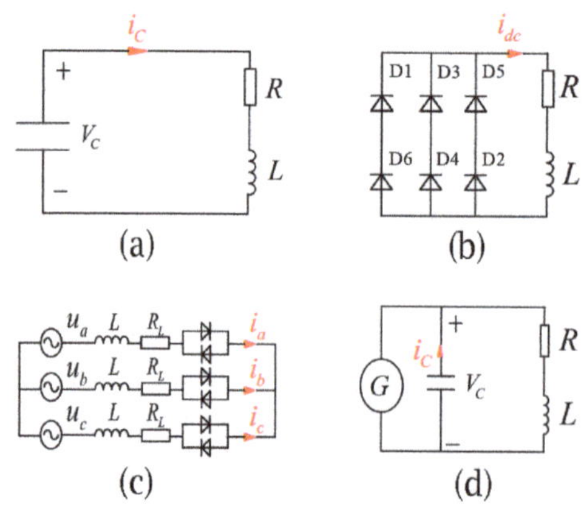

Figure 11. The equivalent circuit diagrams for each stage. (**a**) shows equivalent circuit of capacitor discharge stage; (**b**) shows DC side equivalent circuit of uncontrolled rectifier in initial stage; (**c**) shows AC side equivalent circuit of uncontrolled rectifier in initial stage; (**d**) shows equivalent circuit of uncontrolled rectifier in steady state stage.

After the occurrence of a fault, the short-circuit current provided by the DC side is significantly larger than that of the AC side. Neglecting the AC side's continuation of current flow, the short-circuit loop can be approximated as a second-order RLC discharge circuit composed of resistance, inductance, and DC-side capacitance, as shown in Figure 11a. As the capacitor continues to discharge, the capacitor voltage drops below the AC-side voltage, and the DC side inductance discharges towards the fault point through the circuit formed by the freewheeling diodes of the converter. When the DC voltage is smaller than the peak value of the AC line voltage, the fault circuit enters the uncontrolled rectifier stage, and both the AC side and the capacitor discharge towards the fault point. When the short-circuit impedance is small, the capacitor continues to discharge until the voltage reaches zero. At this moment, the DC side short-circuit reactance accumulates a large amount of energy, and the counter electromotive force on it causes the freewheeling diodes to conduct simultaneously as the capacitor voltage drops to zero, forming an RL first-order free discharge circuit on the DC side. At the same time, the capacitor voltage is clamped by the diodes and remains at zero; the AC side can be considered as experiencing a three-phase short circuit. The AC and DC sides can be decomposed into two relatively independent circuits, as shown in Figure 11b,c. Under the action of the power source, the system gradually reaches a steady state. In the steady state stage, the DC voltage stabilizes at a fixed value, and the short-circuit current remains nearly constant, as shown in Figure 11d.

As the first stage is the main rising phase of the fault current, it is possible to control the fault current of the converter in this stage through control methods. Therefore, the primary focus is on analyzing the impact of the virtual inertia control strategy on the first stage of the fault. According to Figure 11a, assuming the instantaneous DC voltage is U_0 and current is I_0 after the fault occurrence, there is:

$$CL\frac{d^2u_{dc}}{dt^2}+CR\frac{du_{dc}}{dt}+u_{dc}=0 \tag{12}$$

When $R>2\sqrt{\frac{L}{C}}$, the fault circuit is in an over-damped state, and the capacitor voltage will not cross zero. When $R<2\sqrt{\frac{L}{C}}$, the fault circuit is in an under-damped state. However, due to the small equivalent resistance R of the DC network, the fault current in this stage is essentially the capacitor output current. The rate of change of the capacitor current is high, leading to a rapid decrease in the DC bus voltage due to the fast discharge of the capacitor.

If the virtual capacitance parameter is introduced after the fault occurrence, the following can be obtained:

$$(C+C_v)L\frac{d^2u_{dc}}{dt^2}+(C+C_v)R\frac{du_{dc}}{dt}+u_{dc}=0 \tag{13}$$

When $R<2\sqrt{\frac{L}{C+C_v}}$, it corresponds to a second-order underdamped oscillation. Solving this system of equations, there is:

$$\begin{cases} U_{dc}(t)=e^{-at}(\frac{U_0\omega_0}{\omega}\sin(\omega t+\beta)-\frac{I_0}{\omega(C+C_v)}\sin(\omega t)) \\ I_c(t)=e^{-at}(-\frac{I_0\omega_0}{\omega}\sin(\omega t-\beta)+\frac{U_0}{\omega L}\sin(\omega t)) \end{cases} \tag{14}$$

where $\omega=\sqrt{1/L(C+C_v)-(R/2L)^2}$, $\omega_0=\sqrt{\omega^2+a^2}$, $a=R/2L$, $\beta=\arctan(\omega/a)$.

From Equation (14), it can be observed that if the virtual capacitance parameter is maximized within the allowable range, it can effectively limit the magnitude of the fault current. When the virtual capacitance is sufficiently large, it is possible to change the relationship between R and C from under-damped to over-damped, which significantly limits the growth of the fault current, reduces the rate of decrease in the DC bus voltage, and lowers the peak fault current. This provides sufficient time for protective actions to be taken.

4. Stability Analysis

According to Equations (4) and (8), the block diagram of the BGC system control strategy can be obtained, as shown in Figure 12.

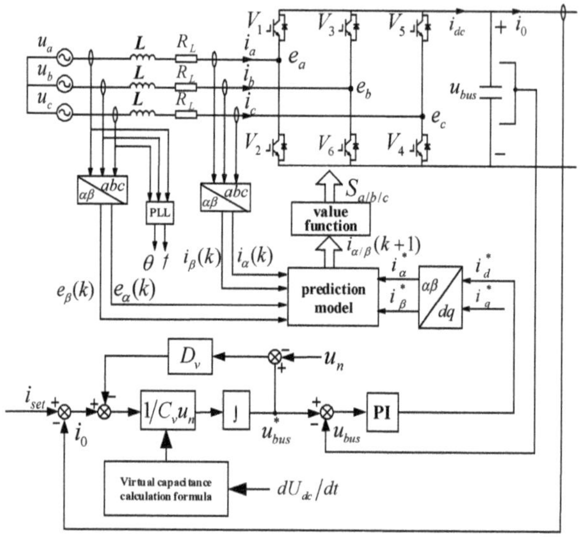

Figure 12. The block diagram of BGC system control strategy.

In order to study the stability of the BGC system after adopting the control strategy proposed in this paper, small signal modeling and analysis of the BGC system are carried out. By writing the state variable in Equation (4) as the sum of steady state values and small disturbance, that is, $i_0 = I_0 + \Delta i_0$, $u_{bus}^* = U_{bus}^* + \Delta u_{bus}^*$, the small signal equation of the virtual inertia control equation can be obtained as:

$$-\Delta i_0 - D_v \Delta u_{bus}^* = C_v u_n \frac{d\Delta u_{bus}^*}{dt} \tag{15}$$

The BGC is set to operate at a unit power factor and does not transmit reactive power to the grid, that is, $i_q = 0$. According to the power balance on both sides of the BGC, there is:

$$\begin{cases} \frac{3}{2} u_d i_d = u_{bus} i_{dc} \\ i_{dc} = i_0 + C\frac{du_{bus}}{dt} \end{cases} \tag{16}$$

The state variable in Equation (16) is written as the sum of steady state value and small disturbance, that is, $i_d = I_d + \Delta i_d$, $u_{bus} = U_{bus} + \Delta u_{bus}$, $i_0 = I_0 + \Delta i_0$, ignoring the disturbance term of power grid voltage and the secondary disturbance term, the small signal equation of Equation (16) is:

$$\frac{3}{2} U_d \Delta i_d = U_{bus}(C\frac{d\Delta u_{bus}}{dt} + \Delta i_0) + \Delta u_{bus} I_0 \tag{17}$$

According to the superposition theorem, ignoring the perturbation term Δi_d and applying Laplace changes to Equation (17), the relationship between Δu_{bus} and Δi_0 is:

$$\frac{\Delta u_{bus}(s)}{\Delta i_0(s)} = -\frac{U_{bus}}{sCU_{bus} + I_0} = G_1(s) \tag{18}$$

Similarly, ignoring the perturbation term Δi_0 and applying Laplace changes to Equation (17), the relationship between Δu_{bus} and Δi_d is:

$$\frac{\Delta u_{bus}(s)}{\Delta i_d(s)} = \frac{3U_d}{2(sCU_{bus} + I_0)} = G_2(s) \tag{19}$$

Thus, the small signal model of the BGC system can be obtained as shown in Figure 13, where $G_v(s) = k_p + k_i/s$. k_p and k_i are proportional constants and integral constants of the PI controller, and the parameters are adjusted by a typical second-order system parameter tuning method.

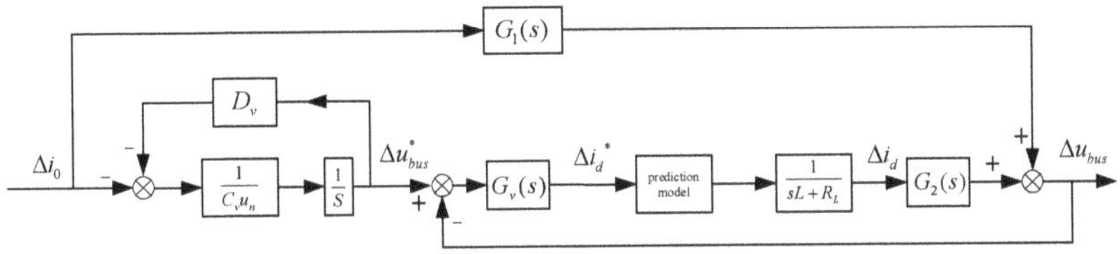

Figure 13. The small signal model of the BGC system.

Since the model prediction algorithm is adopted in the inner loop, considering the tracking accuracy and rapidity of the algorithm, and time delay compensation is introduced, it can be considered that the output current of the converter is equal to the reference current value in real-time [23], that is, the simplified small signal model can be obtained as shown in Figure 14.

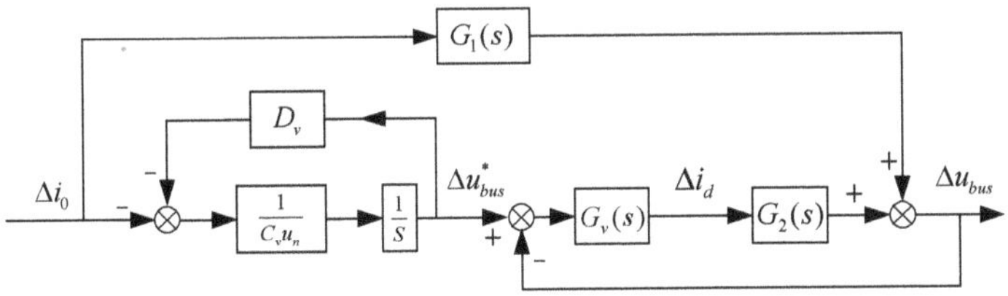

Figure 14. Simplified small signal model of the BGC system.

According to Figure 14, the transfer function between Δu_{bus} and Δi_0 can be obtained:

$$G(s) = -\frac{\Delta u_{bus}(s)}{\Delta i_o(s)} = \frac{a_2 s^2 + a_1 s + a_0}{b_3 s^3 + b_2 s^2 + b_1 s + b_0} \tag{20}$$

where

$$\begin{cases} a_2 = 2C_v U_n U_{bus} \\ a_1 = 2D_v U_{bus} + 3U_d k_p \\ a_0 = 3U_d k_i \\ b_3 = 2CC_v U_n U_{bus} \\ b_2 = 2C_v U_n I_0 + 2CD_v U_{bus} + 3U_d C_v U_n k_p U_{bus} \\ b_1 = 2D_v I_0 + 3U_d D_v k_p + 3U_d C_v U_n k_i \\ b_0 = 3U_d D_v k_p \end{cases} \tag{21}$$

The zero-pole diagram of the BGC control system can be drawn according to Equation (21). Figure 15 shows the dominant pole distribution of the BGC system when different virtual capacitance values C_v are taken under the given voltage damping coefficient D_v.

It can be seen from Figure 15 that under the proposed control strategy, the real part of all changing poles of the BGC system is less than zero, and the poles will gradually approach the imaginary axis but do not cross the imaginary axis into the right half plane. Therefore, when C_v changes within a certain range, the change of C_v will not affect the stability of the BGC system.

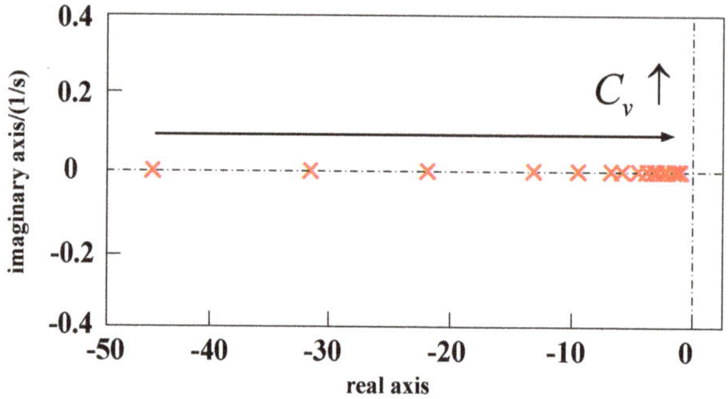

Figure 15. The dominant pole distribution of the BGC system.

5. Simulation Analysis

In order to verify the effectiveness of the control strategy proposed in this paper, the DC microgrid simulation model shown in Figure 1 is built on the Matlab/Simulink simulation platform, and the simulation parameters are shown in Table A3.

5.1. Unit Step Response of Different Virtual Inertia Parameter

Figure 16a shows the simulation results of the unit step response when the damping coefficient D_v is given and the virtual capacitance value C_v is different. It can be seen from Figure 16a that as the value of C_v increases, the magnitude of the DC bus voltage drop gradually decreases, indicating that the inertia of the DC microgrid is enhanced. In addition, the larger the value of C_v, the smaller the amplitude of DC bus voltage drop, the more gentle the change of the DC bus voltage, and the stronger the inertia of the DC microgrid. This is consistent with the previous theoretical analysis. However, too large value of C_v will lead to too long voltage recovery time.

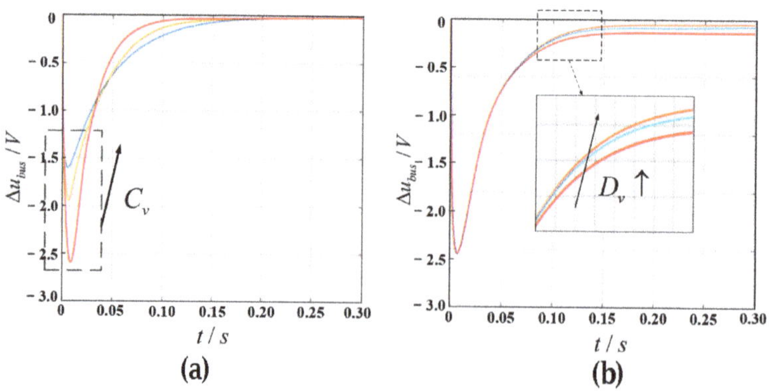

Figure 16. Unit step response of different virtual inertia parameters. (**a**) shows unit step response of different virtual capacitance values; (**b**) shows unit step response of different damping coefficients.

Figure 16b shows the simulation results of the unit step response when the virtual capacitance value C_v is given and the damping coefficient D_v is different. As can be seen from Figure 16b, with the increase of value D_v, the steady-state error of DC bus voltage gradually decreases. It can be seen that D_v mainly affects the steady-state value of DC bus voltage.

5.2. Simulation Comparison of Load Mutation

Figure 17 shows the dynamic response diagram of DC bus when the constant power load surges or drops under three control strategies, namely PI-based parameter fixed AVSG, MPC-based parameter fixed AVSG and MPC -based adaptive AVSG, respectively. Tables A4 and A5 show the corresponding DC bus dynamic response index.

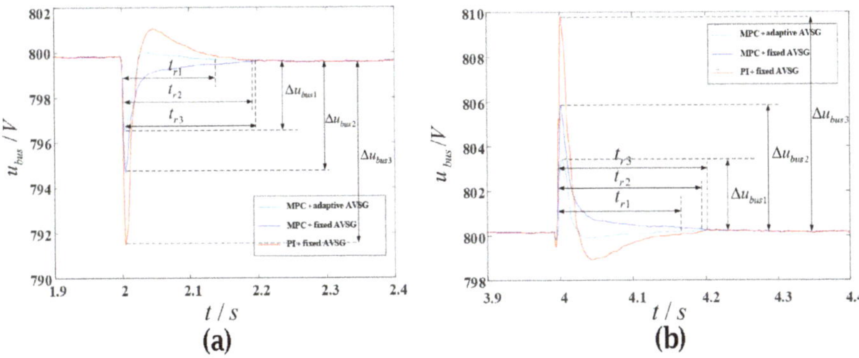

Figure 17. The dynamic response diagram of the DC bus. (**a**) shows the dynamic response diagram of the DC bus when load suddenly increases; (**b**) shows the dynamic response diagram of the DC bus when load suddenly decreases.

During the simulation, the constant power load suddenly increased by 10 kW at $t = 2$ s and then decreased by 10 kW at $t = 4$ s. As can be seen from Tables A4 and A5, under the three control strategies: PI-based parameter fixed AVSG, MPC-based parameter fixed AVSG, and MPC-based adaptive AVSG, when the load suddenly increased, the amplitude of voltage fluctuation is 8.2 V, 5.2 V, and 3.4 V, respectively, and the voltage recovery time is 0.21 s, 0.19 s, and 0.14 s, respectively; when the load suddenly decreased, the amplitude of voltage fluctuation is 9.8 V, 5.9 V, and 3.7 V, respectively, and the voltage recovery time is 0.22 s, 0.19 s, and 0.16 s, respectively.

It can be seen from Figure 17 that among the three control strategies, the adaptive AVSG control based on MPC also has the smallest voltage fluctuation amplitude, the shortest voltage recovery time, and the best dynamic performance.

The above results show that the proposed adaptive AVSG control strategy based on MPC can effectively suppress the DC bus voltage fluctuation when the BGC system is in the network power disturbance. The introduction of adaptive virtual inertia can reduce the voltage fluctuation amplitude, enhance the inertia of the DC microgrid, and smooth the bus voltage change, thus improving the stability of DC bus voltage.

5.3. Grid-Side Power Quality Analysis

Figure 18 shows the current and voltage waveforms at the grid side, as well as the Total Harmonic Distortion (THD) analysis of the grid-side current under three control strategies.

It can be observed that the grid-side voltage and current exhibit good sinusoidal waveforms with minimal distortion. From the THD analysis, it is found in Table A6 that under the three control strategies, the THD is 5.32% for PI-based parameter fixed AVSG control, 3.88% for MPC-based parameter fixed AVSG control, and 2.98% for MPC-based adaptive AVSG control. The MPC-based adaptive AVSG control strategy yields the lowest total harmonic distortion of the grid-side current, indicating the best electrical energy quality at the grid side under this control strategy.

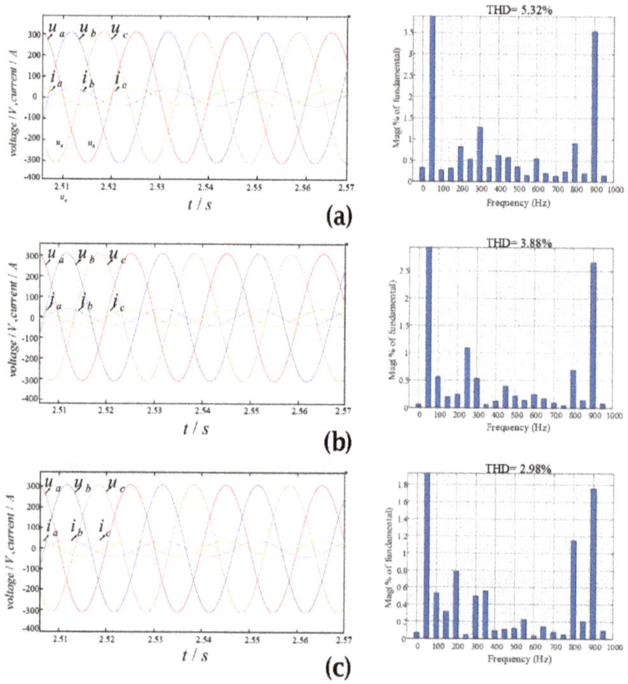

Figure 18. Grid side voltage and current waveform and THD analysis. (**a**) shows the network side waveform and THD analysis under PI+AVSG control strategy; (**b**) shows the network side waveform and THD analysis under MPC + AVSG control strategy; (**c**) shows the network side waveform and THD analysis under MPC + adaptive AVSG control strategy.

5.4. Inter-Pole Short Circuit Fault

When a BGC system experiences an inter-pole fault, the DC bus voltage rapidly drops to zero within a few milliseconds. Under this condition, there is little difference in the performance between the adaptive parameter AVSG and fixed parameter AVSG control strategies. Therefore, for simulation purposes, the fixed parameter approach is sufficient. Figure 19 illustrates the dynamic response comparison of the DC voltage and current before and after applying AVSG control when an inter-pole fault occurs in the BGC system.

From Figure 19a, it is evident that when the BGC system is not utilizing AVSG control, the DC voltage rapidly decreases after a fault occurrence due to the quick discharge of the DC-side capacitor. Within 5.75 ms, the voltage drops to zero. However, when AVSG control is applied, the rate of DC bus voltage decrease slows down, and it takes approximately 8 ms to reach zero. Compared to the scenario without AVSG control, the descent time has been extended by 39.13%. This extended time is beneficial for fault detection in the system, allowing more time for protective actions to be taken.

From Figure 19b, it can be observed that when the BGC system does not employ AVSG control, the fault current rapidly increases after a fault occurrence due to the quick discharge of the DC-side capacitor. The peak fault current reaches 962.55 A. However, when AVSG control is applied, the rate of fault current increase slows down, and the peak current is reduced to 734.74 A. Compared to the scenario without AVSG control, the peak fault current is reduced by 227.81 A. This helps to avoid excessive short-circuit current impacts and prevents device burnout.

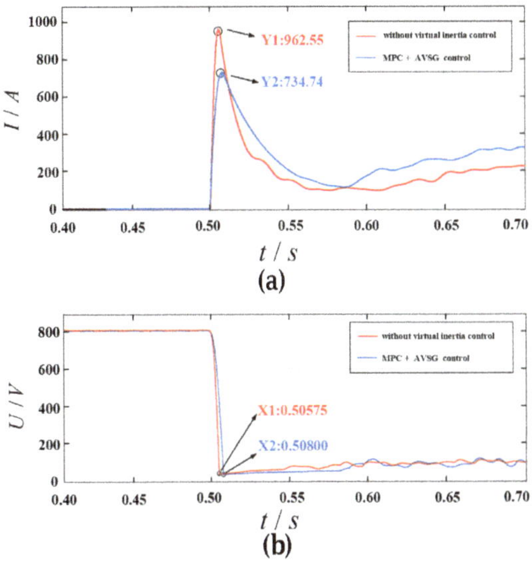

Figure 19. The dynamic response comparison of the DC voltage and current. (**a**) shows the fault current waveform under different control strategies; (**b**) shows the fault voltage waveform under different control strategies.

6. Conclusions

In order to improve the dynamic performance of DC bus voltage, enhance the inertia of DC microgrid, and suppress the drastic fluctuation of DC bus voltage under the power disturbance in the network, this paper improves the traditional virtual inertia control strategy and proposes an adaptive virtual inertia control strategy based on model predictive control, and the following conclusions are obtained:

(1) The model predictive control is used in the inner loop, and the two-step predictive delay compensation is used to realize the fast-tracking of the given current value, eliminating the traditional PI controller and PWM regulator and improving the dynamic performance of the control system.

(2) The adaptive AVSG control is introduced in the outer loop. By combining the inertia coefficient in AVSG with the voltage change rate, the flexible adjustment of the inertia parameters is realized. The BGC system using this control strategy can quickly provide additional power when the power difference occurs in the DC microgrid, thereby enhancing the inertia of the DC microgrid and effectively improving the stability of DC bus voltage and the operation ability of the system under asymmetric conditions.

(3) This study focuses exclusively on the developed four-terminal DC microgrid. When multiple grid-connected units or distributed energy sources are incorporated into the grid, complex systems impose stricter requirements on system stability and coordination among individual units. Further research is still needed in order to address these more stringent demands.

Author Contributions: Conceptualization and methodology, F.Z.; software, validation, and writing—original draft preparation, M.S.; investigation, B.L.; data curation, review, and editing, W.L. All authors have read and agreed to the published version of the manuscript.

Funding: This research was funded by the National Natural Science Foundation of China (52167010).

Data Availability Statement: Data are contained within the article.

Conflicts of Interest: The authors declare no conflict of interest.

Appendix A

Table A1. AC and DC microgrid control variable analogy.

Analogy Term	VSG	AVSG
Droop equation	$\omega - P$	$u - i$
Control objective	ω	u
Output	P	i
Inertia	J	C
Storage capacity	$1/_2\, J\omega^2$	$1/_2\, Cu^2$

Table A2. The relationship between the switching state and the output voltage component.

Switch Status	S_a	S_b	S_c	u_α	u_β
1	0	0	0	0	0
2	0	0	1	0.8165 Udc	0
3	0	1	0	0.4083 Udc	0.7071 Udc
4	0	1	1	0.4083 Udc	0.7071 Udc
5	1	0	0	0.8165 Udc	0
6	1	0	1	0.4083 Udc	0.7071 Udc
7	1	1	0	0.4083 Udc	0.7071 Udc
8	1	1	1	0	0

Table A3. System simulation parameters.

Simulation Parameter	Value
Three − phase grid phase voltage $u_{a,b,c}/V$	220
Rated voltage value U_n/V	800
Filter inductance L/mH	3
Filter resistors R_L/Ω	0.05
DC side capacitance $C/\mu F$	5000
Constant power load R/kW	10
Virtual capacitance value $C_v/\mu F$	1500
Voltage damping factor D_v	5
Proportional factor k_p	10
Integral factor k_i	120

Table A4. DC bus dynamic response index when constant power load suddenly increases.

Control Strategy	Voltage Fluctuations Magnitude/V	Voltage Recovery Time t/s
PI + Fixed AVSG	8.2	0.21
MPC + Fixed AVSG	5.2	0.19
MPC + Adaptive AVSG	3.4	0.14

Table A5. DC bus dynamic response index when constant power load suddenly decreases.

Control Strategy	Voltage Fluctuations Magnitude/V	Voltage Recovery Time t/s
PI + Fixed AVSG	9.8	0.22
MPC + Fixed AVSG	5.9	0.19
MPC + Adaptive AVSG	3.7	0.16

Table A6. THD analysis of the grid-side current under three control strategies.

Control Strategy	THD/%
PI + Fixed AVSG	5.32
MPC + Fixed AVSG	3.88
MPC + Adaptive AVSG	2.98

References

1. Liu, X.; Liu, Y.; Liu, J.; Xiang, Y.; Yuan, X. Optimal planning of AC-DC hybrid transmission and distributed energy resource system: Review and prospects. *CSEE J. Power Energy Syst.* **2019**, *5*, 409–422. [CrossRef]
2. Li, J.; Xu, Z.; Liu, H.; Wang, C.; Wang, L. A Wasserstein Distributionally Robust Planning Model for Renewable Sources and Energy Storage Systems Under Multiple Uncertainties. *IEEE Trans. Sustain. Energy* **2023**, *14*, 1346–1356. [CrossRef]
3. Muhtadi, A.; Pandit, D.; Nguyen, N.; Mitra, J. Distributed Energy Resources Based Microgrid: Review of Architecture, Control, and Reliability. *IEEE Trans. Ind. Appl.* **2021**, *57*, 2223–2235. [CrossRef]
4. Al-Ismail, F.S. DC Microgrid Planning, Operation, and Control: A Comprehensive Review. *IEEE Access.* **2021**, *9*, 36154–36172. [CrossRef]
5. Ansari, S.; Chandel, A.; Tariq, M. Comprehensive Review on Power Converters Control and Control Strategies of AC/DC Microgrid. *IEEE Access.* **2021**, *9*, 17998–18015. [CrossRef]
6. Reddy, Y.; Jithendranath, J.; Chakraborty, A.K.; Guerrero, J.M. Stability Constrained Optimal Operation of Standalone DC Microgrids Considering Load and Solar PV Uncertainties. *IEEE Trans. Power Deliv.* **2023**, *38*, 2673–2681.
7. Zhou, K.; Liu, Z.; Li, J.; Su, M. Stability Analysis in The Stabilization-Free DC Microgrids with CPLs. In Proceedings of the 12th International Conference on Power and Energy Systems (ICPES), Guangzhou, China, 23–25 December 2022; pp. 407–411.
8. Xia, Y.; Yu, M.; Tao, X.; Peng, Y.; Wei, W. Decentralized control for parallel bidirectional power converters of a grid-connected DC microgrid. In Proceedings of the 42nd Annual Conference of the IEEE Industrial Electronics Society, Florence, Italy, 23–27 October 2016; pp. 6085–6090.
9. Wang, L.; Peng, L.; Yang, S. Stability Analysis of a Grid-connected Large-scale DC Microgrid with a Hybrid Wind/Solar Farm. In Proceedings of the 2022 IET International Conference on Engineering Technologies and Applications (IET-ICETA), Changhua, Taiwan, 14–16 October 2022; pp. 1–2.
10. Wang, W.; Lei, X.; Wei, B.; He, K. Research on Adaptive Droop Control Strategy of DC Active Power and Voltage in DC Microgrid. In Proceedings of the 2023 International Conference on Power Energy Systems and Applications (ICoPESA), Nanjing, China, 24–26 February 2023; pp. 627–632.
11. Carnaghi, M.; Cervellini, P.; Judewicz, M.; Retegui, R.G.; Funes, M. Garcia Retegui. Stability analysis of a Networking DC microgrid with distributed droop control and CPLs. *IEEE Lat. Am. Trans.* **2023**, *21*, 966–975. [CrossRef]
12. Unamuno, E.; Barrena, J.A. Design and small-signal stability analysis of a virtual-capacitor control for DC microgrids. In Proceedings of the 19th European Conference on Power Electronics and Applications, Warsaw, Poland, 11–14 September 2017; pp. 1–10.
13. Hosseinipour, A.; Hojabri, H. Virtual inertia control of PV systems for dynamic performance and damping enhancement of DC microgrids with constant power loads. *IET Renew. Power Gener.* **2018**, *12*, 430–438. [CrossRef]
14. Shen, K.; Xue, B.; Zhu, X. Inertia Control Strategy of Direct-driven Wind Generation System in DC Microgrid Based on Analogous Virtual Synchronous Generator. *IET Renew. Power Gener.* **2023**, *49*, 2526–2537.
15. Wen, Z.; Tian, J.; Wang, F. A transient performance improved strategy based on virtual inertia and its allocation for PV-storage DC Microgrid. In Proceedings of the 2022 IEEE International Power Electronics and Application Conference and Exposition (PEAC), Guangzhou, China, 4–7 November 2022; pp. 133–138.
16. Wang, Y.; Wang, C.; Xu, L. Adjustable Inertial Response from the Converter with Adaptive Droop Control in DC Grids. *IEEE Trans. Smart Grid* **2019**, *10*, 3198–3209. [CrossRef]
17. Yan, X.; Xin, H.; Yuan, F. Research on virtual inertial control technology for improving transient stability of DC distribution network. In Proceedings of the 2017 IEEE Conference on Energy Internet and Energy System Integration (EI2), Beijing, China, 26–28 November 2017; pp. 1–5.

18. Wu, W. A Virtual Inertia Control Strategy for DC Microgrids Analogized with Virtual Synchronous Machines. *IEEE Trans. Ind. Electron.* **2017**, *64*, 6005–6016. [CrossRef]
19. Yazdani, S.; Ferdowsi, M.; Shamsi, P. Adaptive-Passive Virtual Inertia Control Based on Energy Balance Between a Synchronous Generator and a Three-Phase Inverter. In Proceedings of the 2020 IEEE Energy Conversion Congress and Exposition (ECCE), Detroit, MI, USA, 11–15 October 2022; pp. 133–138.
20. Liu, J.; Vazquez, S.; Wu, L. Extended State Observer-Based Sliding-Mode Control for Three-Phase Power Converters. *IEEE Trans. Ind. Electron.* **2017**, *64*, 22–31. [CrossRef]
21. Long, B.; Zeng, W.; Rodríguez, J. Voltage Regulation Enhancement of DC-MG Based on Power Accumulator Battery Test System: MPC-Controlled Virtual Inertia Approach. *IEEE Trans. Smart Grid* **2022**, *13*, 71–81. [CrossRef]
22. Zheng, X.; Shi, X.; Li, H. VSG Control Strategy Based on Model Predictive for Islanded Microgrid Inverter. In Proceedings of the IECON 2021–47th Annual Conference of the IEEE Industrial Electronics Society, Toronto, ON, Canada, 13–16 October 2021; pp. 1–7.
23. Xia, C.; Liu, T.; Shi, T.; Song, Z. A Simplified Finite-Control-Set Model-Predictive Control for Power Converters. *IEEE Trans Ind. Inform.* **2014**, *10*, 991–1002.

symmetry

MDPI

Article

Research on Optimal Scheduling Strategy of Microgrid Considering Electric Vehicle Access

Zhimin Wu [1], Yang Zou [1], Feng Zheng [1,*] and Ning Liang [2]

[1] College of Electrical Engineering and Automation, Fuzhou University, Fuzhou 350100, China; mirage12138@126.com (Z.W.); 13275015839@163.com (Y.Z.)
[2] Faculty of Electric Power Engineering, Kunming University of Science and Technology, Kunming 650500, China; ln_kust@163.com
* Correspondence: zf_whu@163.com

Abstract: The random output of renewable energy and the disorderly grid connection of electric vehicles (EV) will pose challenges to the safe and stable operation of the power system. In order to ensure the reliability and symmetry of the microgrid operation, this paper proposes a microgrid optimization scheduling strategy considering the access of EVs. Firstly, in order to reduce the impact of random access to EVs on power system operation, a schedulable model of an EV cluster is constructed based on the Minkowski sum. Then, based on the wavelet neural network (WNN), the renewable energy output is predicted to reduce the influence of its output fluctuation on the operation of the power system. Considering the operation constraints of each unit in the microgrid, the network active power loss and node voltage deviation are taken as the optimization objectives, and the established microgrid model is equivalently transformed via second-order cone relaxation to improve its solution efficiency. Based on network reconfiguration and flexible load participation in demand response, the economy and reliability of system operation are improved. Finally, the feasibility and effectiveness of the proposed method are verified based on the simulation examples.

Keywords: microgrid; electric vehicles; multi-objective optimization; demand response; network reconfiguration

1. Introduction

In order to realize the low-carbon operation of the power system, renewable energy power generation has developed rapidly [1,2]. The increasing demand for electric energy and severe environmental pollution problems have promoted the development of microgrid technology, which is conducive to improving the penetration rate of renewable energy and realizing the on-site production and consumption of energy [3–5]. EVs have great advantages and potential for reducing carbon emissions in the transportation field and alleviating the energy crisis. And EVs are expected to become the main mode of road transportation. The uncertainty of renewable energy output and the disorderly grid connection of EVs have brought great challenges to the safe and stable operation of power systems. To ensure the feasibility and symmetry of the microgrid, it is of great significance to carry out research on the optimal operation of the microgrid considering the access of EVs [6–8].

Due to the volatility and intermittence of renewable energy output, this will aggravate the imbalance between the supply side and the demand side of the power system. Many methods have been applied to reduce the uncertainty of renewable energy output and ensure the safe and stable operation of power systems. In [9], by establishing an energy trading model based on the prediction interval of renewable energy power generation, demand-side flexible resources are used to improve the uncertainty of renewable energy output. By establishing a scenario-based stochastic optimization model, a scenario set considering the error of renewable energy output prediction is generated by sampling in [10,11]. The accuracy of the stochastic optimization method is related to the quantity and quality of the generated scene set. In order to ensure the accuracy and feasibility of the established model, a large number of scenes need to be generated. But this also brings a large computational burden to the solution of the problem. In this paper, the output of renewable energy is predicted with WNN, and the uncertainty of renewable energy output on power systems is alleviated using demand-side flexible resources such as flexible loads and EVs.

As a flexible demand-side resource, microgrid operators can participate in the optimal operation of power systems and improve their operating status by signing charging agreements with EV aggregators. At present, the relevant literature has carried out corresponding research on the problem of EVs participating in microgrid optimal scheduling. In [12], a hierarchical scheduling system was established according to the interest relationship between EV charging and discharging stations and microgrid operators. The time-of-use electricity price was formulated based on different working conditions to guide EVs to charge and discharge orderly, and the mixed integer linear programming algorithm was used to calculate and solve the proposed model. In [13], the minimum operating cost of microgrids was considered the objective function, and the influence of EV charging load uncertainty on the optimal scheduling of microgrids was also considered. The disordered and ordered charging models were established to solve the optimal scheduling scheme in the worst scenario. In [14], considering the uncertainty of EV users' default, the first and second stage scheduling models were established with the minimum operating cost of microgrid and the minimum difference of EV users' income, respectively. The proposed models can effectively reduce the operating cost of the system and deal with EV users' default. Most of the existing research focuses on the uncertainty of microgrid operation scheduling caused by EV users' disorderly charging and does not fully consider the high-dimensional computational burden caused by the direct participation of large-scale EVs in the optimal scheduling of power systems [15]. In this paper, by establishing an EV cluster schedulable model, the computational pressure of problem solving is reduced while ensuring an accurate description of the charging behavior of EVs.

In order to reduce the impact of distributed power output fluctuation on the operation stability of microgrid systems, this paper proposes a microgrid optimal scheduling strategy considering the schedulability of EV clusters. In order to characterize the travel habits of EVs and explore the load–storage capacity of EV clusters, the load–storage schedulable capacity domain of EV clusters is constructed based on the Minkowski sum. Then, based on the WNN, the output of the wind turbine is predicted to reduce the influence of its volatility and intermittence on the day-ahead optimal scheduling scheme of the microgrid. Considering the operation constraints of each unit in the power system, a microgrid optimization model with the minimum active network loss and node voltage deviation as the optimization objective is established, and the model is equivalently transformed based on second-order cone relaxation. Through network reconfiguration and flexible load participation in demand response, the power flow distribution and operation of microgrid systems are optimized, and the operation reliability and economy of microgrid systems are improved. Finally, the feasibility and effectiveness of the proposed method are verified based on simulation examples. The main contributions are as follows:

(1) In order to reduce the challenge and influence of the disorderly grid connection of EVs on the safe and stable operation of the power system, a dispatchable capacity model of EV set load storage is established based on Minkowski sum. In order to reduce the influence of renewable energy output uncertainty on the safe and stable operation of power systems, the renewable energy output is predicted based on WNN;
(2) To improve the overall voltage quality and economic benefits of the system, taking into account the reduction of active network loss and the reduction of voltage deviation as the optimization objectives, the load fluctuation of the microgrid system operation can be effectively suppressed, and the operation cost of the system can be significantly reduced by means of network reconfiguration and flexible load participation in demand response.

The rest of this paper is organized as follows: Section 2 presents the schedulable model of the EV cluster, the day-ahead optimal scheduling model of the microgrid, and the transformation and deformation of the model. Section 3 introduces the established simulation test system parameters. In Section 4, the effectiveness and feasibility of the proposed method are verified using a comparative analysis of simulation examples. Finally, Section 5 concludes this work.

2. Theoretical Analysis

The research object of this paper is the superior power grid and microgrid managed using different operators. It relies on the information interaction between the two to make decisions on the optimal scheduling scheme of the microgrid and optimize the operation status of the microgrid system. As shown in Figure 1, the microgrid system structure diagram considering EV access is established in this paper.

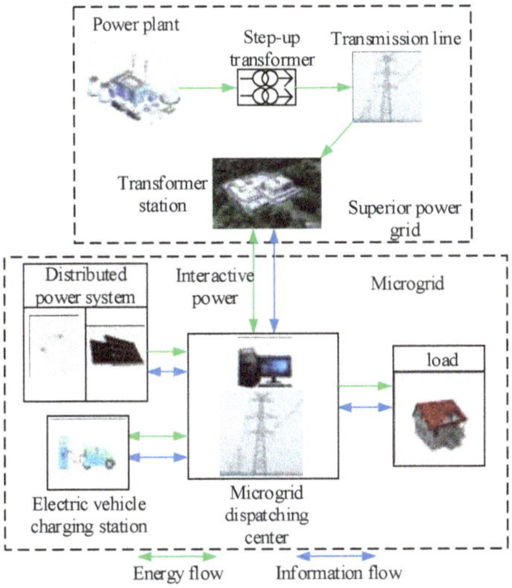

Figure 1. Structure diagram of microgrid system considering EV access.

The upper power grid and the microgrid realize power interaction through the tie line. The microgrid dispatching center formulates an optimized dispatching plan by collecting relevant data on the equipment and lines. It maximizes the overall efficiency of operation under the premise of ensuring the safety and stability of the microgrid systems. The model established in this paper is multi-period day-ahead optimization scheduling. It can be written in the following compact form as follows:

$$\begin{cases} \min F & = \min \sum_{i=1}^{n} F_i(x_t), t \in \mathrm{T} \\ s.t. & h(x_t) = 0 \\ & g(x_t) \leq 0 \end{cases} \tag{1}$$

where F is the objective function of the optimization model; x_t is the decision variable of the optimization model; $F_i(x_t)$ is the objective function of the ith optimization model; and $h(x_t)$ and $g(x_t)$ are the equality constraints and inequality constraints of the optimization model.

Figure 2 shows the optimal microgrid scheduling framework considering the schedulability of the EV cluster. Firstly, the influence of the direct grid connection of individual EVs on the safe and stable operation of the power grid is reduced by establishing a schedulable model of EV clusters. Secondly, based on the WNN, the renewable energy output is predicted, and the system operation state is improved by the flexible load participating in the demand response on the demand side. Based on the network reconfiguration, the reliability of the power grid is improved by improving the power flow distribution of the system. Finally, by setting different operating scenarios for comparative analysis, the feasibility and effectiveness of the proposed method are verified.

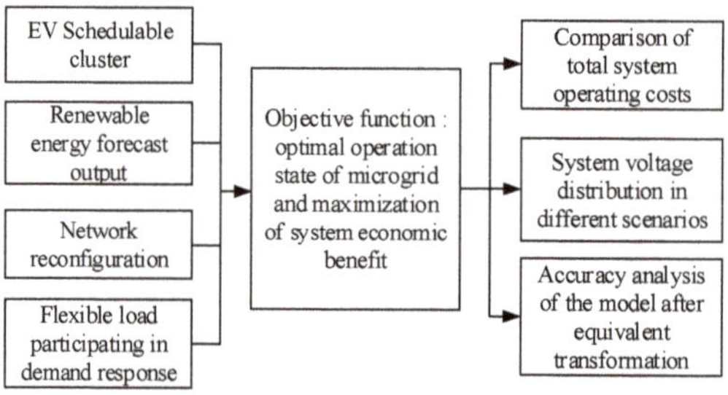

Figure 2. Microgrid optimal scheduling framework considering schedulability of EV cluster.

2.1. Schedulable Charging and Discharging Model of EV Cluster

In view of the challenges brought by the large-scale disorderly grid-connection of EVs to the safe and stable operation of the power system, this paper is based on Minkowski sum equivalents of the individual EVs with large prediction randomness to obtain the charging and discharging model of the EV cluster so that it can flexibly and controllably participate in the optimal scheduling of the power system.

In order to establish an accurate and feasible schedulable charging and discharging model of the EV cluster based on a large number of historical data of the charging and discharging behaviors of EV users, this paper uses the Gaussian mixture model to model the arrival/departure times of EV users:

$$\begin{cases} f_k^{\mathrm{a}}(t_{\mathrm{a}}) = \sum\limits_{l=1}^{L} \rho_l \frac{1}{\sqrt{2\pi}\sigma_{\mathrm{a},l}} \exp(-\frac{t_{\mathrm{a},l}-\mu_{\mathrm{a},l}}{2\sigma_{\mathrm{a},l}^2}), t_{\mathrm{a}} \in T \\ f_k^{\mathrm{d}}(t_{\mathrm{d}}) = \sum\limits_{l=1}^{L} \rho_l \frac{1}{\sqrt{2\pi}\sigma_{\mathrm{d},l}} \exp(-\frac{t_{\mathrm{d},l}-\mu_{\mathrm{d},l}}{2\sigma_{\mathrm{d},l}^2}), t_{\mathrm{d}} \in T \end{cases} \tag{2}$$

where t_{a} and t_{d} are the times for EV users to arrive and leave the charging and discharging station; ρ_l is the corresponding distribution proportion coefficient of the Gaussian mixture model; $\mu_{\mathrm{a},l}$ and $\sigma_{\mathrm{a},l}$ are the expectation and variance of arrival time; and $\mu_{\mathrm{d},l}$ and $\sigma_{\mathrm{d},l}$ are the expectation and variance of departure time.

Figure 3 shows the Gaussian mixture model of EV arrival time and departure time.

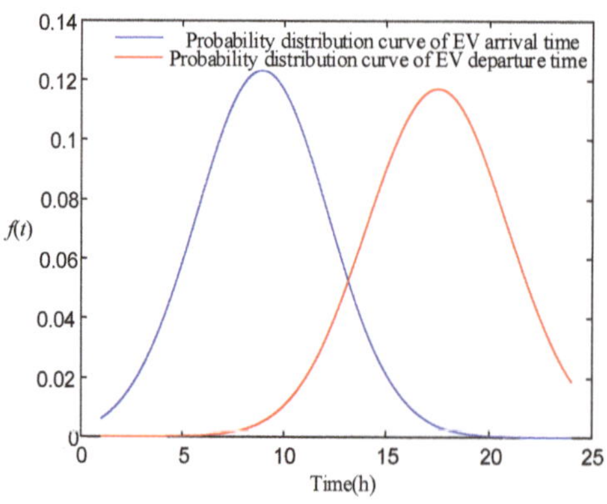

Figure 3. Gaussian mixture model of EV arrival time and departure time.

Due to the randomness of the charging and discharging behavior of individual EV users, it is not guaranteed to accurately predict the travel time distribution of each EV. Therefore, the EV cluster is considered a research object to reduce the randomness of individual EVs and improve the applicability of the established model. The grid-connected scale of EVs at each charging and discharging station can be expressed as follows:

$$\begin{cases} N_{k,t}^{\mathrm{CS}} = N_{k,t-1}^{\mathrm{CS}} + N_{k,t}^{\mathrm{A}} - N_{k,t}^{\mathrm{D}} \\ N_{k,t}^{\mathrm{A}} = (F_k^{\mathrm{a}}(t+1) - F_k^{\mathrm{a}}(t)) \cdot N_k^{\mathrm{EV}} \\ N_{k,t}^{\mathrm{D}} = (F_k^{\mathrm{d}}(t+1) - F_k^{\mathrm{d}}(t)) \cdot N_k^{\mathrm{EV}} \end{cases} \tag{3}$$

where $N_{k,t}^{\mathrm{CS}}$ is the number of EVs in the charging and discharging station of unit k at hour t; $N_{k,t}^{\mathrm{A}}$ and $N_{k,t}^{\mathrm{D}}$ are the number of EVs arriving and departing at the charging and discharging station of unit k at hour t; $F_k^{\mathrm{a}}(t)$ and $F_k^{\mathrm{d}}(t)$ are the cumulative density functions of Equation (2); and N_k^{EV} is the total number of EVs that reach the charging and discharging station in one day of unit k.

By introducing Boolean variables to define the grid-connected and off-grid states of EV users, the EV individuals in the same scheduling period can use Minkowski sum to characterize the schedulability of EV clusters, and the multi-dimensional decision variables of the original EV individuals are equivalently transformed into the single-dimensional decision variables of the EV cluster, alleviating the pressure of the model calculation to solve the dimension:

$$\begin{cases} P_{k,t}^{\mathrm{CS,c}} = \sum\limits_{n \in I_k^{\mathrm{EV}}} u_{n,t} P_{n,t}^{\mathrm{ev,c}}; P_{k,t}^{\mathrm{CS,d}} = \sum\limits_{n \in I_k^{\mathrm{EV}}} u_{n,t} P_{n,t}^{\mathrm{ev,d}} \\ E_{k,t}^{\mathrm{CS,max}} = \sum\limits_{n \in I_k^{\mathrm{EV}}} u_{n,t} E_n^{\mathrm{max}}; E_{k,t}^{\mathrm{CS,min}} = \sum\limits_{n \in I_k^{\mathrm{EV}}} u_{n,t} E_n^{\mathrm{min}} \end{cases} \tag{4}$$

where $P_{k,t}^{\mathrm{CS,c}}$ and $P_{k,t}^{\mathrm{CS,d}}$ are the charging and discharging power of the charging and discharging station of unit k at hour t; $u_{n,t}$ is the on-grid and off-grid state of EV of unit n at hour t; I_k^{EV} is the number of EV clusters at the charging and discharging station of unit k; $P_{n,t}^{\mathrm{EV,c}}$ and $P_{n,t}^{\mathrm{EV,d}}$ are the charging and discharging power of the EV of unit n at hour t; $E_{k,t}^{\mathrm{CS,max}}$ and $E_{k,t}^{\mathrm{CS,min}}$ are the upper and lower limits of the amount of electricity of the charging and discharging station of unit k at hour t; and E_n^{max} and E_n^{min} are the upper and lower limits of the power of the EV of unit n.

2.2. Day-Ahead Optimal Scheduling Model of Microgrid

The output of distributed generation has the characteristics of intermittence and fluctuation. In order to reduce the impact on the operation and scheduling of microgrids, it is necessary to predict the output of distributed generation. WNN is based on a BP neural network, and the transfer function of hidden layer nodes is a wavelet basis function. WNN takes into account the forward propagation of signals and the back propagation of errors. The program flow chart of the WNN algorithm is shown in Figure 4. The overall structure of WNN is shown in Figure 5.

The hidden layer output $h(j)$ and predictive output $y(k)$ of the model are expressed as follows:

$$h(j) = h_j \left(\frac{\sum\limits_{i=1}^{k} w_{ij} x_i - b_j}{a_j} \right), j = 1, 2, \ldots, l \tag{5}$$

$$y(k) = \sum_{i=1}^{l} w_{ik} h(i), k = 1, 2, \ldots m \tag{6}$$

where w_{ij} is the connection weight of unit ij; w_{ik} is the connection weight of unit ik.

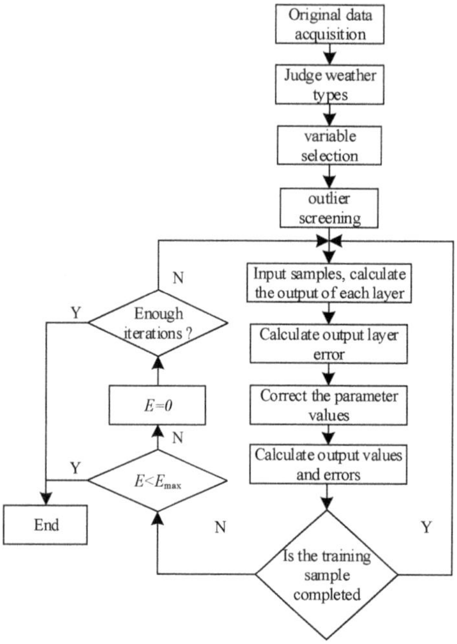

Figure 4. WNN algorithm flow chart.

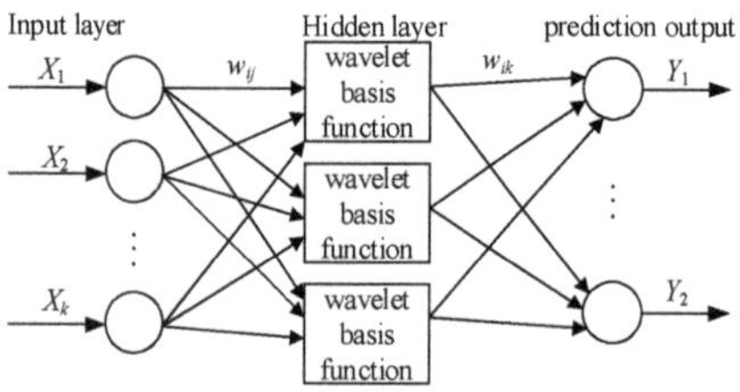

Figure 5. WNN structure diagram.

The connection weight and network parameters of WNN are modified using the gradient correction method so that the predicted output of the model gradually approaches the ideal target value. In this paper, the Morlet wavelet basis function is selected as follows:

$$y = \cos(1.75x)e^{-\frac{x^2}{2}} \tag{7}$$

In order to reduce the influence of distributed power access on the operation state of microgrids, this paper selects two indexes of network active power loss F_{loss} and node voltage deviation F_{u} to quantitatively analyze the effectiveness and feasibility of a microgrid optimal scheduling strategy, which can be shown as follows:

$$\begin{cases} F_{\text{loss}} = \sum_{t=1}^{T}\sum_{ij=1}^{M} I_{ij,t}^2 r_{ij} \\ F_{\text{u}} = \sum_{t=1}^{T}\sum_{i=1}^{N} \frac{|U_{i,t}-U_{\text{e}}|}{U_{\text{e}}} \end{cases} \tag{8}$$

where $I_{ij,t}$ is the line current of unit ij at hour t; r_{ij} is the line resistance of unit ij; $U_{i,t}$ is the node voltage of unit i at hour t; and U_{e} is the node-rated voltage.

Due to the different quantization units of the objective function in the established multi-objective optimization problem, it is difficult to directly and objectively judge which kind of power grid operation state is better. It needs to be normalized to comprehensively evaluate the operating state of the power system. F_{loss} and F_{u} are used as the objective functions to solve the problem, respectively. The two sets of values correspond to the operating boundary values of the two objective functions. Then, F_{loss} and F_{u} are normalized according to the operating boundary values. The auxiliary variable ξ represents the minimum value of each objective function after normalization. The auxiliary variable ξ is used to comprehensively evaluate the operating state of the power system: the larger the ξ is, the better the operation state of the power grid is. The original multi-objective optimization problem is transformed into a single-objective problem for solving the maximum value of ξ:

$$f_i^* = \frac{F_i^{\max} - F_i}{F_i^{\max} - F_i^{\min}}, i = 1, 2 \tag{9}$$

$$\xi = \min(f_1^*, f_2^*) \tag{10}$$

where f_i^* is the normalized value of the ith objective function F_i; $F_i^{\max}$ and $F_i^{\min}$ are the maximum and minimum values of F_i calculated when another objective function is used as the solution object; and ξ is the minimum value of the normalization results of each objective function.

In order to ensure the feasibility and effectiveness of the established microgrid optimal scheduling model considering the access of EVs, the system unit operation constraints considered in this paper are as follows:

(1) Power flow balance constraints:

$$\begin{cases} p_{j,t} = P_{ij,t} - r_{ij} I_{ij,t}^2 - \sum\limits_{k:j \to k} P_{jk,t}, |p_{j,t}| \geq \delta \\ q_{j,t} = Q_{ij,t} - x_{ij} I_{ij,t}^2 - \sum\limits_{k:j \to k} Q_{jk,t}, |q_{j,t}| \geq \delta \\ I_{ij,t}^2 = \frac{P_{ij,t}^2 + Q_{ij,t}^2}{U_{i,t}^2} \end{cases} \tag{11}$$

where $p_{j,t}$ and $q_{j,t}$ are the node active power and reactive power of unit j at hour t; $P_{ij,t}$ and $Q_{ij,t}$ are the branch active power and reactive power of unit ij at hour t; x_{ij} is the line reactance of unit ij; and δ is a very small positive number;

(2) Safe operation constraints:

$$\begin{cases} U_{i,\min} \leq U_{i,t} \leq U_{i,\max} \\ I_{ij,\min} \leq I_{ij,t} \leq I_{ij,\max} \end{cases} \tag{12}$$

where $U_{i,\max}$ and $U_{i,\min}$ are the upper and lower limits of voltage amplitude of unit i; $I_{ij,\max}$ and $I_{ij,\min}$ are the upper and lower limits of branch current of unit ij;

(3) EV cluster operation constraints [16]:

$$\begin{cases} 0 \leq P_{k,t}^{\text{CS,c}} \leq u_{k,t}^{\text{CS}} P_{k,\max}^{\text{CS,c}} \\ 0 \leq P_{k,t}^{\text{CS,d}} \leq (1 - u_{k,t}^{\text{CS}}) P_{k,\max}^{\text{CS,d}} \\ E_k^{\text{CS,min}} \leq E_{k,t}^{\text{CS}} \leq E_k^{\text{CS,max}} \\ E_{k,t}^{\text{CS}} = E_{k,t-1}^{\text{CS}} + \eta_{\text{c}} P_{k,t}^{\text{CS,c}} - P_{k,t}^{\text{CS,d}} / \eta_{\text{d}} \end{cases} \tag{13}$$

where $u_{k,t}^{\text{CS}}$ is the Boolean variable of the equivalent charging and discharging state of EV cluster of unit k at hour t; $P_{k,\max}^{\text{CS,c}}$ and $P_{k,\max}^{\text{CS,d}}$ are the upper limits of equivalent charging and discharging power of EV cluster of unit k; $P_{k,t}^{\text{CS}}$ is the equivalent grid-connected power of EV cluster in charging station of unit k at hour t; $E_{k,t}^{\mathrm{CS}}$ is the equivalent battery capacity of unit k at hour t; $E_{k}^{\mathrm{CS,max}}$ and $E_{k}^{\mathrm{CS,min}}$ are the upper and lower limits of battery capacity of EV cluster in charging station; and η_{c} and η_{d} are the charging and discharging coefficient of EV;

(4) Flexible load operation constraints [17]:

$$\begin{cases} P_{i,t}^{\mathrm{load}} = P_{i,t}^{\mathrm{load0}} + P_{i,t}^{\mathrm{fle}} = P_{i,t}^{\mathrm{load0}} + P_{i,t}^{\mathrm{stop}} + P_{i,t}^{\mathrm{tran}} \\ P_{i,t}^{\mathrm{stop}} = P_{i,t}^{\mathrm{stop0}} - \Delta P_{i,t}^{\mathrm{stop}}, \Delta P_{i,t}^{\mathrm{stop}} \geq 0 \\ P_{i,t}^{\mathrm{tran}} = P_{i,t}^{\mathrm{tran0}} + P_{i,t}^{\mathrm{add}} - P_{i,t}^{\mathrm{sub}}, P_{i,t}^{\mathrm{add}} \geq 0, P_{i,t}^{\mathrm{sub}} \geq 0 \\ \sum\limits_{t=1}^{T} P_{i,t}^{\mathrm{tran}} = \sum\limits_{t=1}^{T} P_{i,t}^{\mathrm{tran0}} \\ P_{i,t}^{\mathrm{tran,min}} \leq P_{i,t}^{\mathrm{tran}} \leq P_{i,t}^{\mathrm{tran,max}} \end{cases} \tag{14}$$

where $P_{i,t}^{\mathrm{load}}$ is the total load power of unit i at hour t; $P_{i,t}^{\mathrm{load0}}$ and $P_{i,t}^{\mathrm{fle}}$ are the fixed load power and flexible load power of unit i at hour t; $P_{i,t}^{\mathrm{stop}}$ is the interruptible load power of unit i at hour t; $P_{i,t}^{\mathrm{stop0}}$ and $\Delta P_{i,t}^{\mathrm{stop}}$ are the initial interruptible load power and load interruption involved in demand response of unit i at hour t; $P_{i,t}^{\mathrm{tran}}$ and $P_{i,t}^{\mathrm{tran0}}$ are the transferable load power and initial transferable load power of unit i at hour t; $P_{i,t}^{\mathrm{add}}$ and $P_{i,t}^{\mathrm{sub}}$ are the increment and reduction of transferable load power of unit i at hour t; and $P_{i,t}^{\mathrm{tran,max}}$ and $P_{i,t}^{\mathrm{tran,min}}$ are the upper and lower limits of a transferable load participating in demand response;

(5) Interaction power constraints between main network and microgrid:

$$-P_{\mathrm{change,max}} \leq P_{\mathrm{change},t} \leq P_{\mathrm{change,max}} \tag{15}$$

where $P_{\mathrm{change},t}$ and $P_{\mathrm{change,max}}$ are the interaction powers between the main network and microgrid at hour t and maximum interaction power;

(6) Network reconfiguration constraints:

$$\begin{cases} \sum\limits_{(ij)\in B} z_{ij} = n_{\mathrm{b}} - n_{\mathrm{s}} \\ m_{ij} = (1 - z_{ij}) \cdot \mathrm{M} \\ U_{j,t}^{2} \geq U_{i,t}^{2} - m_{ij} - 2(r_{ij}P_{ij} + x_{ij}Q_{ij}) + (r_{ij}^{2} + x_{ij}^{2})I_{ij,t}^{2} \\ U_{j,t}^{2} \leq U_{i,t}^{2} + m_{ij} - 2(r_{ij}P_{ij} + x_{ij}Q_{ij}) + (r_{ij}^{2} + x_{ij}^{2})I_{ij,t}^{2} \end{cases} \tag{16}$$

where z_{ij} is the Boolean variable of line breaking state; n_{b} and n_{s} are the total network node number and network root node number; and M is a sufficiently large positive number.

2.3. *Transformation and Deformation of the Model*

The constraints of the microgrid day-ahead optimal scheduling model established above include quadratic terms and integer terms. It belongs to a mixed-integer nonlinear non-convex mathematical problem, which is a polynomial complexity non-deterministic (NP-hard) problem. Figure 6 shows the schematic figure of second-order cone relaxation. Based on the second-order cone relaxation method, this paper transforms the original non-convex and nonlinear problem into a second-order cone optimization problem that can be solved more efficiently by relaxing the quadratic equality constraint into a second-order cone constraint.

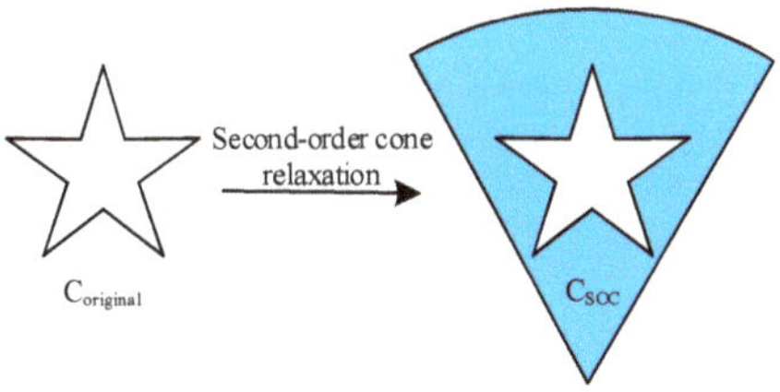

Figure 6. Schematic figure of second-order cone relaxation.

Let $\tilde{I}_{ij,t} = I_{ij,t}^2$ and $\tilde{U}_{i,t} = U_{i,t}^2$, then the safe operation constraints can be written as follows:

$$\begin{cases} I_{ij,\min}^2 \leq \tilde{I}_{ij,t} \leq I_{ij,\max}^2 \\ U_{i,\min}^2 \leq \tilde{U}_{i,t} \leq U_{i,\max}^2 \end{cases} \tag{17}$$

The power flow balance constraints can be written as follows:

$$\begin{cases} p_{j,t} = P_{ij,t} - r_{ij}\tilde{I}_{ij,t} - \sum\limits_{k:j\to k} P_{jk,t}, |p_{j,t}| \geq \delta \\ q_{j,t} = Q_{ij,t} - x_{ij}\tilde{I}_{ij,t} - \sum\limits_{k:j\to k} Q_{jk,t}, |q_{j,t}| \geq \delta \\ \tilde{I}_{ij,t} = \frac{P_{ij,t}^2 + Q_{ij,t}^2}{\tilde{U}_{i,t}} \end{cases} \tag{18}$$

After equivalent transformation and relaxation, the standard second-order cone form can be obtained as follows:

$$\left\| \begin{matrix} 2P_{ij,t} \\ 2Q_{ij,t} \\ \tilde{I}_{ij,t} - \tilde{U}_{i,t} \end{matrix} \right\|_2 \leq \tilde{I}_{ij,t} + \tilde{U}_{i,t}, \forall ij \in B \tag{19}$$

3. Model Building

In order to further verify the feasibility and effectiveness of the microgrid optimal scheduling model considering the access of EVs, based on the Matlab 2021b platform, the model is modeled and the CPLEX solver is called to solve the problem. This paper conducts research and analysis based on the improved IEEE33 node system, and the node location distribution is shown in Figure 7. The basic parameters of the IEEE33 node network are detailed in [18].

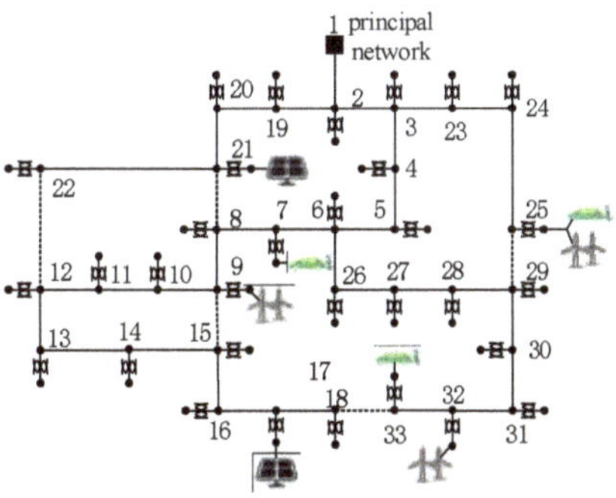

Figure 7. The extended IEEE33 node test system.

The rated voltage level of the test network system is 12.66 kV, and the voltage amplitude is ±1.05 pu. The maximum branch current is 500 A. The interactive power between the main network and the microgrid is 750 kW. It is assumed that the connected EVs are the same type, the capacity of each EV is 24 kWh, and the access power range is −3~3 kW. The charging and discharging coefficients of EVs are 0.95. The operating costs of the established microgrid system are shown in Table A1. The time-of-use electricity price is shown in Table A2. The load curve of the microgrid is shown in Figure 8.

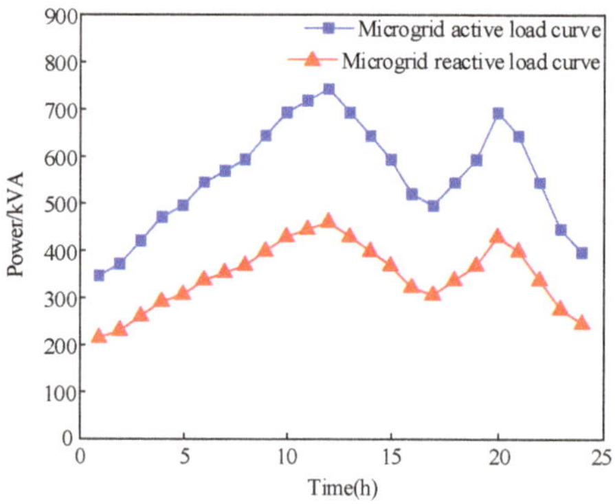

Figure 8. Microgrid load curve.

4. Result Analysis

In order to reduce the influence of fluctuation and intermittence in distributed power output on microgrid operation, this paper predicts the output of distributed power based on WNN. Taking photovoltaic power generation as an example, the output prediction results are shown in Figure 9. It can be seen from the results in the figure that the photovoltaic output prediction based on WNN has high prediction accuracy. It can provide original data support for the formulation of a day-ahead optimal scheduling scheme for microgrids.

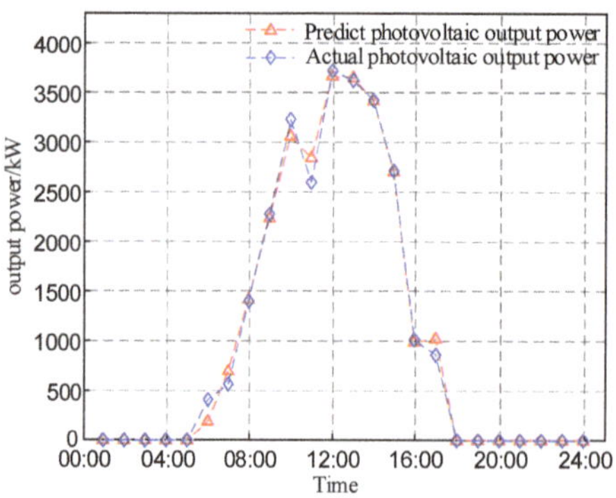

Figure 9. Photovoltaic output prediction based on WNN.

Considering the access of EVs, the microgrid is randomly optimized. Multiple sets of examples are set to compare and analyze the effectiveness and feasibility of the scheduling results. In order to verify the superiority of the proposed microgrid optimal scheduling scheme considering the access to EVs, the following four cases are set up for analysis in Table A3. Table A4 shows the microgrid operation results under different scenarios. The results show that the optimal scheduling of microgrids with EV access, considering network reconfiguration and demand response, can significantly reduce the network loss and node voltage deviation of microgrids and improve the economic benefits of microgrid operation.

Figure 10 shows the microgrid voltage levels under different operating scenarios. The voltage amplitudes of Case 1 and Case 2 are higher than the rated voltage amplitude due to the lack of network reconfiguration and demand response, as shown in Figures 10a and 10b, respectively. The extended IEEE33 node test systems after network reconfiguration are shown in Figure 11a,b. Network reconfiguration can effectively improve the power flow distribution of the microgrid system. Figure 10c shows that the voltage distribution of Case 3 has been significantly improved. As shown in Figure 10d, Case 4 considers the influence of demand response on the basis of Case 3, which further improves the voltage quality of the system.

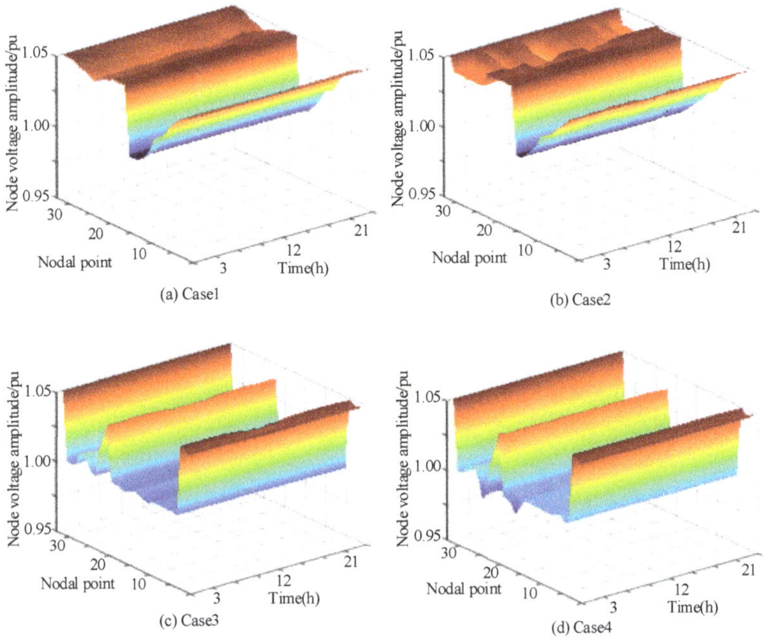

Figure 10. The microgrid voltage levels under different operating scenarios.

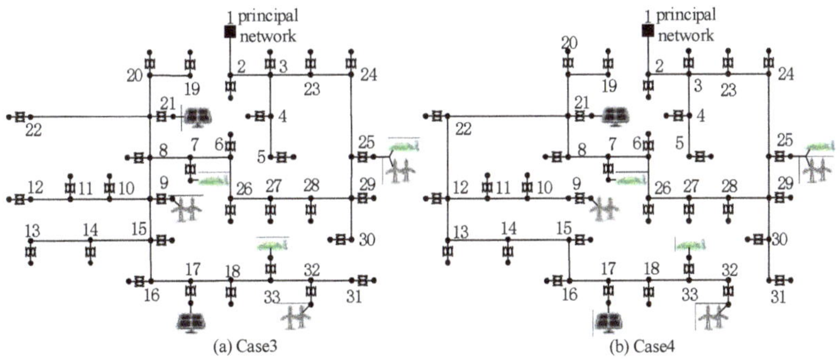

Figure 11. The extended IEEE33 node test system after network reconfiguration.

Figure 12 shows the comparison of flexible loads before and after participating in demand response. Without considering the participation of flexible load in demand response, the load peak-valley difference of the microgrid system is 396.26 kW. Considering the flexible load participating in the demand response, the load peak-valley difference of the microgrid system is 308.12 kW, which is 88.14 kW lower than before. The simulation results show that the participation of flexible load in demand response can effectively suppress the load fluctuation of system operation and improve the operation reliability of microgrid systems.

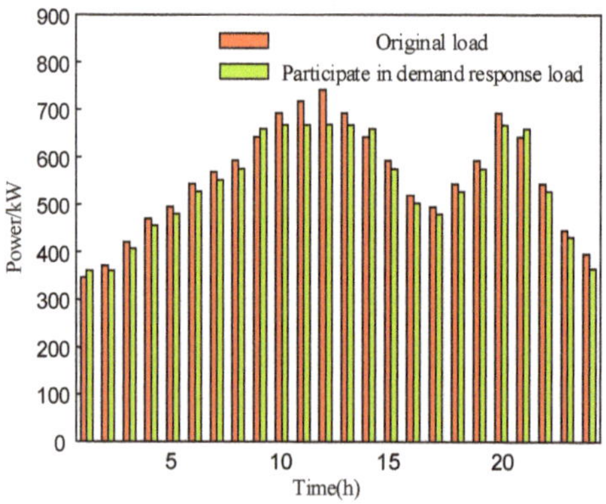

Figure 12. Comparison of flexible load before and after participating in demand response.

In order to improve the calculation speed of the established model and ensure the accuracy of the solution, this paper equivalently transforms the established model based on second-order cone relaxation and sets the model accuracy analysis index $\Delta_{ij,t}$ to verify that the model after second-order cone relaxation is accurate and feasible:

$$\Delta_{ij,t} = \left| \widetilde{U}_{j,t} \widetilde{I}_{ij,t} - (P_{ij,t}^2 + Q_{ij,t}^2) \right| \tag{20}$$

Figure 13 shows the error analysis histogram of the equivalent model. The different colors represent different sizes of error values. Taking Case 4 as an example for verification analysis, it can be seen from Figure 13 that the order of magnitude of the solution error of the model after the second-order cone relaxation equivalent transformation is 10^{-5}. The solution error of the model established in this paper is within the allowable range. The second-order cone relaxation can improve the calculation speed of the problem while ensuring the accuracy of the solution, and the optimal scheduling model for microgrids considering EV access after second-order cone transformation is effective.

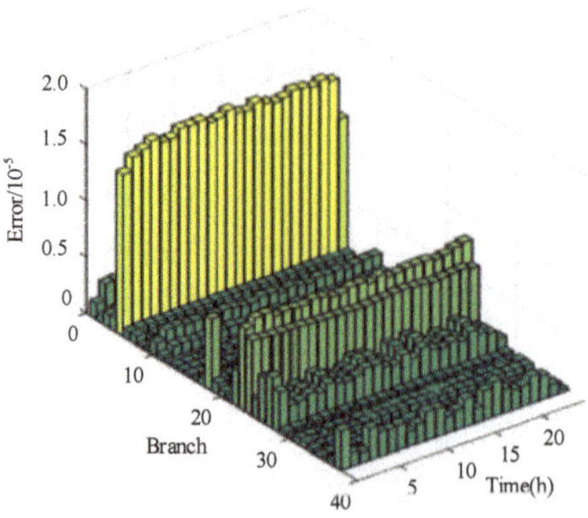

Figure 13. The error analysis histogram of the equivalent model.

5. Conclusions

In this paper, considering the random output of distributed generation and the operation constraints of microgrid units, a microgrid optimization scheduling model considering the access of EVs is established to ensure the feasibility and symmetry of the microgrid. The construction of an EV cluster schedulable model based on Minkowski sum can effectively improve the flexibility and accuracy of large-scale EV cluster pre-scheduling. The prediction of renewable energy output based on WNN can effectively reduce the influence of its output uncertainty and volatility on the safe and stable operation of power systems. Considering the operation constraints of each unit in the microgrid system and taking into account the objective function of the network active power loss and node voltage deviation as the optimization objects, the operation state of the microgrid system is optimized based on network reconfiguration and flexible load participation demand response. This method can significantly improve the operation reliability and economy of the microgrid system. Within the allowable calculation error range, the established model is equivalently transformed based on the second-order cone relaxation, which improves the solution speed of the model while ensuring its accuracy.

The microgrid optimal scheduling model considering EV access established in this paper is a day-ahead optimal scheduling model. By predicting the output of renewable energy and establishing the scheduling model of the EV cluster, the influence of these uncertain factors on the safe and stable operation of the power system is reduced, and the operation state of power grid is improved through demand-side flexible resources, but the influence of prediction errors on the formulation of the optimal scheduling strategy is not fully considered. In the future research process, the influence of prediction error on power system operation will be fully considered to make it more in line with the actual operation of the power grid.

Author Contributions: Conceptualization and methodology, Z.W.; software, validation, and writing—original draft preparation, Y.Z.; investigation, F.Z.; data curation, review, and editing, N.L. All authors have read and agreed to the published version of the manuscript.

Funding: This research is supported by the National Natural Science Foundation of China (Research on optimal scheduling method of elastic power grid considering surplus electric energy gasification under the background of 'carbon neutrality', 52167010).

Data Availability Statement: Not applicable.

Conflicts of Interest: The authors declare no conflict of interest.

Appendix A

Table A1. Operation cost of microgrid system.

Parameter	Value(USD/kWh)	Parameter	Value(USD/kWh)
The unit cost of active power loss	0.1158	The unit compensation cost of transferable load	0.0073
The unit cost of interactive power	0.093	The unit compensation cost of interruptible load	0.0577
The unit output cost of wind turbine	0.0435	The unit compensation cost of EV users	0.0791

Table A2. Time-of-use electricity price.

Time	Electricity Sales Price (USD/kWh)	Electricity Purchase Price (USD/kWh)
1:00–7:00(Valley time)	0.0724	0.0556
8:00–10:00, 19:00–24:00 (Peak time)	0.1951	0.1503
11:00–18:00(Usual time)	0.1302	0.1003

Table A3. Operating condition of cases.

Scene Mode	The Operation Considered
Case 1	None
Case 2	The impact of EV access
Case 3	The impact of EV access and network reconfiguration
Case 4	The impact of EV access, network reconfiguration, and demand response

Table A4. Microgrid operation results under different scenarios.

Scene Mode	f_{loss}(MW)	f_{u}(pu)	F (Normalization)	Microgrid Operation Cost (USD)
Case 1	8.2672	58.3628	0.7972	2807.91
Case 2	7.0688	57.4321	0.8205	2621.39
Case 3	2.9829	18.1345	0.8516	1983.57
Case 4	2.4405	17.7196	0.9665	2177.79

References

1. Kusakana, K. Optimal Economic Dispatch of Grid-Interactive Renewable Prosumers with Hybrid Storage and Peer to Peer Energy Sharing Capabilities. *Int. J. Electr. Electron. Eng. Telecommun.* **2021**, *10*, 209–216. [CrossRef]
2. Hong, T.; Cao, J.; Zhao, W.; Lu, M. Electric Vehicle Charging Scheduling Algorithm Based on Online Multi-objective Optimization. In Proceedings of the 2021 International Wireless Communications and Mobile Computing (IWCMC), Harbin, China, 28 June–2 July 2021; pp. 1141–1146.
3. Jiang, L.; Zhang, Y.; Xiao, C.; Zhang, J.; Li, J. Optimal Scheduling of Electric Vehicle Clusters Considering Uncertainty of User Demand Response. In Proceedings of the 2022 IEEE 6th Conference on Energy Internet and Energy System Integration (EI2), Chengdu, China, 28–30 October 2022; pp. 2927–2930.
4. Zhang, K.; Gao, B.; Han, Y.; Dong, Z. Optimization Scheduling of Electric Vehicle Charging Load Based on Improved PSO. In Proceedings of the 2022 IEEE 2nd International Conference on Mobile Networks and Wireless Communications (ICMNWC), Tumkur, India, 2–3 December 2022; pp. 1–6.
5. Abdalla, M.A.A.; Min, W.; Haroun, A.H.G.; Elhindi, M. Optimal Energy Scheduling Strategy for Smart Charging of Electric Vehicles from Grid-Connected Photovoltaic System. In Proceedings of the 2021 7th International Conference on Electrical, Electronics and Information Engineering (ICEEIE), Malang, Indonesia, 2 October 2021; pp. 37–42.
6. Srilakshmi, E.; Singh, S.P. Energy regulation of EV using MILP for optimal operation of incentive based prosumer microgrid with uncertainty modelling. *Int. J. Electr. Power Energy Syst.* **2022**, *134*, 107353. [CrossRef]
7. Jin, H.; Sangkeum, L.; Hussain, N.S.; Dongsoo, H. Development of Charging/Discharging Scheduling Algorithm for Economical and Energy-Efficient Operation of Multi-EV Charging Station. *Appl. Sci.* **2022**, *12*, 4786. [CrossRef]
8. Zhang, X.; Farajian, H.; Wang, X.; Latifi, M.; Ohshima, K. Scheduling of renewable energy and plug-in hybrid electric vehicles based microgrid using hybrid crow-Pattern search method. *J. Energy Storage* **2022**, *47*, 103605.
9. Jia, Y.; Wan, C.; Cui, W.; Song, Y.; Ju, P. Peer-to-Peer Energy Trading Using Prediction Intervals of Renewable Energy Generation. *IEEE Trans. Smart Grid* **2023**, *14*, 1454–1465. [CrossRef]
10. Antoniadou-Plytaria, K.; Steen, D.; Tuan, L.A.; Carlson, O.; Mohandes, B.; Ghazvini, M.A.F. Scenario-Based Stochastic Optimization for Energy and Flexibility Dispatch of a Microgrid. *IEEE Trans. Smart Grid* **2022**, *13*, 3328–3341. [CrossRef]
11. Yu, D.; Ebadi, A.G.; Jermsittiparsert, K.; Jabarullah, N.H.; Vasiljeva, M.V.; Nojavan, S. Risk-Constrained Stochastic Optimization of a Concentrating Solar Power Plant. *IEEE Trans. Sustain. Energy* **2020**, *11*, 1464–1472. [CrossRef]

12. Xiao, Z.; Zhang, K.; Feng, J. Hierarchical optimal dispatching of wind/PV/diesel islanded microgrid with EVs charging station. *J. Tianjing Univ.* **2022**, *41*, 61–74.
13. Shao, S.; Ma, X.; Yuan, W.; Zhang, K.; Fu, X.; Huang, C. Robust optimal dispatching method for uncertain microgrid including electric vehicles. *J. Electr. Eng.* **2023**, *18*, 201–209.
14. Li, C.; Xu, M.; Cai, S. Two-stage optimal scheduling strategy for micro-grid considering EV default uncertainty. *Trans. China Electrotech. Soc.* **2023**, *38*, 1838–1851.
15. Zhou, M.; Wu, Z.; Wang, J.; Li, G. Forming Dispatchable Region of Electric Vehicle Aggregation in Microgrid Bidding. *IEEE Trans. Ind. Inform.* **2021**, *17*, 4755–4765. [CrossRef]
16. Waseem, K.M.; Wang, J. Multi-agents based optimal energy scheduling technique for electric vehicles aggregator in microgrids. *Int. J. Electr. Power Energy Syst.* **2022**, *134*, 107346.
17. Mostafa, B.; Mohamed, E.; Abdelfatah, A.; Shaaban, M.F.; Baseem, K.; Salah, K. Optimal day-ahead scheduling in micro-grid with renewable based DGs and smart charging station of EVs using an enhanced manta-ray foraging optimisation. *IET Renew. Power Gener.* **2022**, *16*, 2413–2428.
18. Baran, M.E.; Wu, F.F. Network reconfiguration in distribution systems for loss reduction and load balancing. *IEEE Trans. Power Deliv.* **1989**, *4*, 1401–1407. [CrossRef]

symmetry

MDPI

Article

Optimal Location and Sizing of Photovoltaic-Based Distributed Generations to Improve the Efficiency and Symmetry of a Distribution Network by Handling Random Constraints of Particle Swarm Optimization Algorithm

Muhammad Abid Ali [1], Abdul Rauf Bhatti [1,*], Akhtar Rasool [2,*], Muhammad Farhan [1] and Ebenezer Esenogho [2]

[1] Department of Electrical Engineering and Technology, Government College University Faisalabad, Faisalabad 38000, Pakistan; muhammadabidali@gcuf.edu.pk (M.A.A.); mfarhan@gcuf.edu.pk (M.F.)

[2] Department of Electrical Engineering, University of Botswana, Gaborone UB0061, Botswana; esenoghoe@ub.ac.bw or drebenic4real@gmail.com

* Correspondence: bhatti_abdulrauf@gcuf.edu.pk (A.R.B.); akhtar@sabanciuniv.edu (A.R.)

Abstract: Distributed generators (DGs) are increasingly employed in radial distribution systems owing to their ability to reduce electrical energy losses, better voltage levels, and increased dependability of the power supply. This research paper deals with the utilization of a Particle Swarm Optimization algorithm by handling its random constraints to determine the most appropriate size and location of photovoltaic-based DG (PVDG) to keep the asymmetries of the phases minimal in the grid. It is thus expected that this algorithm will provide an efficient and consistent solution to improve the overall performance of the power system. The placement and sizing of the DG are done in a way that minimizes power losses, enhances the voltage profile, i.e., bringing symmetry in the voltage profile of the system, and provides maximum cost savings. The model has been tested on an IEEE 33-bus radial distribution system using MATLAB software, in both conditions, i.e., with and without PVDG. The simulation results were successful, indicating the viability of the proposed model. The proposed PSO-based PVDG model further reduced active power losses as compared to the models based on the teaching–learning artificial bee colony algorithm (TLABC), pathfinder algorithm (PFA), and ant lion optimization algorithm (ALOA). With the proposed model, active power losses have reduced to 17.50%, 17.48%, and 8.82% compared to the losses found in the case of TLABC, PFA, and ALOA, respectively. Similarly, the proposed solution lessens the reactive power losses compared to the losses found through existing TLABC, PFA, and ALOA techniques by an extent of 23.06%, 23%, and 23.08%, respectively. Moreover, this work shows cost saving of 15.21% and 6.70% more than TLABC and ALOA, respectively. Additionally, it improves the voltage profile by 3.48% of the power distribution system.

Keywords: distributed generator; PVDG; PSO algorithm; voltage profile improvement; cost savings; power losses; radial distribution network; constraints handling

1. Introduction

Electrical energy demand is increasing because of the world's rising population and the usage of more electrical-based appliances in human life. It is required to generate more electrical power to fulfill these demands. Electrical power should be generated from sources that are neither depleting nor causing global warming. Thus, renewable-based electrical power generation (PV, wind turbines, microturbines, biomass, etc.) is one of the preferred choices. Among renewable-based electrical power generation, electrical power generation from solar power-based DGs is on a rising trend.

Currently, the integration of DGs has become an attractive choice for technical, economic, and environmental benefits in power distribution networks [1,2]. The distributed generator is used to generate electrical power on a small scale (1 kW to 50 MW) and is generally embedded in the electrical power distribution networks [3]. The DG unit assists in improving the efficiency of the electrical system by decreasing power losses, stabilizing system oscillation, and improving voltage profile, reliability, and security. This is accomplished by injecting active and reactive power into the load side, as reported in several research studies [4–6]. The advantages of integrating DGs at the distribution side are clear; they require less time and money to install than traditional centralized generators [7–9]. This research aims to use a biogeography-based optimization algorithm to join multiple power voltage distributed generators (PVDGs) in the power distribution system. It is expected to reduce power losses and total harmonics distortion while improving the efficiency of the system [10]. The integration of PVDG in RDS can yield a range of advantages, such as decreased power losses, improved voltage profile, and increased voltage stability index. This integration can also reduce global warming by decreasing the greenhouse effect. In [11], the FFA algorithm was used to identify the optimal position and size of DG in the RDS. In [12], the GA technique was used to determine the optimal location and size of solar-based DG in RDS for minimizing power losses. Similarly, [13] proposed a PSO technique for integrating PVDG at the optimal position and size to reduce actual power losses and improve the voltage profile. The Whale optimization algorithm for multi-objective optimization is proposed for inserting distributed generators in RDS [14]. The renewable-based DGs are in sub-transmission and distribution systems to compare their impacts on Voltage profile and power losses [14]. The enactment of renewable-based DGs into the distribution network has been incorporated by using multi-state modeling through probability density function [15]. A hybrid technique known as TLABC has been employed to determine the most suitable position and size of PV and wind DGs in RDS, to reduce power losses and reduce costs. This technique is based on active power loss reduction [16]. The paper introduces an intelligent augmented social network seeking power dispatch (ORPD) in energy networks. It outperforms the social network seek (ASNS) algorithm for the best reactive (SNS) algorithm with the aid of attaining as much as a 22% power loss discount and up to a 93% development in voltage profiles on tested IEEE fashionable grids [17]. This article offers an improved primarily Heap-based optimizer with a Deeper Exploitative development (HODEI) set of rules for power distribution feeder reconfiguration (PDFR) and allotted generator (DG) allocation; it outperforms conventional techniques in voltage profiles and health metrics [18]. The paper offers an advanced hybrid evolutionary algorithm (PODESCA) and a primarily sensitivity-based decision-making technique for the optimal planning of shunt capacitors in radial distribution structures, reaching higher effects than preceding techniques [19]. This paper introduces a unique mixed-evolutionary technique, the quasi-oppositional differential evolution Lévy flights method (QODELFM), for solving the ideal making plans of distribution generators in radial distribution networks, demonstrating its superiority over existing techniques phrases of robustness and efficiency [20]. This paper proposes an international framework for short-time collection modeling with a rolling mechanism, gray model, and meta-heuristic algorithms. It outperforms popular models and enhances the accuracy and speed of complex structure prediction. Dragonfly and whale optimization boost performance [21]. This paper introduces the Quasi-opposition-based studying and Q-learning-based Marine Predators set of rules (QQLMPA) to beautify the overall performance of the traditional Marine Predators algorithm (MPA) for solving optimization troubles. Q-learning enables better utilization of beyond iteration facts, at the same time as quasi-opposition-based studying improves populace diversity, decreasing convergence to neighbored optima [22]. This study introduces QLADIFA, a novel optimization algorithm combining Q-learning with the adaptive logarithmic spiral-Levy flight firefly algorithm. QLADIFA leverages fireflies' environmental awareness and memory, leading to improved performance compared to existing methods. Numerical experiments validate its effectiveness on benchmark functions and various engineering problems [23]. This paper examines a 150.7 kW grid-connected PV system at GCU Faisalabad. the use of PVSyst 7.4 and Metronome, it carried out an average yearly PR of 79.64%, with a peak of 85.4% in January. The PV array produced 218.12 MWh of DC electricity, injecting 211.70 MWh of AC power into the grid yearly [24].

Incorrect siting and sizing of Distributed Generation (DG) can have a negative effect on the existing system's technical balance. To improve the radial distribution system performance, a single or hybrid technique for optimal siting and sizing of DG is essential. Recently, a combination of Real Power Loss Sensitivity Index (RPLSI) and Artificial Ecosystem-based Optimization (AEO) was proposed to identify the optimal placement of photovoltaic and wind-powered DG units in a radial distribution system, to improve the voltage profile and reduce power losses [25]. The accurate PV array-based DGs siting in RDS by using a differential evolution (DE) algorithm is presented in [26] to obtain the optimal reduction in actual power losses and voltage improvement. Recently, an innovative pathfinder algorithm (PFA) has been developed to identify the best possible locations for incorporating solar-based distributed energy resources (DERs) in a radial distribution system (RDS) [27]. This algorithm leverages a backtracking search optimization technique to reduce active power losses [28]. Moreover, an ALOA algorithm has been proposed to identify the most suitable size and position of photovoltaic (PV) and wind-based DERs, which would ultimately reduce power losses, enhance the voltage profile, and improve voltage stability, thus maximizing cost savings [29].

Previous research has highlighted the capacity for the reduction in power losses and improvement in the voltage profile, execution time, and cost savings. This provides an opportunity for further reduction in real power losses, execution time, cost savings, and DG size. To this end, the integration of solar-based DG using Particle Swarm Optimization (PSO) has not been addressed in detail. This study used the Backward Forward Sweep Method (BFSM) to compare the power losses and voltage profile in the IEEE 33-bus system with and without PVDG. Moreover, PSO was deployed to identify the most suitable location and size for photovoltaic-based distributed generators in a radial distribution network. Simulation of the proposed optimized algorithm in MATLAB has been used to generate the results.

The PSO algorithm was first introduced in 1995. Meanwhile, it has been used as a robust technique for solving optimization issues in a wide variety of applications. It is becoming very popular for its simplicity of implementation and also for its ability to quickly converge to a good solution. It requires no information about the gradient of the function to be optimized and uses only primitive mathematical operators. Compared to other optimization methods, it is faster, cheaper, and more efficient [30–32]. In addition, there are a few parameters to adjust in PSO. Thus, PSO is well suited to solving non-linear, non-convex, continuous, discrete, and integer variable problems. On the other hand, this algorithm does not always work well and there is still room for development. In comparison to other optimization techniques, along with Genetic Algorithms (GA), Differential Evolution (DE), or Simulated Annealing (SA), PSO frequently reveals faster convergence, superior international exploration capabilities, and ease of implementation. But the choice of optimization technique relies upon on the unique characteristics and complexity of the hassle, and in a few cases, other algorithms may additionally outperform PSO below positive situations. Consequently, it is far more crucial to remember the problem's nature and necessities earlier than deciding on the maximum appropriate optimization method for a given radial disbursed strength machine.

This research's main aims are to compute the optimal size and position of single and multiple PVDG units for reducing the real power losses, boosting the voltage profile, and maximizing cost savings by using the PSO algorithm. In this work, the PSO is utilized by handling the random constraints of the original PSO algorithm to improve the efficiency and symmetry of a distribution network. Moreover, Improvement in the voltage profile, Reduction in Active Power Loss (%), Reduction in Reactive Power Loss (%), Execution Time (Sec), and Maximum Cost Savings (USD) of the radial distribution system have been achieved through proposed work by handling the random constraints. On the other hand, no existing published work has achieved all five above-mentioned improvements simultaneously. The main contributions of this paper are listed below precisely.

i. Photovoltaic PV distributed generation, as well as constant load, is all factored into the RDG sizing and allocation problem.
ii. The stochastic characteristics are achieved by using appropriate probability density functions (PDFs).
iii. The Particle Swarm optimization algorithm (PSO), a metaheuristic algorithm, is used to determine the optimal solution with high exploitation potential and exploration aptitude.
iv. The FBSM load flow approach is used to calculate the number of power losses and voltage profiles or symmetry/asymmetry in the voltages.
v. PVDG is injected into the RDS at its optimal location and sizing to minimize the active power loss, reactive power loss, cost savings, and improve the voltage profile.
vi. To show the effectiveness and performance of the proposed model, an IEEE 33 RDS is considered.
vii. The simulation results of the proposed technique are compared with those of recently available algorithms in the literature.

This paper is structured in a way to cover the research work in its entirety. Section 2 talks about the problem and the relevant constraints for optimal PVDG placement. Section 3 explains the proposed optimization technique for the placement and sizing of the PVDG. Section 4 evaluates the simulation results obtained from the procedure. Finally, Section 5 sums up the article with the appropriate conclusion.

2. Methodology

Solar-based DGs at their optimal size and location in the radial distribution system are shown in Figure 1. This figure illustrates that all the data are provided to the control system, which decides the optimal location and size of DGs by using PSO. Additionally, it calculates the voltage profile, active and reactive power losses, and annual cost savings in radial distribution systems.

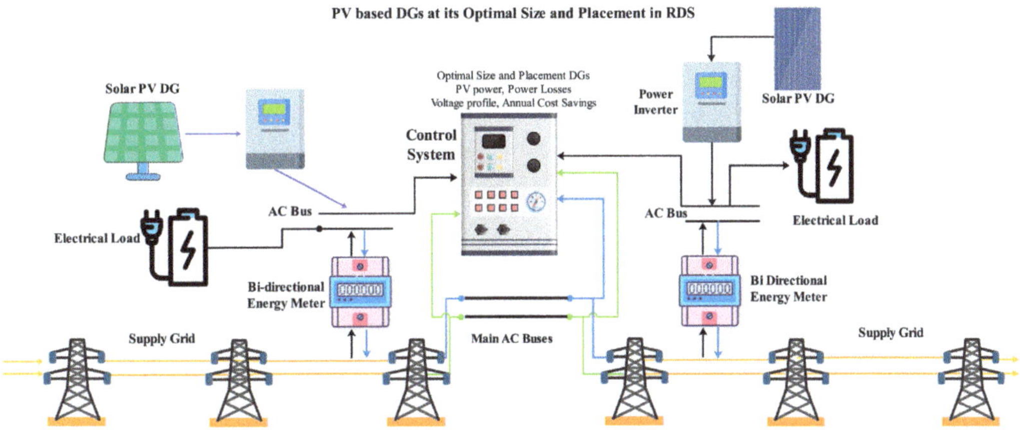

Figure 1. Electric power distribution system with PVDGS.

This proposed work is dedicated to inserting the DG at its optimal location and sizing in the radial distribution network and improving its technical and economic benefits.

The proposed system involves the layout and development of a model using the PSO approach in MATLAB software. Initially, the existing system is studied, and various technical and economic values can be calculated. Eventually, the particle swarm optimization approach could be applied to decide the optimal location and size of PVDG. Once the optimal places and sizes in its RDS are calculated. For making sure the accuracy and reliability of results obtained from MATLAB, a comparison among the numerous techniques has been accomplished, theory cross-checking the validity of the findings.

3. Problem Formulation

The objective of this study is to optimize the active power losses, reactive power losses, voltage profile improvement, and cost savings. Additionally, the minimum and maximum voltage magnitudes and power balance were used as constraints to ensure the desired outcome of the problem.

3.1. Objective Function

The statistics presented in [33] have shown that the distribution system contains about 13% of electrical power losses from the total power generation. The main purpose of the optimal position and size of PVDGs in a power distribution network is to achieve the maximum possible benefits by increasing the efficiency of the system in terms of reduction in power losses, improvement in the voltage profile, and cost savings. BFSM has been used to compute electrical power losses and voltage profiles [34]. The objectives of minimizing the active (P_L) and reactive (Q_L) power losses have been mathematically formulated as in Equations (1) and (2) [35].

$$\text{Minimize } P_L = \sum_{i=1}^{N} P_{loss} = \sum_{i=1}^{N} I_{br,i}{}^2 \times R_i \; for \; i = 1,2 \ldots N \tag{1}$$

where $I_{br,i}$ and R_i is the *ith* branch current and the branch resistance, respectively.

$$\text{Minimize } Q_L = \sum_{i=1}^{N} Q_{loss} = \sum_{i=1}^{N} I_{br,i}{}^2 \times X_i \;\; for \; i = 1,\ 2 \ldots N \tag{2}$$

where $I_{br,i}$ and X_i is the *ith* branch current and the branch impedance, respectively.

The voltage profile problem of the distribution network is again related to power quality. This is normally less important than the power losses from the utility point of view. However, in the recent era, it looks like due to the penetration of highly intermittent natural renewable-based DGs in power distribution systems, the interest in voltage profiles at the distribution level is increasing. The voltage at different nodes may differ due to sudden changes in load and generation requirements.

$$V_{Profile} = \sum_{i=1}^{n_i} (V_i - V_{rated}) \; where \; i = 1,2 \ldots n \tag{3}$$

V_i is the voltage at bus i and V_{rated} is a rated voltage of the distribution system and selected as 1 p.u. in this study.

where ΔV = total change in voltage profile

$$\Delta V = 1.05p.u. \leq v \geq 0.95p.u. \tag{4}$$

3.2. Constraints

There are two types of constraints: equality constraints, and inequality constraints.

3.2.1. Equality Constraints

The following operating conditions must be fulfilled during the optimization process.

$$P_{Grid} + P_{DG} = P_{Loss} + P_{Load} \quad (5)$$

$$Q_{Grid} + Q_{DG} = Q_{Loss} + Q_{Load} \quad (6)$$

where, P_{Grid} and Q_{Grid} are the total active and reactive power inserted by the grid into the system. P_{DG} and Q_{DG} are active and reactive power injected by the distributed generator. P_{Loss} and Q_{Loss} are active and reactive power losses. P_{Load} and Q_{Load} are active and reactive power consumed, respectively.

3.2.2. Inequality Constraints

- Voltage Limitation.
- For keeping a proper stable voltage magnitude or voltage symmetry of the whole IEEE 33-bus network, the absolute voltage value at all nodes of the distribution system should meet the defined constraints.

$$V_{min} \leq |V_i| \leq V_{max} \quad (7)$$

- Current Limitation.
- For keeping a proper current flow in all branches, it should not exceed the rated limit. The absolute value of the current at all nodes of the RDS should meet the defined constraints.

$$|I_{ij}| \leq |I_{ij}|^{max} \quad (8)$$

- Thermal line restriction:
- The thermal line restriction condition is mentioned in Equation (9).

$$|S_{li}| \leq |I_{li}|^{max} \quad (9)$$

3.3. *Photovoltaic (PV)-Based DG Model*

Solar-based DGs or PV modules convert sunlight directly into electrical power. The amount of power generation is directly dependent on the sunlight intensity. This power generation is in the form of DC and the demand side consumes power in the form of AC. An inverter is a device that is used to transform direct current (DC) power from a solar generator into alternating current (AC). The converter would provide compatible AC output power with the AC utility distribution system. According to a paper [36], the output power capacity (P_{cpv}) of the PVDG is as follows.

$$P_{cpv} \rightarrow f(A_{sp}, I_{solar}, \mu_{sp}) \quad (10)$$

where A_{sp} is the area of solar panels; I_{solar} is solar irradiance, which is the function of time; and μ_{sp} is solar cells' efficiency in the PVDG. The calculation of the $P_{cpv}(\Delta t)$ at a time instance using the equation is given below.

$$P_{cpv}(\Delta t) = A_{sp} \times I_{solar} \times \mu_{sp}(\Delta t) \quad (11)$$

Therefore, the power generated from solar photovoltaic panels can be considered power generated from a non-dispatchable source. Another important feature of this source is that it provides active and reactive power (either stable or unity power factor depending on the usage of the converter). If this source needs to provide power with a constant power factor, then a static electronic converter is used. The PVDG model is generally considered a constant power factor model. The maximum power evaluation of the PVDG (Pmax) has been computed using the equation given below.

$$P_{max} = \frac{1}{m \to n} \sum_{m=1}^{n=24} P_{cpv}(\Delta t)_{mn} \tag{12}$$

4. Particle Swarm Optimization (PSO) Algorithm

The Particle Swarm Optimization algorithm is a powerful tool for solving optimization problems in a stochastic manner. It mimics the behavior of animals that search for food in groups, such as a school of fish or a flock of birds. This technique is useful in finding the optimal solution in each search space. Many researchers have considered the use of this technique due to its substantiated strength, ease of implementation, and universal examination ability in many applications. This optimization technique was introduced by Kennedy and Ebert in 1995, in which a group of the swarm (named population) was randomly created. Every particle inside the search space had an individual momentum and speed in correlation with the object. This speed and direction would be adjusted based on the particle's history of the best experiences and the collective best experiences of its surroundings. Due to this, the particle has the tendency to move in a particular direction toward the desired goal in the search region [30–32]. Each particle moves in an N-dimensional search space with the position and velocity of a particle could be updated by using Equations (13) and (14) as given.

$$\mathrm{V}_{\mathrm{p}}^{\mathrm{k}+1} = \omega \mathrm{V}_{\mathrm{p}}^{\mathrm{k}} + \mathrm{c}_1 * \mathrm{rand}_1 * \left(\mathrm{p}_{\mathrm{best}} - T_{\mathrm{p}}^{\mathrm{k}}\right) + \mathrm{c}_2 * \mathrm{rand}_2 * \left(\mathrm{g}_{\mathrm{best}} - T_{\mathrm{p}}^{\mathrm{k}}\right) \tag{13}$$

$$T_p^{k+1} = T_p^k + \gamma * V_p^{k+1} \tag{14}$$

- T^k is the present search point and Tk+1 is the changed search point.
- V^{k} is the present velocity and Vk+1 is the changed velocity.
- c_1 and c_2 are weighing coefficients.
- $rand_1$ and $rand_2$ are random numbers [0, 1]; $\mathrm{c}_1 = \mathrm{c}_2 = 2$; inertia weight is $\omega = \omega_{max} - k(\omega_{max} - \omega_{min})/k_{max}$ and $\omega_{min} = 0.4, \omega_{max} = 0.9$ [37]. K and k_{max} are present and the maximum iteration number, respectively.

The proposed model is illustrated in Figure 2, which utilizes the Particle Swarm Optimization (PSO) technique to identify the best location and size of PVDG. This method enables the model to achieve an optimal solution. The algorithm begins by setting the input parameters and selecting the line and bus data of the IEEE 33-bus system. The FBSM is used to evaluate the number of power losses and voltage profiles (symmetric or asymmetric voltages) before the integration of the DG (Distributed Generation). The PSO (Particle Swarm Optimization) algorithm is then applied to identify the appropriate placement and size of the PVDG (Photovoltaic Distributed Generation). In each round, the FBSM is again used for computing the voltage profiles and power losses. The proposed model obtains the best position and size of PVDG, which lessens power losses, reduces cost, and improves the voltage profile for the target issues. The conforming DG fitness value represents the improvement for the mentioned problems.

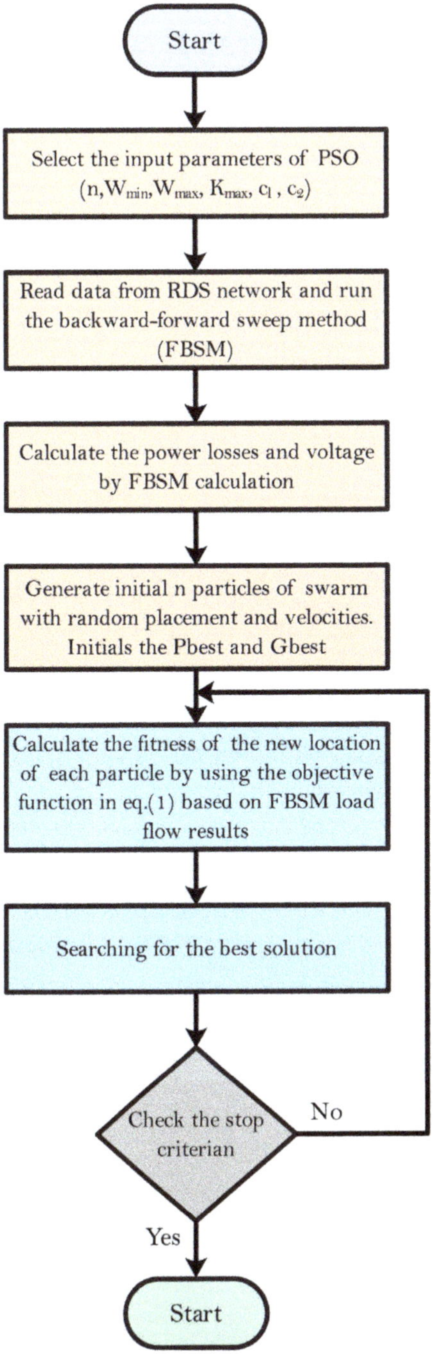

Figure 2. Flowchart of proposed PSO algorithm.

5. Results and Discussion

In the first stage, a base distribution system without DG was considered for analysis. Hereafter, a comparative study of the RDS with and without PVDG was considered. The proposed metaheuristic method was executed in MATLAB 2018a software. The simulation was been performed on a laptop with specifications of Intel® Core (TM) i7-3720 QM CPU@2.60 GHz and 4 GB RAM.

5.1. Performance Analysis of IEEE 33-Bus Power System without PVDG

The efficiency of the suggested approach was validated by using the IEEE 33-bus system, which is the benchmark for testing the performance of any system. The following performance parameters have been selected for comparison, i.e., active (Ploss), reactive (Qloss) power losses, and voltage profile (V). The 33 kV bus network was selected without PVDG and the input data of the proposed model was taken from the line and load data of the IEEE 33-bus system.

The IEEE 33-bus RDS, shown in Figure 3 [38], is composed of 33 buses and 32 lines. It is a standard type of network and is widely used in power sector research. The impedance of each line has different values, and this power distribution system is connected to a centralized power grid system. Different power sources like hydro, coal, nuclear, ocean, wind, PV, and geothermal power plants are connected to the grid as centralized power sources. The maximum and minimum voltage limits have been considered at ±5 for all buses of the network. The voltage level of all buses is 12.66 kV the load of the total active power is 3.715 MW, and the load of total reactive power is 2.3 MVAR.

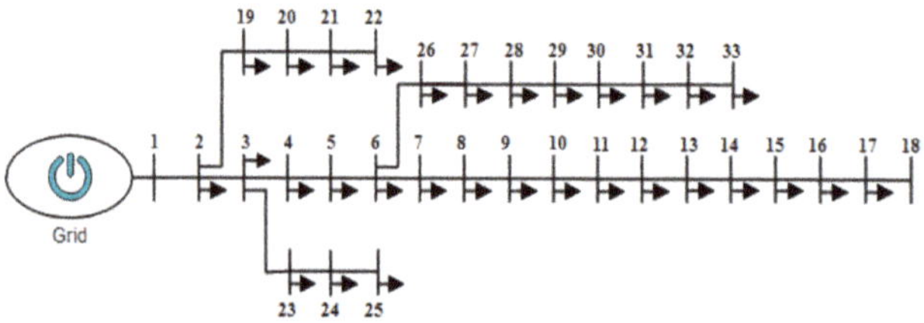

Figure 3. IEEE 33 radial distribution network.

The real power of the system with the load can be seen in Figure 4. It is noted that the initial active (Pi) and reactive (Qi) powers were zero at bus one and they changed from bus two to bus thirty-three. The total active power load was 3715 kW, and the total reactive power load was 2300 kVAR. The highest active power values were found at buses twenty-four and twenty-five, both having 420 kW, and the highest reactive power was at bus thirty with 600 kVAR. The minimum active and reactive powers were both zero at bus one.

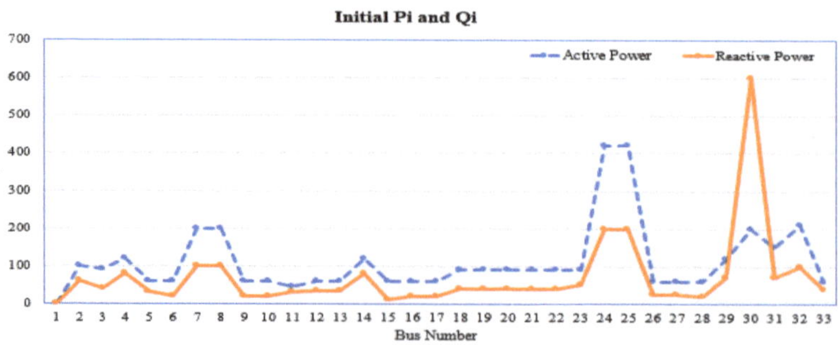

Figure 4. A load data plot (active and reactive power) for IEEE 33-bus system.

The impedance of the bus system is demonstrated in Figure 5, which was derived from the IEEE 33-bus network. This figure provides a graphical representation of the resistance and reactance of the line. The maximum resistance is observed at 1.542 ohms at line 19 and reactance is observed at 1.7210 ohms at line 16. The minimum resistance is 0.0922 ohms and a reactance of 0.0470 ohms has been observed at line one.

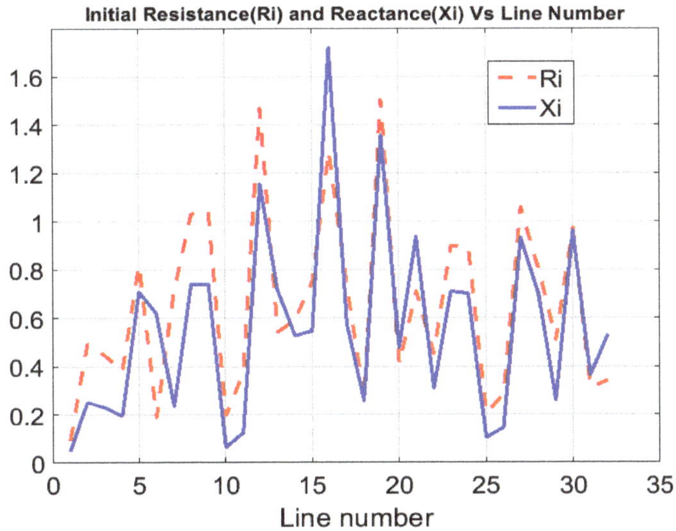

Figure 5. A plot of take-out load data (Resistance and reactance) for the IEEE 33-bus network.

The computation of the base impedance (Z_b) has been performed by assuming the kV and MVA of the IEEE 33-bus network. Base impedance is calculated using Equation (15).

$$Z_b = \frac{V^2}{MVA} \tag{15}$$

Furthermore, the per-unit value (p.u.) of resistance "$(R)_{p.u}$" and reactance "$(X)_{p.u}$" of each line is calculated as given in Equations (16) and (17).

$$(R)_{p.u} = \frac{R_i}{Z_b} \tag{16}$$

where R_i represents the preliminary value of resistance obtained from the line database of the network.

$$(X)_{p.u} = \frac{X_i}{Z_b} \tag{17}$$

where X_i represents the preliminary value of reactance also obtained from the line database of the network.

Figure 6 shows a graph of the obtained value of resistance $(R_{p.u})$ and reactance $(X)_{p.u}$ against thirty-two lines of the 33-bus system. It is observed in the plot that the highest resistance is 0.93850849 shown in line number 19 and the highest value of reactance is 1.073775 shown in line 16. In line number 1, the lowest resistance of 0.057525912 along with the reactance value of 0.029324 is observed.

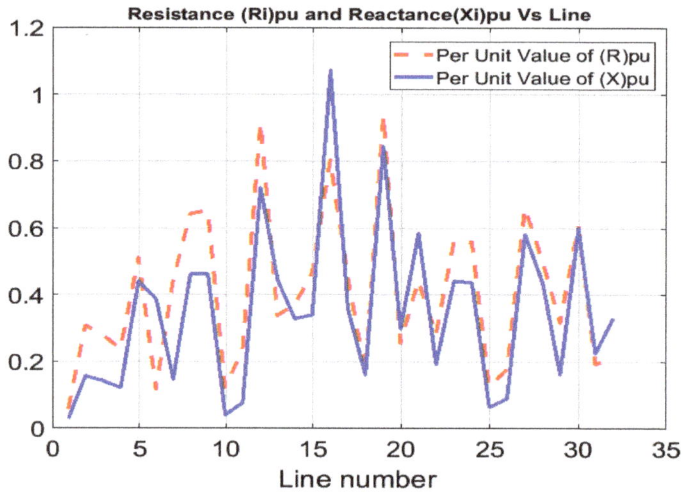

Figure 6. IEEE 33-bus network (per unit value of resistance and reactance for each line).

Figure 7 demonstrates the graph for the analysis of active and reactive power loss without a solar-based DG unit. It is indicated that bus 33 has peak active power losses of 206.95 kW and reactive power losses of 137.46 kVAR.

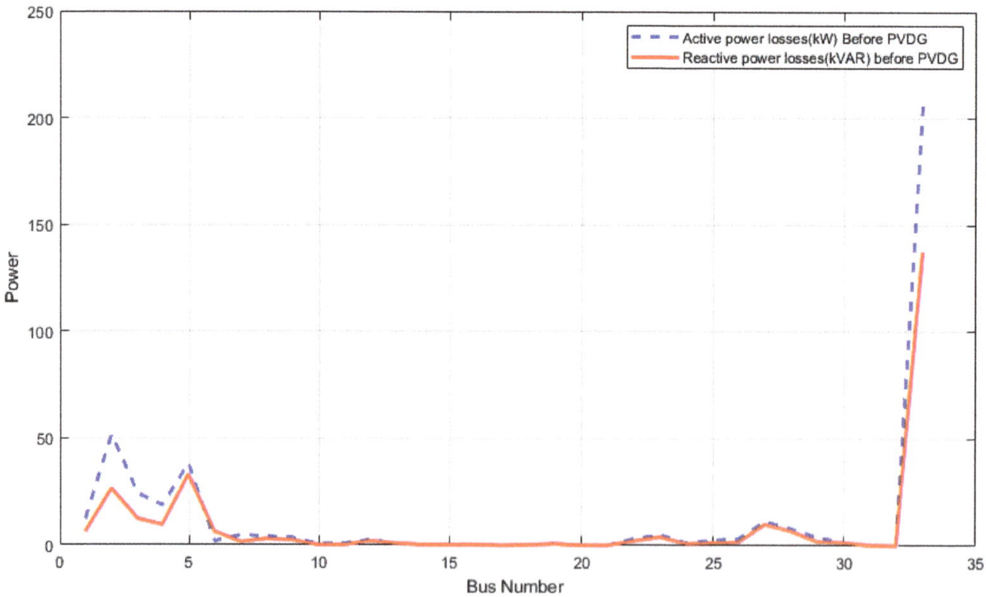

Figure 7. Active and reactive power losses without PVDG.

5.2. *Performance Analysis of IEEE 33-Bus Power System with PVDG*

This section investigates the effect of the optimum placement and size of PVDG in a radial distribution system. All the bus bars of the network are taken into consideration as possible candidates for the integration of PVDG, apart from bus number 1 which is regarded as a slack bus to relate to an external grid utility.

Figure 8 illustrates the optimal location and size of a solar-based distributed generator in a 33-bus system. The RDS system is connected to a centralized power utility with conventional and renewable power sources. When the proposed model is executed, it is noticed that minimum electrical power losses and improved voltage levels are observed at the optimal placement (node number 9) and size (2440 kW) of PVDG.

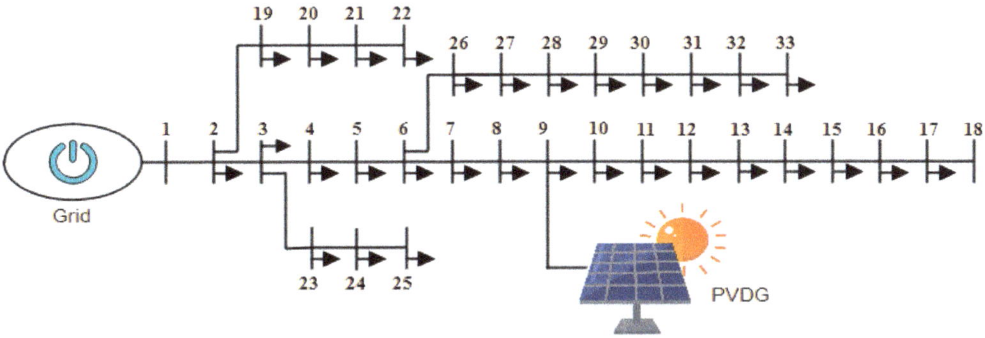

Figure 8. IEEE 33 radial distribution system with PVDG.

The optimized location and size of the PVDG have enabled a significant decrease in active and reactive power losses. The active power loss decreased from 206.95 kW to 91.75 kW, and the reactive power loss decreased from 137.46 kVAR to 64.79 kVAR, as demonstrated in Figure 9.

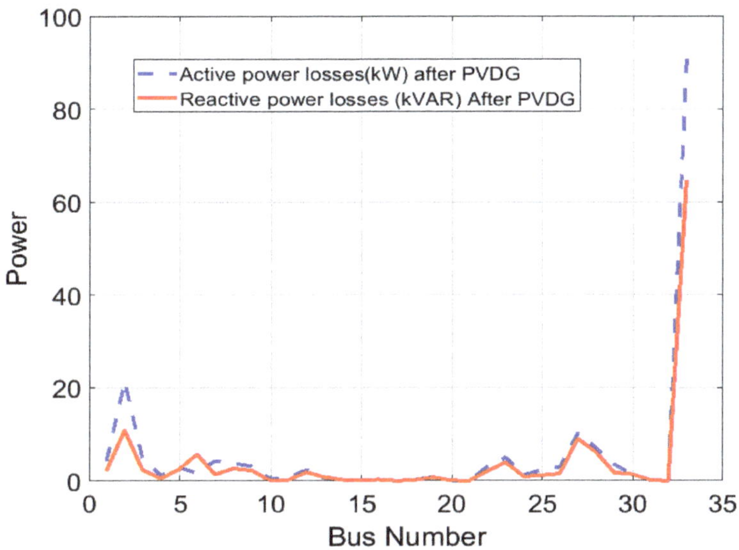

Figure 9. Active and reactive power losses with PVDG.

The effect of the PVDG unit on the voltage profile of a system is shown in Figure 10. The comparison between the system voltage profile with and without the PVDG system is visible. It is observed that the voltage profile improved when solar-based DG is integrated into the 33-bus system. At node 18, the voltage was observed to be 0.9116 per unit, which was improved to 0.9575 per unit when the PVDG system was incorporated. The highest voltage was observed at node one, which was 1 per unit. Incorporating the PVDG system, the overall voltage profile of the system was improved by 3.48%, providing better performance.

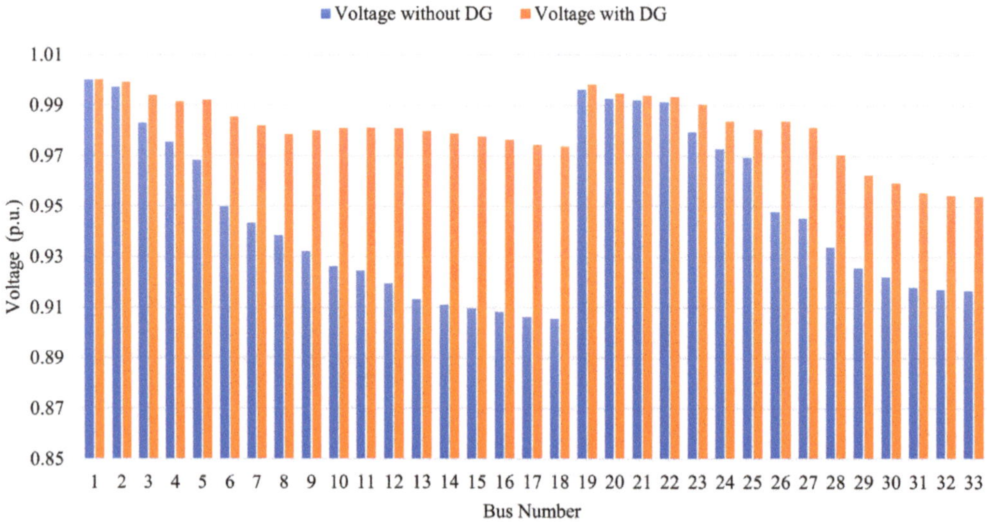

Figure 10. Voltage profile without PVDG and with PVDG.

The total electrical power losses in the RDS before the integration of the solar-based distributed generators were 206.95 kW of active power and 137.46 kVAR of reactive power. Table 1 provides the results of the model, which point out that total active and reactive power losses decreased by 55.66% and 52.85%, respectively after the integration of a 2440 kW PVDG unit. Additionally, the voltage level increased from 0.9116 p.u. to 0.9575 p.u., providing better overall performance. The execution time is just 3.254238 s.

Table 1. Main results found by the proposed model (IEEE 33-bus system).

Subject	without PVDG	with PVDG
Total Active Power loss (kW)	206.95	91.75
Total Reactive Power loss (kVAR)	137.46	64.79
Loss decrement in Pi	-	55.66%
Loss decrement in Qi	-	52.85%
Minimum Voltage V (p.u.)	0.9116@bus 18	0.9575@bus 18
Maximum Voltage V (p.u.)	0.9970@bus 2	0.9985@bus 2
Cost of losses ($)	108,772.92	48,223.4
Saving ($/year)	-	60,549.12
Total DG (Size@Location)	-	2440 kW@bus 9
Execution time (s)	-	3.254238

The proposed system energy cost savings results are shown in Figure 11, which shows that power losses annual saving cost has been increased. If the cost of electric power energy is taken at $0.06, the annual energy saving cost is $60,527.12, which is higher than the annual cost savings presented in [12,16,38].

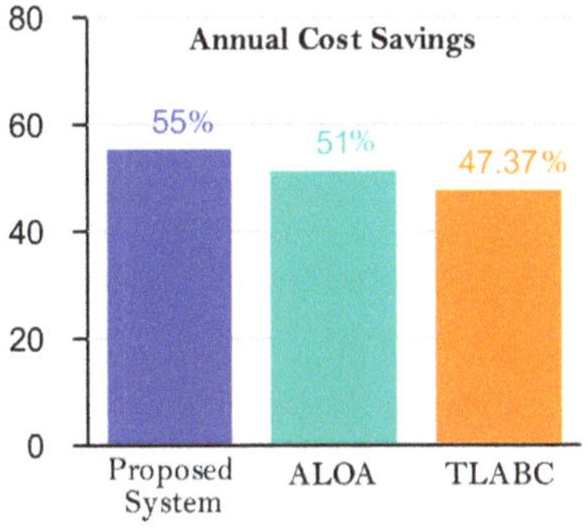

Figure 11. Comparison of annual saving cost of power losses.

The proposed metaheuristic algorithm improves convergence features using less computation time in addition the execution period for the proposed model system is 8.62 s to complete and get optimized results This study has established that the PVDG unit is most effective when placed at the 9th bus in the system with a capacity of 2.44 MW. This significant decrease in power losses has been noted, with a drop of 55.66% in active power losses and 52.78% in reactive power losses. This is an impressive achievement. The proposed results have less active power losses in RDS as compared with the firefly technique presented in [11] as shown above in Table 2. The results comparison has also been carried out with the GA technique [12], in which DG's size is 2.89 MW with a power loss reduction of 46.65%. The proposed model provides better results for active and reactive power when compared with [16,25–27,38] as given in Table 2. To evaluate the proposed research, a comparison was made with a PSO-based optimization technique [13]. This technique only considered the optimal sizing and location in terms of reduced active power loss and improved voltage profile. On the other hand, the proposed model considers additional parameters as listed in Section 2.

Table 2. Comparison of the proposed technique with existing control algorithms.

Author	Year	Control Algorithm	Min. Voltage Improved@Bus	Reduction in Active Power Loss (%)	Reduction in Reactive Power Loss (%)	Execution Time (s)	Maximum Cost Savings (USD)
Remha et al. [11]	2017	FFA	0.9412@18	47.39%	-	-	-
T. Matlokosti [12]	2017	GA	0.9175@18	46.65%	-	-	-
E.S. Ali [38]	2017	ALOA	0.9503@18	51.15%	42.88%	-	$56,726.5
M. Khasanov [16]	2019	TLABC	0.94237@18	47.37%	42.891%	-	$52,536.3
M. Khasanov [25]	2020	AEO	0.94237@18	47.37%	-	-	-
V Janamala [27]	2020	PFA	0.9424@18	47.38%	42.89%	25.342	-
J. Urinby [26]	2021	DE	0.95836@18	47.38%	-	-	-
Rekha C. M. [13]	2022	PSO	0.9180@17	49.28%	32.38%	-	-
Proposed System	2023	PSO	0.9575@18	55.66%	52.78%	3.254238	$60,527.12

6. Conclusions

Particle Swarm Optimization (PSO) is employed to effectively position and adjust the size of a solar-based Distributed Generator (DG) to minimize power losses and improve the voltage profile. To assess the performance of this model, the IEEE 33-bus system was used. The results show that power losses were minimized, the voltage profile improved, and cost savings were maximized when using PVDG units. The results show that the PVDG PSO-based model offers less active power losses as compared to the non-PVDG one. The proposed solution has proved to be superior to other techniques, as it can accurately pinpoint the optimal location and size of the PVDG. This makes it an invaluable tool for the purpose. Furthermore, the insertion of the solar PVDG into a 33-bus system can lessen energy (active and reactive power) losses to 55.66% and 52.78% when compared with the base case. So, it can be concluded that solar-based DGs with PSO algorithm can be a better choice for decreasing electrical power loss, improving in voltage profile, and increasing cost savings in the power RDS system. Additionally, the overall efficiency of the RDS network has been improved. The proposed model can ease the dependency of the utility system during the load demand. PVDG units can be installed in the area where sunlight intensity is adequate.

Future research could focus on the effects of combining PV and wind turbines in RDS along with the addition of energy storing systems. Additionally, the development of a wind-based DG model and the comparison of solar and wind DG results could be explored. Finally, future efforts should be devoted to addressing the uncertainty in load requirements.

Author Contributions: Conceptualization, A.R.B. and M.A.A.; methodology, M.A.A.; software, M.A.A.; validation, A.R.B., A.R., M.F. and E.E.; formal analysis, M.A.A.; investigation, M.A.A.; resources, A.R.B., A.R. and E.E.; data curation, A.R.B.; writing—original draft preparation, M.A.A. and M.F.; writing—review and editing, A.R.B. and A.R.; visualization, A.R.B.; supervision, A.R.B.; project administration, A.R.B. All authors have read and agreed to the published version of the manuscript.

Funding: This research received no external funding.

Data Availability Statement: IEEE 33-bus data was utilized in this study and is publicly available on the IEEE website.

Acknowledgments: The authors acknowledge the support of IEEE for the provision of the 33 bus data publicly utilized for research purposes.

Conflicts of Interest: The authors declare no conflict of interest.

References

1. Chiradeja, P.; Ramakumar, R. An approach to quantify the technical benefits of distribution system. *IEEE Trans. Energy Convers.* **2004**, *19*, 764–773.
2. Pepermans, G.; Driesen, J.; Haeseldonckx, D.; D'haeseleer, W.; Belmans, R. Distribution generation: Definition, benefits and issues. *Energy Policy* **2005**, *33*, 787–798.
3. Mohammedi, R.D.; Hellal, A.; Arif, S.; Mosbah, M. Optimal DG Placement and Sizing in Radial Distribution Systems Using NSGA-II for Power Loss Minimization and Voltage Stability Enhancement. *Int. Rev. Electr. Eng.* **2013**, *8*, 6.
4. Sun, K.; Yao, W.; Fang, J.; Ai, X.; Wen, J.; Cheng, S. Impedance modeling and stability analysis of grid-connected DFIG-based wind farm with a VSC-HVDC. *IEEE J. Emerg. Sel. Top. Power Electron.* **2019**, *8*, 1375–1390.
5. Layadi, T.M.; Champenois, G.; Mostefai, M. Economic and ecological optimization of multi-source systems under the variability in the cost of fuel. *Energy Convers. Manag.* **2018**, *177*, 161–175.
6. Zhang, X.; Tan, T.; Zhou, B.; Yu, T.; Yang, B.; Huang, X. Adaptive distributed auction-based algorithm for optimal mileage based AGC dispatch with high participation of renewable energy. *Int. J. Electr. Power Energy Syst.* **2020**, *124*, 106371.
7. Wang, Q.; Yao, W.; Fang, J.; Ai, X.; Wen, J.; Yang, X.; Xie, H.; Huang, X. Dynamic modeling and small signal stability analysis of distributed photovoltaic grid-connected system with large scale of panel level DC optimizers. *Appl. Energy* **2019**, *259*, 114132.
8. Barik, S.; Das, D. A novel Q-PQV bus pair method of biomass DGs placement in distribution networks to maintain the voltage of remotely located buses. *Energy* **2020**, *194*, 116880.
9. Zhang, X.S.; Yang, B.; Yu, T.; Jiang, L. Dynamic Surrogate Model Based Optimization for MPPT of Centralized Thermoelectric Generation Systems Under Heterogeneous Temperature Difference. *IEEE Trans. Energy Convers.* **2020**, *35*, 966–976.
10. Duong, M.Q.; Pham, T.D.; Nguyen, T.T.; Doan, A.T.; Van Tran, H. Determination of Optimal Location and Sizing of Solar Photovoltaic Distribution Generation Units in Radial Distribution Systems. *Energies* **2019**, *12*, 174. [CrossRef]
11. Remha, S.; Chettih, S.; Arif, S. Optimal DG location and sizing for minimum active power loss in radial distribution system using firefly algorithm. *Int. J. Energetica* **2017**, *4*, 6–10.
12. Matlokotsi, T.; Chowdhury, S. Optimal Placement and Sizing of Renewable Distributed Generation in Electricity Networks Considering Different Load Models. *IEEE Conf.* **2017**, *1*, 1–6. [CrossRef]
13. Rekha, C.M.; Shankaralingappa, C.B. Optimal placement of PSO based Photovotaic (PV) distributed generator (DG) in a distribution system. *J. Tianjin Univ. Sci. Technol.* **2022**, *55*, 493–2137.
14. Prakash, D.; Lakshminarayana, C. Multiple DG placements in radial distribution system for multi objectives using Whale Optimization Algorithm. *Alex. Eng. J.* **2018**, *57*, 2797–2806. [CrossRef]

15. Kumar, M.; Nallagownden, P.; Elamvazuthi, I. Optimal Placement and Sizing of Renewable Distributed Generations and Capacitor Banks into Radial Distribution Systems. *Energies* **2017**, *10*, 811. [CrossRef]
16. Khasanov, M.; Kamel, S.; Xie, K.; Zhou, P.; Li, B. Allocation of Distributed Generation in Radial Distribution Networks Using an Efficient Hybrid Optimization Algorithm. In Proceedings of the 2019 IEEE Innovative Smart Grid Technologies-Asia (ISGT Asia), Chengdu, China, 21–24 May 2019; pp. 1300–1305.
17. Sarhan, S.; Shaheen, A.; El-Sehiemy, R.; Gafar, M. An Augmented Social Network Search Algorithm for Optimal Reactive Power Dispatch Problem. *Mathematics* **2023**, *11*, 1236.
18. Shaheen, A.M.; Elsayed, A.M.; Ginidi, A.R.; El-Sehiemy, R.A.; Elattar, E. A heap-based algorithm with deeper exploitative feature for optimal allocations of distributed generations with feeder reconfiguration in power distribution networks. *Knowl. -Based Syst.* **2022**, *241*, 108269.
19. Mahfoud, R.J.; Alkayem, N.F.; Sun, Y.; Haes Alhelou, H.; Siano, P.; Parente, M. Improved Hybridization of Evolutionary Algorithms with a Sensitivity-Based Decision-Making Technique for the Optimal Planning of Shunt Capacitors in Radial Distribution Systems. *Appl. Sci.* **2020**, *10*, 1384.
20. Jamil Mahfoud, R.; Sun, Y.; Faisal Alkayem, N.; Haes Alhelou, H.; Siano, P.; Shafie-Khah, M. A Novel Combined Evolutionary Algorithm for Optimal Planning of Distributed Generators in Radial Distribution Systems section. *Appl. Sci.* **2019**, *9*, 3394.
21. Cui, Z.; Wu, J.; Ding, Z.; Duan, Q.; Lian, W.; Yang, Y.; Cao, T. A hybrid rolling grey framework for short time series modelling. *Neural Comput. Appl.* **2021**, *33*, 11339–11353.
22. Zhao, S.; Wu, Y.; Tan, S.; Wu, J.; Cui, Z.; Wang, Y.G. QQLMPA: A quasi-opposition learning, and Q-learning based marine predators' algorithm. *Expert Syst. Appl.* **2023**, *213*, 119246.
23. Tan, S.; Zhao, S.; Wu, J. QL-ADIFA: Hybrid optimization using Q-learning and an adaptive logarithmic spiral-levy firefly algorithm. *Math. Biosci. Eng.* **2023**, *20*, 13542–13561.
24. Tamoor, M.; Bhatti, A.R.; Farhan, M.; Zaka, M.A.; ZakaUllah, P. Solar Energy Capacity Assessment and Performance Evaluation of Designed Grid-Connected Photovoltaic Systems. *Appl. Sci.* **2023**, *37*, 39. [CrossRef]
25. Khasanov, M.; Kamel, S.; Tostado-Véliz, M.; Jurado, F. Allocation of Photovoltaic and Wind Turbine Based DG Units Using Artificial Ecosystem-based Optimization. In Proceedings of the 2020 IEEE International Conference on Environment and Electrical Engineering and 2020 IEEE Industrial and Commercial Power Systems Europe, Madrid, Spain, 9–12 June 2020; pp. 1–5.
26. Urinboy, J.; Hasanov, M. Improvement Performance of Radial Distribution System by Optimal Placement of Photovoltaic Array. *Int. J. Eng. Inf. Syst.* **2021**, *5*, 157–159.
27. Janamala, V. A new meta-heuristic pathfinder algorithm for solving optimal allocation of solar photovoltaic system in multi-lateral distribution system for improving resilience. *SN Appl. Sci.* **2021**, *3*, 118.
28. El-Fergany, A. Optimal allocation of multi-type distributed generators using backtracking search optimization algorithm. *Int. J. Electr. Power Energy Syst.* **2015**, *64*, 1197–1205.
29. Ali, E.S.; Abd Elazim, S.M.; Abdelaziz, A.Y. Ant Lion optimization algorithm for optimal location and sizing of renewable distrib-uted generations. *Renew Energy* **2017**, *101*, 1311–1324.
30. Atwa, Y.M.; El-Saadany, E.F.; Salama, M.M.A.; Seethapathy, R. Optimal Renewable Resources Mix for Distribution System Energy Loss Minimization. *IEEE Trans. Power Syst.* **2009**, *25*, 360–370.
31. Kennedy, J.; Eberhart, R. Particle Swarm Optimization. In Proceedings of the ICNN'95—International Conference on Neural Networks, Perth, Australia, 27 November–1 December 1995; Volume 4, pp. 1942–1948. [CrossRef]
32. Eberhart, R.; Kennedy, J. A new optimizer using particle swarm theory. In Proceedings of the Sixth International Symposium on Micro Machine and Human Science, Nagoya, Japan, 4–6 October 1995; pp. 39–43. [CrossRef]
33. Bohre, A.K.; Agnihotri, G.; Dubey, M. Hybrid butterfly-based particle swarm optimization for optimization problems. In Proceedings of the First International Conference on Networks and Soft Computing (ICNSC), Guntur, India, 19–20 August 2014; pp. 172–177. [CrossRef]
34. Zeinalzadeh, A.; Mohammadi, Y.; Moradi, M.H. Optimal multi objective placement and sizing of multiple DGs and shunt capacitor banks simultaneously considering load uncertainty via MOPSO approach. *Int. J. Electr. Power Energy Syst.* **2015**, *67*, 336–349.
35. Haque, M. Efficient load flow method for distribution systems with radial or mesh configuration. *IEE Proc. Gener. Transm. Distrib.* **1996**, *143*, 33–38.
36. Salkuti, S.R.; Battu, N.R. An effective network reconfiguration approach of radial distribution system for loss minimization and voltage profile improvement. *Bull. Electr. Eng. Informatics* **2021**, *10*, 1819–1827. [CrossRef]
37. Eberhart, R.C.; Shi, Y. Comparing inertial weights and constriction factor in particle swarm optimization. In Proceedings of the International Congress on Evaluating Computation, La Jolla, CA, USA, 16–19 July 2000; pp. 84–89.
38. Vita, V. Development of a Decision-Making Algorithm for the Optimum Size and Placement of Distributed Generation Units in Distribution Networks. *Energies* **2017**, *10*, 1433. [CrossRef]

symmetry

MDPI

Article

Utilizing Full Degrees of Freedom of Control in Voltage Source Inverters to Support Micro-Grid with Symmetric and Asymmetric Voltage Requirements

Akhtar Rasool [1,*], Fiaz Ahmad [2], Muhammad Salman Fakhar [3], Syed Abdul Rahman Kashif [3] and Edwin Matlotse [1]

[1] Department of Electrical Engineering, University of Botswana, Gaborone UB 0061, Botswana
[2] Department of Electrical and Computer Engineering, Air University, Islamabad 44000, Pakistan
[3] Department of Electrical Engineering, University of Engineering and Technology, Lahore 54000, Pakistan
* Correspondence: rasoola@ub.ac.bw or akhtar@alumni.sabanciuniv.edu

Abstract: This article proposes a novel equivalent control method for voltage source inverters (VSI) with disturbance observers (DOB) to support the symmetric and asymmetric voltage requirements of a micro-grid (MG) while also matching the MG output power requirements. The method leverages the degrees of freedom (DOF) of the VSI under symmetric and asymmetric MG voltage conditions by utilizing the mean-point voltage of the MG, which is often overlooked in literature studies due to this being grounded. The method enables the three-phase inverter to generate voltages as needed by the MG inconsistently due to changing loads in the MG circuits or phases. The method is also insensitive to disturbances because of the DOB, being part of the controller. The proposed method is validated under both the balance and imbalance voltage demands of the MG. The mean voltage of the MG is used as a set-point to be corroborated as a mean voltage at the inverter's output, in addition to active-reactive power references. The novel model is developed by augmenting the new, mean-point voltage as part of the system dynamics. The proposed method is simulated in MATLAB/Simulink® and is verified for its hardiness and effectiveness.

Keywords: voltage source inverter control; disturbance estimation; voltage sag; voltage swell; distributed generator; micro-grid

1. Introduction

With the advent of modern applications and the expanding populations of cities, the demand for electricity is increasing unchecked. This demand is bringing a great deal of pressure on the already functioning power system's infrastructure. Not restricted to the infrastructural issues only, power generation is also being constrained by the limited fossil fuels and changing trends towards clean and renewable energy options. The renewable energy sources (RES) have their own challenges because of their fluctuating nature, thus leading to open problems to be solved by the research and development community [1,2].

The smart grid (SG) is a modern paradigm of power systems, compared to its traditional version, due to having the state of the art of all engineering and scientific fields merged into one system. SG is a novel framework due to its bidirectional power flows, smart data acquisition methods, remote data operations, smart diagnostics, desirable monitoring, smart communications, distributed controllers, smart energy sharing, consumer participation, renewable energy integration, re-configurable capability and smart conversion functions [1–3]. SG is realized by the integration of MG units, while also connecting it to the upstream power network. MG is a local power system architecture and autonomously controllable power unit, which can function in isolation or in grid-connected modes. MG is generally formed by the integration of RES or by the integration of distributed generators (DGs), where every DG is either based on fuel cells, wind or solar energy. Every DG also functions in a self-controllable manner so as to optimize the energy generation and to meet the needs of the MG [4]. The structure of the MG composed of DGs, depending upon VSI, is given below in Figure 1 [5].

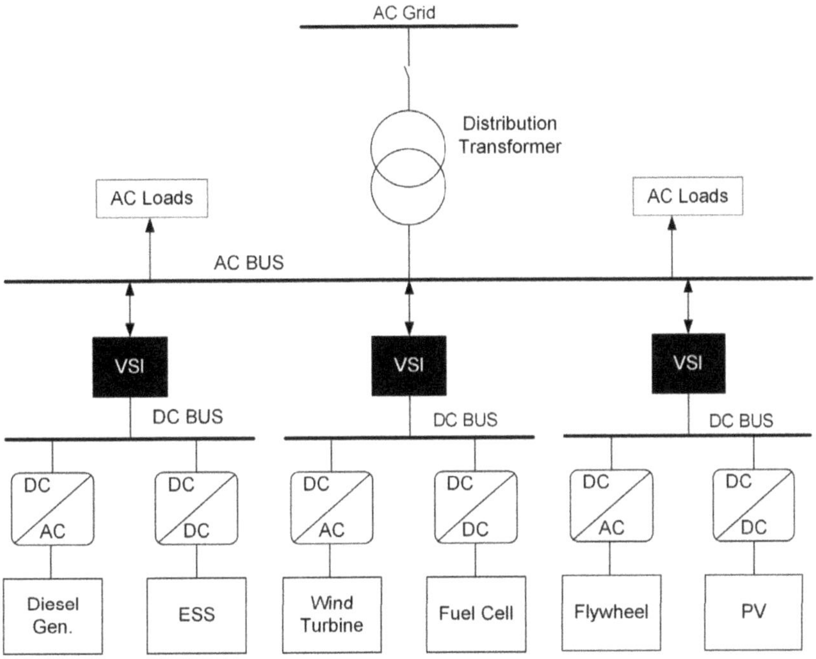

Figure 1. The structure of the MG composed of DGs based on VSI [5].

DG's main causes of tripping from the MG are voltage variations (swell or sag), which are either due to fault conditions or impedance mismatches among the phase circuits of the MG. This tripping of DGs can cause a serious imbalance in the MG circuit, leading to total blackouts. For this reason, grid codes (GCs) suggest three demands [6]:

- The DG must stay connected to the MG for up to 150 ms, even if the MG voltage drops to zero.
- The DG must support voltage recovery by injecting a reactive current into the MG.
- The DG must ramp up the active power to normal operation immediately after clearing the imbalance or fault.

Therefore, DGs are required by the GCs to maintain the MG voltage or MG frequency within the required range and is a matter of converter control. The role of power converters is very unique since the converters are the interfaces enabling the synchronization, voltage control, current control and power flow control in the MG and SG [5,7,8]. A number of studies can be found in the literature regarding the control of three-phase VSI, specifically in view of controlling the imbalance of power in the MG [9–13]. All of these established works offer complexity, since they are designed to maintain pure sinusoidal voltages as the output of the VSI, whose conception dates back to the 1980s due to the stringent supply requirements of the three-phase motors [14–17].

In order to supplement the compensations for the imbalances produced on the grid side, a number of research and control formulations have been applied to VSI. The sliding mode control (SMC) method has also been applied, considering that the converters have switching devices, which in nature are discontinuous and quite consistent with the adopted control technique [18,19]. However, all of these studies are found employing two-dimensional control, which results in the under-utilization of the capabilities of three-phase converters [20–23].

The unknown variables or the control constituents in any system are known as the degrees of freedom (DOF), which are two in the case of three-phase converters in the synchronous reference frame, meaning that the third degree is free for utilization [22]. The random nature of the loads in the distribution system connected to the MG introduces a demand for imbalance voltage generation or imbalance current compensation from the VSI. This asymmetric voltage generation demand from the VSI requires it to employ the full capabilities (DOF) of its switching matrix [24]. All three-phase converters have three independent control inputs but, in general, in all techniques, the desired outputs are defined either as control of current (orthogonal dq-/$\alpha\beta$-frames) or control of power (active and reactive, power frames), resulting in 2D control requirements. Hence, the target is to use the third DOF to provide a much needed and desired output in addition to two orthogonal currents or powers.

In this article, the superfluous DOF available in the VSI's switching matrix is used for the voltages' asymmetry as an additional requirement to bring a balance between the MG (with loads) and the VSI source. The organization of the paper follows in a logical manner. First, the system's novel model is developed, and then the novel control formulation is established, and finally the simulation results are described to demonstrate the method's robust and vigorous performance.

2. System's Modeling

The renewable energy system's output can be modeled using a DC link capacitor, which, through a VSI, a 3-phase filter, and a 3-phase inductance (equivalence of transmission line), is connected to a 3-phase MG. The grid is considered to be working either in balance or in imbalance conditions due to the contingent nature of the real-time loads or the occurrence of faults. A generalized diagram of the VSI source connected to the three-phase MG is given in Figure 2.

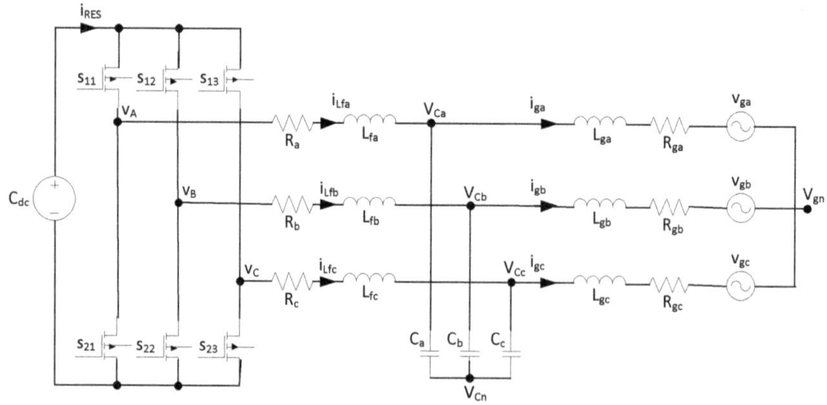

Figure 2. Interconnecting structure of RES and the MG [24].

The active and reactive or orthogonal powers can be evaluated at the MG end using expression (1).

$$\begin{bmatrix} P_g \\ Q_g \end{bmatrix} = \begin{bmatrix} \mathbf{v}_g^T \cdot \mathbf{i}_g \\ \mathbf{v}_{g\perp}^T \cdot \mathbf{i}_g \end{bmatrix} = \begin{bmatrix} \mathbf{v}_g^T \mathbf{i}_g \\ \mathbf{v}_{g\perp}^T \mathbf{i}_g \end{bmatrix} \quad (1)$$

where

$$\begin{aligned}\mathbf{v}_{\mathbf{g}\perp}^{\mathbf{T}} &= \text{vector orthogonal to MG voltage vector}(\mathbf{v}_{\mathbf{g}}^{\mathbf{T}})\\ \mathbf{v}_{\mathbf{g}}^{\mathbf{T}} &= \begin{bmatrix} v_{ga} & v_{gb} & v_{gc} \end{bmatrix} = \text{voltage vector of MG}\\ \mathbf{i}_{\mathbf{g}}^{\mathbf{T}} &= \begin{bmatrix} i_{ga} & i_{gb} & i_{gc} \end{bmatrix} = \text{current vector of MG.}\end{aligned}$$

Taking the derivatives of power (1) yields

$$\begin{bmatrix} \dot{P}g \\ \dot{Q}g \end{bmatrix} = \begin{bmatrix} \dot{\mathbf{v}}_{\mathbf{g}}^{\mathbf{T}}\mathbf{i}_{\mathbf{g}} + \mathbf{v}_{\mathbf{g}}^{\mathbf{T}}\dot{\mathbf{i}}_{\mathbf{g}} \\ \dot{\mathbf{v}}_{\mathbf{g}\perp}^{\mathbf{T}}\mathbf{i}_{\mathbf{g}} + \mathbf{v}_{\mathbf{g}\perp}^{\mathbf{T}}\dot{\mathbf{i}}_{\mathbf{g}} \end{bmatrix} \tag{2}$$

Considering the L_g values to be zero in Figure 2, Kirchhoff's voltage law gives the grid current expression to be

$$\mathbf{L}_{\mathbf{g}}\frac{d\mathbf{i}_{\mathbf{g}}}{dt} = \mathbf{v}_{\mathbf{c}} - \mathbf{v}_{\mathbf{g}} - \mathbf{R}_{\mathbf{g}}\mathbf{i}_{\mathbf{g}} \tag{3}$$

Now, substituting the grid current dynamics into Equation (2), the following dynamics of active/reactive powers are obtained.

$$\begin{bmatrix} \dot{P}g \\ \dot{Q}g \end{bmatrix} = \begin{bmatrix} \dot{\mathbf{v}}_{\mathbf{g}}^{\mathbf{T}}\mathbf{i}_{\mathbf{g}} - \left(\mathbf{L}_{\mathbf{g}}^{-1}\mathbf{v}_{\mathbf{g}}^{\mathbf{T}}\mathbf{v}_{\mathbf{g}} + \mathbf{L}_{\mathbf{g}}^{-1}\mathbf{R}_{\mathbf{g}}\mathbf{v}_{\mathbf{g}}^{\mathbf{T}}\mathbf{i}_{\mathbf{g}}\right) \\ \dot{\mathbf{v}}_{\mathbf{g}\perp}^{\mathbf{T}}\mathbf{i}_{\mathbf{g}} - \left(\mathbf{L}_{\mathbf{g}}^{-1}\mathbf{v}_{\mathbf{g}\perp}^{\mathbf{T}}\mathbf{v}_{\mathbf{g}} + \mathbf{L}_{\mathbf{g}}^{-1}\mathbf{R}_{\mathbf{g}}\mathbf{v}_{\mathbf{g}\perp}^{\mathbf{T}}\mathbf{i}_{\mathbf{g}}\right) \end{bmatrix} + \mathbf{L}_{\mathbf{g}}^{-1}\begin{bmatrix} \mathbf{v}_{\mathbf{g}}^{\mathbf{T}} \\ \mathbf{v}_{\mathbf{g}\perp}^{\mathbf{T}} \end{bmatrix}\mathbf{v}_{\mathbf{c}} \tag{4}$$

If the errors in the orthogonal powers' exchange are denoted by e_{Pg}, e_{Qg} and the set-points for these powers are represented by P_g^{ref}, Q_g^{ref}, then the dynamics of errors for these powers' $(\dot{e}_{Pg}, \dot{e}_{Qg})$ exchange can be given as below [24]:

$$\begin{bmatrix} \dot{e}_{Pg} \\ \dot{e}_{Qg} \end{bmatrix} = \begin{bmatrix} \dot{P}_g^{ref} - \left(\dot{\mathbf{v}}_{\mathbf{g}}^{\mathbf{T}}\mathbf{i}_{\mathbf{g}} - \mathbf{L}_{\mathbf{g}}^{-1}\mathbf{v}_{\mathbf{g}}^{\mathbf{T}}\mathbf{v}_{\mathbf{g}} - \mathbf{L}_{\mathbf{g}}^{-1}\mathbf{R}_{\mathbf{g}}\mathbf{v}_{\mathbf{g}}^{\mathbf{T}}\mathbf{i}_{\mathbf{g}}\right) \\ \dot{Q}_g^{ref} - \left(\dot{\mathbf{v}}_{\mathbf{g}\perp}^{\mathbf{T}}\mathbf{i}_{\mathbf{g}} - \mathbf{L}_{\mathbf{g}}^{-1}\mathbf{v}_{\mathbf{g}\perp}^{\mathbf{T}}\mathbf{v}_{\mathbf{g}} - \mathbf{L}_{\mathbf{g}}^{-1}\mathbf{R}_{\mathbf{g}}\mathbf{v}_{\mathbf{g}\perp}^{\mathbf{T}}\mathbf{i}_{\mathbf{g}}\right) \end{bmatrix} - \mathbf{L}_{\mathbf{g}}^{-1}\begin{bmatrix} \mathbf{v}_{\mathbf{g}}^{\mathbf{T}} \\ \mathbf{v}_{\mathbf{g}\perp}^{\mathbf{T}} \end{bmatrix}\mathbf{v}_{\mathbf{c}} \tag{5}$$

where

$$\begin{aligned}\mathbf{L}_{\mathbf{g}} &= \text{diagonal MG inductance matrix of order } 3 \times 3\\ \mathbf{R}_{\mathbf{g}} &= \text{diagonal MG resistance matrix of order } 3 \times 3\\ \mathbf{v}_{\mathbf{c}}^{\mathbf{T}} &= \begin{bmatrix} v_{ca} & v_{cb} & v_{cc} \end{bmatrix} = \text{converter voltage vector.}\end{aligned}$$

It is obvious from Equation (5) that the converter output voltage components $(\mathbf{v}_{\mathbf{s}}^{\mathbf{T}})$ are independently affecting the orthogonal power exchange between the RES and the MG, since both the powers are reliant on the collinear vector of the MG voltage $(\mathbf{v}_{\mathbf{g}}^{\mathbf{T}})$ and the orthogonal vector of the MG voltage $(\mathbf{v}_{\mathbf{g}\perp}^{\mathbf{T}})$. As seen in Equation (5), the vector of control $(\mathbf{v}_{\mathbf{c}}^{\mathbf{T}})$ has 3 constituents, while the vector of control error $(\mathbf{e}_{\mathbf{PQ}}^{\mathbf{T}})$ has 2 constituents. Consequently, there is a possibility to use the third available superfluous degree in the vector of control $(\mathbf{v}_{\mathbf{c}}^{\mathbf{T}})$ so as to enforce or satisfy a new requirement of the system, which will accomplish the utilization of the full control capabilities of the VSI. Writing Equation (5) in a compact form yields

$$\begin{bmatrix} \dot{e}_{Pg} \\ \dot{e}_{Qg} \end{bmatrix} = \begin{bmatrix} \dot{P}_g^{ref} - f_P \\ \dot{Q}_g^{ref} - f_Q \end{bmatrix} - \mathbf{L}_{\mathbf{g}}^{-1}\begin{bmatrix} \mathbf{v}_{\mathbf{g}}^{\mathbf{T}} \\ \mathbf{v}_{\mathbf{g}\perp}^{\mathbf{T}} \end{bmatrix}\mathbf{v}_{\mathbf{c}} \tag{6}$$

where

$$\begin{aligned}f_P &= \text{disturbance terms of active power of order } 1 \times 1\\ f_Q &= \text{disturbance terms of reactive power of order } 1 \times 1.\end{aligned}$$

Now, it is important to mention that in the case of a balanced or symmetric MG, the average of the 3-phase voltages equals zero, as denominated by point (v_{gn}) in Figure 2. However, in the case of an imbalanced or asymmetric MG, the average of the 3-phase voltages results in an instantaneous, non-zero value due to the stochastic nature of the real-time loads connected to the MG. It is important to highlight again that this MG's neutral value (v_{gn}) is generally considered to be grounded in the literature [20–23] and negative sequence control schemes are used to compensate for the imbalances appearing in the form of a negative sequence in the MG. Contrarily, in this article, the mean voltage (v_{gn}) of the MG is kept floating and also it is used as an additional requirement of the MG, which is reflected towards the generation side (or the RES-based VSI source side) as a desired additional reference value. In order to obtain the desired results, it is specified in the error dynamics form as given below in Equation (7):

$$\dot{\eta} = f_\eta - \mathbf{L_g}^{-1}\mathbf{b}^\mathbf{T}\mathbf{v_c} = v_{gn} - v_{cn} \tag{7}$$

where

$$\begin{aligned} f_\eta &= \text{disturbance terms of mean MG voltage of order } 1 \times 1 \\ v_{gn} = \tfrac{1}{3}(v_{ga} + v_{gb} + v_{gc}) &= \text{mean voltage of MG of order } 1 \times 1 \\ v_{cn} = \tfrac{1}{3}(v_{ca} + v_{cb} + v_{cc}) &= \text{mean voltage of VSI of order } 1 \times 1 \\ \mathbf{b}^\mathbf{T} &= \begin{bmatrix} \tfrac{1}{3} & \tfrac{1}{3} & \tfrac{1}{3} \end{bmatrix} = \text{mean voltage magnitude vector of order } 1 \times 3. \end{aligned}$$

where the novel variable dynamics ($\dot{\eta}$) show the rate of change for this contemporary voltage requirement (η) of the MG to be produced by the VSI.

After augmenting the newly introduced dynamics (7) of the new requirement into the power control error dynamics (6), the final and complete control error dynamics are found to be

$$\begin{bmatrix} \dot{e}_{Pg} \\ \dot{e}_{Qg} \\ \dot{\eta} \end{bmatrix} = \begin{bmatrix} \dot{P}_g^{ref} - f_P \\ \dot{Q}_g^{ref} - f_Q \\ f_\eta \end{bmatrix} - \mathbf{L_g}^{-1} \begin{bmatrix} \mathbf{v_g^T} \\ \mathbf{v_{g\perp}^T} \\ \mathbf{b^T} \end{bmatrix} \mathbf{v_c} \tag{8}$$

Further, in a more compact form, the errors of dynamical system (8) can be mentioned in matrix form as below:

$$\dot{\mathbf{e}}_{PQ\eta} = \mathbf{f}_{PQ\eta} + \mathbf{B}_{PQ\eta}\mathbf{v_c} \tag{9}$$

In Equations (8) and (9), the dimensions of the error vector ($\dot{\mathbf{e}}_{PQ\eta}$) are matched with those of the control vector ($\mathbf{v_c}$) and, according to the basic control theory, the system (9) is now fully controllable because the distribution matrix of control ($\mathbf{B}_{PQ\eta}$) has full rank. Further, as the system (9) is not rank-deficient, a unique transformation ($\mathbf{x} = \mathbf{T}\,\mathbf{e}_{PQ\eta}$) is now possible in such a way that the new system ($\dot{\mathbf{x}} = \mathbf{f_x} + \mathbf{B_x}\mathbf{v_c}$) has a diagonal control distribution matrix ($\mathbf{B_x}$). Hence, the new converted system is a dynamically decoupled system and has three first-order, decoupled sub-systems. One sub-system corresponds to the active power, the second to the reactive power and the third sub-system corresponds to the mean-point voltage of the MG, meaning that the system has been transformed into a one-on-one function system.

Now, the non-singularity of the control distribution matrix ($\mathbf{B}_{PQ\eta}$) completely depends upon the selection of the third row vector ($\mathbf{b^T}$), since the first and second rows already have vectors, which are orthogonal. It means that the variable (η) has to be selected carefully so that it meets the requirements as defined above. It is now important to highlight that the zero-sequence voltage or zero-sequence current cannot exist in a balanced, 3-phase system, but these sequences do exist in an unbalanced, 3-phase system. Accordingly, the variable ($\dot{\eta}$) has been formulated in such a way as to make the zero-sequence voltage of the VSI output $\left(v_{cn} = \frac{1}{3}(v_{ca} + v_{cb} + v_{cc})\right)$ to track the grid's zero-sequence voltage $\left(v_{gn} = \frac{1}{3}(v_{ga} + v_{gb} + v_{gc})\right)$, and the three-phase output voltages ($\mathbf{v_c^T}$) of the inverter are generated so as to meet the zero-sequence voltage (v_{cn}) of the VSI. Hence, in the system (8), selecting the third variable as $\dot{\eta} = v_{gn} - v_{cn}$ will achieve the desired output voltages for the MG.

3. Control Formulation

As found in the above section, the variable η's selection is constrained by the need that the distribution matrix of control ($\mathbf{B}_{PQ\eta}$) must not be rank-deficient. As per the literature, the third variable (η) can help to generate balance voltages in converters for ac machines by using any non-linear control technique [18]. On the other hand, the zero-sequence voltage or current exists for the imbalance scenario, i.e., the reference for η does exist, which means that the zero-sequence voltage of the MG (v_{gn}) can be tracked by the zero-sequence voltage of the inverter ($v_{cn} = \frac{1}{3}(v_{c1} + v_{c2} + v_{c3})$). This can be achieved by selecting $\dot{\eta} = v_{gn} - v_{Cn}$. Hence, it leads the control formulation to lie within the (P, Q, η)-frame of reference (FOR), in contrast to $(P, Q)/(d, q)$-FOR present in most of the literature [20,21].

Accordingly, the choice of the control variable from (9) can be made as below:

$$\mathbf{v_c} = -\mathbf{B}_{PQ\eta}^{-1}\left(\hat{\mathbf{f}}_{PQ\eta} + \mathbf{K}_{PQ\eta}\mathbf{e}_{PQ\eta}\right) \tag{10}$$

where $\hat{\mathbf{f}}_{PQ\eta}$ depicts the estimation of the $\mathbf{f}_{PQ\eta}$, yielding closed-loop error dynamics to be

$$\dot{\mathbf{e}}_{PQ\eta} + \mathbf{K}_{PQ\eta}\mathbf{e}_{PQ\eta} = \left(\mathbf{f}_{PQ\eta} - \hat{\mathbf{f}}_{PQ\eta}\right) \tag{11}$$

It is obvious from (11) that the control errors will diminish to zero and the references will be tracked provided that the estimation of $\mathbf{f}_{PQ\eta}$ is such that $\left(\hat{\mathbf{f}}_{PQ\eta} \to \mathbf{f}_{PQ\eta}\right), \forall\, \mathbf{K}_{PQ\eta} > 0$.

Considering the pair $\left(\mathbf{e}_{PQ\eta}, \mathbf{v_c}\right)$ calculable, having $(\mathbf{f}_{PQ\eta})$ as an input, which is undecided but fulfilling $(\hat{\mathbf{f}}_{PQ\eta} = 0)$, then the dynamics of the new system $(\mathbf{z} = \mathbf{f}_{PQ\eta} - \mathbf{L}\mathbf{e}_{PQ\eta})$, where $\mathbf{L} > \mathbf{0}$ can be mentioned as $\dot{\mathbf{z}} = -\mathbf{L}\left(\mathbf{z} + \mathbf{L}\mathbf{e}_{PQ\eta} + \mathbf{B}_{PQ\eta}\mathbf{v_c}\right)$. This system helps to calculate $\mathbf{z}$, thus estimating the unknown input as $\hat{\mathbf{f}}_{PQ\eta} = \mathbf{z} + \mathbf{L}\mathbf{e}_{PQ\eta}$. Then, taking suitable values of $\mathbf{L} > \mathbf{0}$ will separate the dynamics of the observer from the closed-loop dynamics, resulting in $\left(\hat{\mathbf{f}}_{PQ\eta} \to \mathbf{f}_{PQ\eta}\right)$ and thus $\left(\mathbf{e}_{PQ\eta} \to 0\right)$. Hence, the selection $\mathbf{v_c} = -\mathbf{B}_{PQ\eta}^{-1}\left(\hat{\mathbf{f}}_{PQ\eta} + \mathbf{K}_{PQ\eta}\mathbf{e}_{PQ\eta}\right)$ confirms the power control and maintains the additional, novel requirement $\eta \to 0$. Notice that $\mathbf{v_c}$ is continuous.

Further, in order to complete the control design, the kinetics of the inductor currents and capacitor voltages may be evaluated from Figure 2 as given in the equations below:

$$\frac{d}{dt}\begin{bmatrix} i_{Lfa} \\ i_{Lfb} \\ i_{Lfc} \end{bmatrix} = -\mathbf{L_f^{-1}}\begin{bmatrix} v_{ca} \\ v_{cb} \\ v_{cc} \end{bmatrix} + \mathbf{L_f^{-1}}\begin{bmatrix} v_A \\ v_B \\ v_C \end{bmatrix} \tag{12}$$

$$\frac{d}{dt}\begin{bmatrix} v_{ca} \\ v_{cb} \\ v_{cc} \end{bmatrix} = -\mathbf{C_f^{-1}}\begin{bmatrix} i_{ga} \\ i_{gb} \\ i_{gc} \end{bmatrix} + \mathbf{C_f^{-1}}\begin{bmatrix} i_{Lfa} \\ i_{Lfb} \\ i_{Lfc} \end{bmatrix} \tag{13}$$

$$\frac{d}{dt}[v_{dc}] = -C_{dc}^{-1}\left[i_{RES} - \mathbf{i_{Lf}^T s}\right];\ \mathbf{s^T} = \begin{bmatrix} s_{11} & s_{12} & s_{13} \end{bmatrix} \tag{14}$$

$$\begin{cases} v_A = v_{dc}\, s_{11} \\ v_B = v_{dc}\, s_{12} \\ v_C = v_{dc}\, s_{13} \end{cases} \tag{15}$$

$$s_{1k} = \begin{cases} 1, & \text{if the switch } (s_{1k}) \text{ is closed and the switch } (s_{2k}) \text{ is open} \\ 0, & \text{if the switch } (s_{1k}) \text{ is open and the switch } (s_{2k}) \text{ is closed} \end{cases} \tag{16}$$

Here,

i_{RES} = current to/from RES
v_{dc} = DC-bus voltage
C_{dc} = DC-bus capacitor
$\mathbf{L_f}$ = filter inductance diagonal matrix
$\mathbf{C_f}$ = filter capacitance diagonal matrix
$\mathbf{i_{Lf}^T} = [i_{Lfa} \; i_{Lfb} \; i_{Lfc}]$ = inductor/converter o/p current vector
$\mathbf{i_g^T} = [i_{ga} \; i_{gb} \; i_{gc}]$ = converter o/p current vector supplied to grid
$\mathbf{v_c^T} = [v_{ca} \; v_{cb} \; v_{cc}]$ = o/p voltage vector at PCC
$\mathbf{v_s^T} = [v_A \; v_B \; v_C]$ = voltage vector at o/p of switching matrix
$\mathbf{s^T} = [s_{11} \; s_{12} \; s_{13}]$ = converter switching vector

$\mathbf{s^T}$ is the switching vector, which is used to determine the condition of the switches within the switching matrix. If the voltages (v_A, v_B, v_C) are taken as control inputs, then the dynamics (10)–(16) constitute discontinuous inputs of control, and the voltages (v_{ca}, v_{cb}, v_{cc}) are controlled outputs. As given in Equation (10), the already evaluated voltage ($\mathbf{v_c}$) is to be used to maintain the wanted power flow, so it can be employed to determine the set of three-phase voltages (v_A, v_B, v_C). Though the controller design for three-phase systems may be formulated in different FORs subject to the skills of designers; however, it is advantageous to carry this out in a synchronous frame, mainly due to enabling smaller control gains' selection, possible without any additional benefit.

To attain the desired output voltages ($\mathbf{v_c}$) based on the converter's filter dynamics (12)–(16), the switching pattern for the VSI can either be determined by applying sliding-mode control directly or by averaging the system and using any suitable PWM technique. The second approach is actually a well-known and practical approach for such systems compared to the sliding-mode control. Further, in order to complete the design, a cascaded layout with an exterior loop to evaluate the currents required to adjust the needed voltages ($\mathbf{v_c}$) and in the internal loop to compute the voltages inevitable to maintain the desired currents have been adopted. Now, the switching pattern is determined by using a PWM method.

By using Equation (10) as a reference for the output voltage of the converter, the kinetics of the control error of the inverter's voltage ($\mathbf{e_{vC}} = \mathbf{v_c^{ref}} - \mathbf{v_c}$) can be given as below in Equations (17) and (18):

$$\begin{bmatrix} \dot{e}_{vCa} \\ \dot{e}_{vCb} \\ \dot{e}_{vCc} \end{bmatrix} = \begin{bmatrix} \dot{v}_{Ca}^{ref} \\ \dot{v}_{Cb}^{ref} \\ \dot{v}_{Cc}^{ref} \end{bmatrix} - \mathbf{C_f^{-1}} \begin{bmatrix} i_{ga} \\ i_{gb} \\ i_{gc} \end{bmatrix} - \mathbf{C_f^{-1}} \begin{bmatrix} i_{Lfa} \\ i_{Lfb} \\ i_{Lfc} \end{bmatrix} \tag{17}$$

In a compact form,

$$\mathbf{\dot{e}_{vC}} = \mathbf{f_{vC}} - \mathbf{C_f^{-1} i_{Lf}} \tag{18}$$

Similar to the selection of (10), here, the choice of the control variable ($\mathbf{i_{Lf}}$) can be made as follows:

$$\mathbf{i_{Lf}} = \mathbf{C_f}(\mathbf{\hat{f}_{vC}} + \mathbf{K_{vC} e_{vC}}); \mathbf{K_{vC}} > \mathbf{0} \tag{19}$$

Hence, the error dynamics have the form,

$$\mathbf{\dot{e}_{vC}} + \mathbf{K_{vC} e_{vC}} = (\mathbf{f_{vC}} - \mathbf{\hat{f}_{vC}}) \tag{20}$$

Here, the error in control converges to zero at a rate set by the gains $\mathbf{K_{vC}}$ provided that a suitable observer design is shaped for an unknown input ($\mathbf{f_{vC}}$). After having found the reference for inductor currents ($\mathbf{i_{Lf}^{ref}} = \begin{bmatrix} i_{LP}^{ref} & i_{LQ}^{ref} & i_{L\eta}^{ref} \end{bmatrix}^T$), the converter o/p voltage ($\mathbf{v_s}$) and switching pattern of the converter can be based on the dynamics (12). Further projecting these dynamics (12) into the FOR (P, Q, η) and making use of the transformations

$(x_P = \frac{\mathbf{v}_g^T}{\mathbf{v}_g^T}\mathbf{x}_{abc}; x_Q = \frac{\mathbf{v}_{g\perp}^T}{\mathbf{v}_g^T}\mathbf{x}_{abc}; x_\eta = \frac{\mathbf{b}^T}{\mathbf{v}_g^T}\mathbf{x}_{abc})$, the current control error ($\mathbf{e}_{iLf} = \mathbf{i}_{Lf}^{ref} - \mathbf{i}_{Lf}$) dynamics can be evaluated by

$$\frac{d}{dt}\begin{bmatrix} e_{LP} \\ e_{LQ} \\ e_{L\eta} \end{bmatrix} = \frac{d}{dt}\begin{bmatrix} i_{LP}^{ref} \\ i_{LQ}^{ref} \\ i_{L\eta}^{ref} \end{bmatrix} + \mathbf{L}_g^{-1}\mathbf{C}_f^{-1}\mathbf{v}_g\begin{bmatrix} v_{CP} \\ v_{CQ} \\ v_{C\eta} \end{bmatrix} - \mathbf{L}_g^{-1}\mathbf{C}_f^{-1}\mathbf{v}_g\begin{bmatrix} v_{SP} \\ v_{SQ} \\ v_{S\eta} \end{bmatrix} \tag{21}$$

and in a compact form,

$$\dot{\mathbf{e}}_{iLf} = \mathbf{f}_{iLf} - \mathbf{L}_g^{-1}\mathbf{C}_f^{-1}\mathbf{v}_g^T\mathbf{v}_s \tag{22}$$

where

$$\begin{aligned} \mathbf{v}_g^T &= \begin{bmatrix} v_{gP} & v_{gQ} & v_{g\eta} \end{bmatrix}^T = \text{voltage vector of MG} \\ \mathbf{v}_s^T &= \begin{bmatrix} v_{sP} & v_{sQ} & v_{s\eta} \end{bmatrix}^T = \text{output vector of switching matrix} \end{aligned}$$

It was mentioned earlier that the the distribution matrix of control is diagonal; hence, the mathematical model of the system is constituted of three systems of the first order. Two orthogonal set-points of powers with the mean MG voltage are used for the PW modulation. This design is based on the combination of applying elementary disturbance observers with the basic controllers. The observers' design in the first-order systems is a simple estimation of inputs, which are undetermined. This enables the utilization of nested loops and similar structures for controllers. The added advantage of the model formulation and controller design selection lies in choosing the supplemental control requirement to be able to use the full DOF of the switching matrix to compensate for the asymmetries of the MG-connected system. The proposed control structure for the VSI connected to the MG is given below in Figure 3.

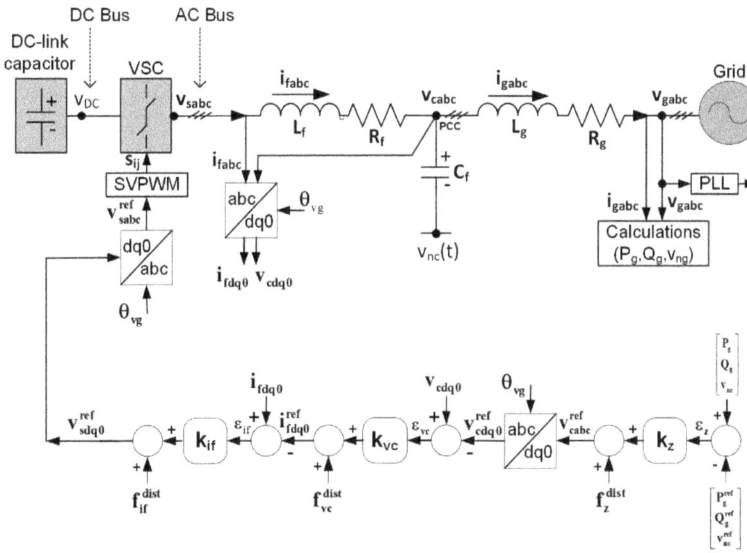

Figure 3. Novel control structure of VSI supported MG.

4. Simulation Results

The proposed control method's formulation is ascertained with the help of simulating the proposed arrangement of the grid-connected VSI-based source. Two of the grid cases, the balanced grid and the imbalanced grid, have been taken into account one-by-one for simulation in order to obtain the results in the following subsection.

4.1. Case A: VSI Source Results with the MG in Balance Condition

The grid voltage for each phase of the balanced grid is set at 100 Volts, as seen in Figure 4a. Figure 4b depicts the respective current injections to the grid as per the requirements of active (or) and reactive powers seen in Figure 4c,d. Figure 4e shows the average voltage (or the neutral voltage) requirement of the MG and the corresponding mean voltage of the VSI source. For the symmetric or balanced grid case, the designed controller is able to keep the required mean voltage to zero at the output of the converter. It is quite evident that the designed control procedure is maintaining the desired injections of orthogonal powers to the MG, as seen in Figure 4c,d.

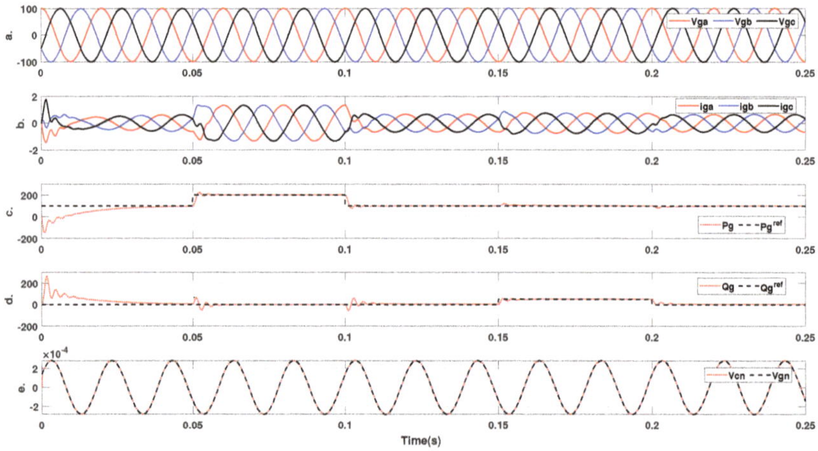

Figure 4. Power step responses of the symmetric MG connected VSI source. (**a**) Grid voltages ($v_g[V]$), (**b**) Grid currents ($i_g[A]$), (**c**) Active power ($P_g[W]$), (**d**) Reactive power ($Q_g[VAR]$), (**e**) Neutral/mean voltages ($v_n[V]$).

Further, the locus plots of the MG voltages, inverter's output voltages and MG currents are all shown in Figure 5, which validates the hardiness of the control structure for the balanced MG circuit.

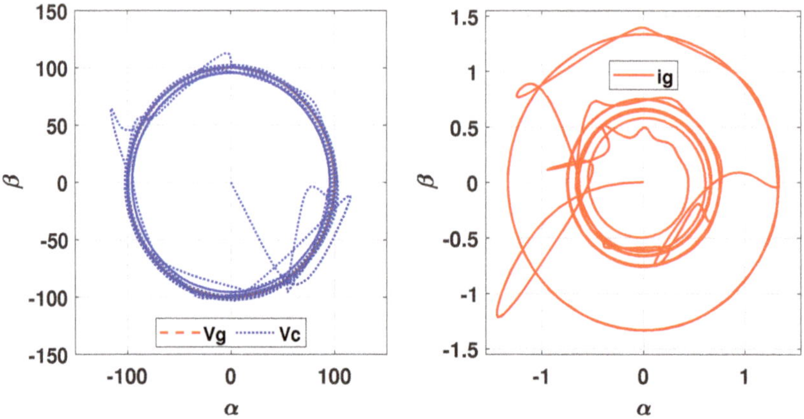

Figure 5. Locus plots of Grid voltages ($v_g[V]$), Converter output voltages ($v_c[V]$), and Grid currents ($i_g[A]$) for power step response with symmetric MG conditions.

Finally, the disturbance estimation results are given in Figure 6, showing the results of the output of the controller (or the inverter's voltage outputs), which are clearly in symmetric form as per the requirement of the balanced MG. The changes are visible around 0.05 s, 0.1 s, 0.15 s and 0.2 s since these are the instants where the power's set-points are rendered. Figure 6c shows the respective disturbance estimations for the imbalanced MG voltage conditions.

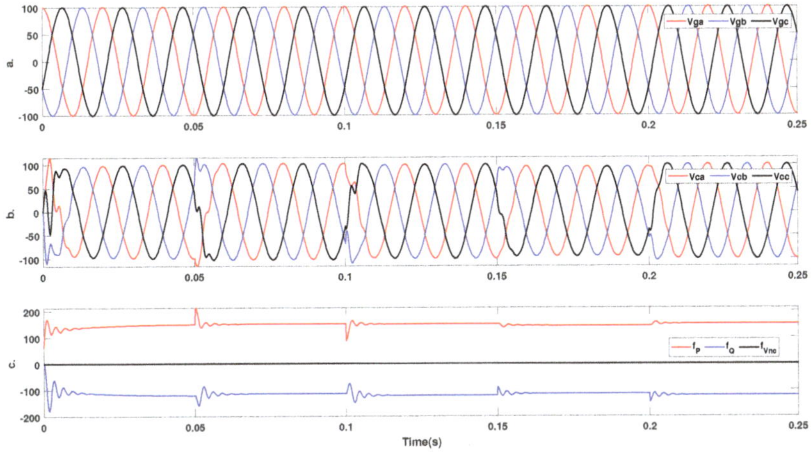

Figure 6. Disturbance estimation results for Pg, Qg and v_{nc} with symmetric MG. (**a**) Grid voltage ($v_g[V]$), (**b**) VSI output voltages ($v_c[V]$) and (**c**) Disturbance estimations ($\hat{f}_{PQ\eta}[V]$).

4.2. Case B: VSI Source Results with the MG in Imbalanced Condition

In the case of imbalanced MG requirements, the converter's controlled results are given in Figures 7–9. The step-wise response of the orthogonal powers and the corresponding voltages and currents of the MG are given in Figure 7. A deliberate 10% voltage reduction in phase b is introduced in the MG for the entire duration of the simulation to depict the converter's controller response for the case of the imbalanced MG.

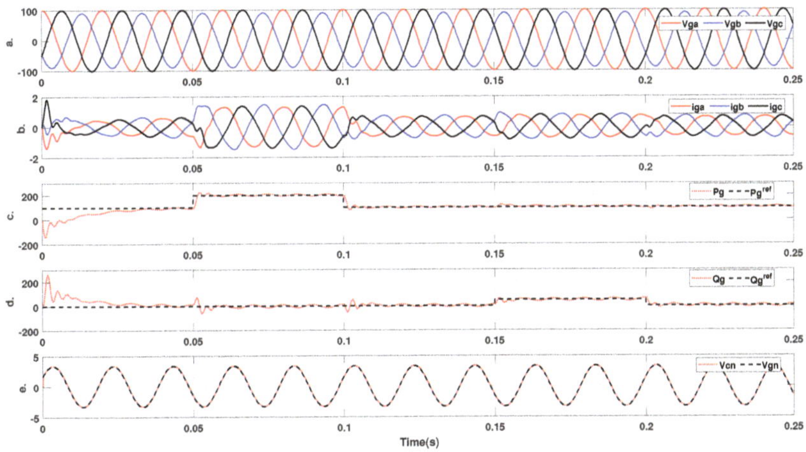

Figure 7. Power step responses of the asymmetric MG connected VSI source. (**a**) Grid voltages ($v_g[V]$), (**b**) Grid currents ($i_g[A]$), (**c**) Active power ($P_g[W]$), (**d**) Reactive power ($Q_g[VAR]$), (**e**) Neutral/mean voltages ($v_n[V]$).

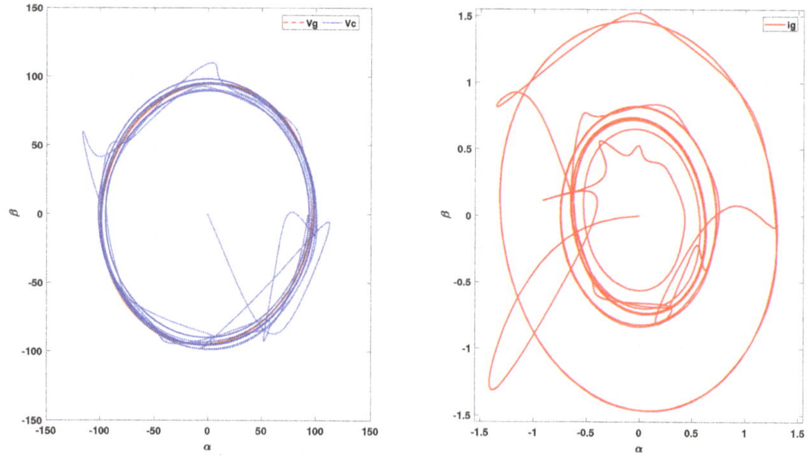

Figure 8. Locus plots of Grid voltages ($v_g[V]$), Converter output voltages ($v_c[V]$), and Grid currents ($i_g[A]$) for power step response with asymmetric MG conditions.

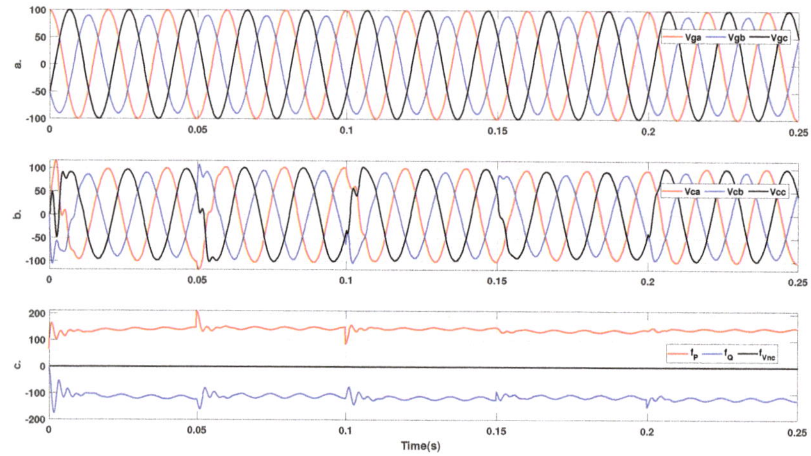

Figure 9. Disturbance estimation results for Pg, Qg and v_{nc} with asymmetric MG. (**a**) Grid voltage ($v_g[V]$), (**b**) VSI output voltages ($v_c[V]$) and (**c**) Disturbance estimations ($\hat{f}_{PQ\eta}[V]$).

In Figure 7, the first row shows the imbalance grid voltages, the second row shows the corresponding grid current requirements, the third and fourth rows show the power's step response requirements of the grid and the last row shows the corresponding average/neutral voltages, with all the average voltages displaying clearly non-zero values corresponding to the grid's imbalance demands. During intervals 0.05–0.1 s and 0.15–0.2 s, a step variation in the reference powers (Pg, Qg) was introduced and the corresponding power injections by the three-phase inverter can be seen to be in line with the reference values, with very few oscillations compared to the balanced MG case, as in Figure 4.

The VSI source results with the imbalanced grid from Figure 7 have also been depicted by corresponding locus plots in Figure 8. The locus plots of the output voltages, the MG voltages and the MG currents all have been plotted alongside each other to highlight the efficacy of the controller's capability to enforce the desired output voltages of the inverter, which are seen to be oval, confirming the matching with the imbalance requirements.

Finally, Figure 9 shows the results of the controller output (or the inverter's output voltages), which are obviously imbalanced as per the requirement of the imbalanced grid. The transients can be noticed around 0.05 s, 0.1 s, 0.15 s and 0.2 s since these are the instants where the power's set-points are rendered. Figure 9c shows the respective disturbance estimations for the imbalanced MG voltage conditions.

The values used for different parameters and control gains in the simulation are tabulated in the following Table 1.

Table 1. Parameter values and control gains for VSI-based MG.

Quantity (Symbol)	Magnitude Units
Grid Voltage (v_g)	100-Volts
Grid Inductance (L_g)	50-mH
Filter Inductance (L_f)	22-mH
Filter Capacitance (C_f)	220-µF
Grid Resistance (R_g)	100-mΩ
Control Gains (K_P, K_Q)	55
Observer Gains (L)	1200

5. Conclusions

The simulation results have proven the capability of both the proposed mathematical model of the MG-connected DG structure by augmenting the MG mean error voltage, and the relevant disturbance observer-based controller design, in the effective utilization of the full DOF of VSI under imbalanced MG voltage conditions. The proposed method has proven to be a generic method to satisfy both the symmetric and asymmetric conditions of the MG. The proposed method has successfully achieved this goal by enabling the VSI filter's mean voltage to track the MG's mean voltage and the switching inverter to generate output voltages to satisfy the VSI's filter's mean voltage. The proven controller can also be named the (P,Q,0) or (d,q,0) controller, due to mean voltage or zero sequence being part of the novel mathematical model and controller design. The results have also proven that the proposed methodology enables the three-phase inverter to generate the voltages and currents necessary to meet the required active and reactive powers at the end of the MG, irrespective of the symmetric and asymmetric MG voltage requirements. The proposed structure also works effectively irrespective of the level of disturbance, due to the DOB included as part of the control structure. The future plan is to augment the mean voltage of the MG in the MG-connected VSI model in the double synchronous frame or the model based on symmetrical components so as to explore whether it can provide even better results in terms of a fast response to imbalance requirements.

Author Contributions: Conceptualization, methodology, software simulation, formal analysis, investigation, writing—original draft preparation, writing—review and editing, A.R. and F.A.; writing—review and editing, S.A.R.K., M.S.F. and E.M.; visualization, supervision, project administration, funding acquisition, A.R. All authors have read and agreed to the published version of the manuscript.

Funding: This research received no external funding and the APC was partially funded by the University of Botswana, https://www.ub.ac.bw.

Data Availability Statement: Data sharing is not applicable to this article.

Conflicts of Interest: The authors declare no conflict of interest.

Abbreviations

The following abbreviations are used in this manuscript:

VSI	Voltage Source Inverter
DG	Distributed Generator
MG	Micro-Grid

SG	Smart Grid
DOF	Degree of Freedom
RES	Renewable Energy Systems
PCC	Point of Common Coupling
PWM	Pulse Width Modulation
SMC	Sliding-Mode Control
FOR	Frame of Reference
RF	Reference Frame

References

1. Bharothu, J.N.; Sridhar, M.; Rao, R.S. A literature survey report on Smart Grid technologies. In Proceedings of the IEEE International Conference on Smart Electric Grid (ISEG), Guntur, India, 19–20 September 2014; pp. 1–8.
2. Ekanayake, J.B.; Jenkins, N.; Liyanage, K.; Wu, J.; Yokoyama, A. Smart grid: Technology and applications. In *Smart Grid: Technology and Applications*; John Wiley & Sons: Hoboken, NJ, USA, 2012.
3. Ahmad, F.; Rasool, A.; Ozsoy, E.; Sekar, R.; Sabanovic, A.; Elitaş, M. Distribution system state estimation-A step towards smart grid. *Renew. Sustain. Energy Rev.* **2018**, *81*, 2659–2671. [CrossRef]
4. Kaviri, S.M.; Pahlevani, M.; Jain, P.; Bakhshai, A. A review of AC microgrid control methods. In Proceedings of the 8th IEEE International Symposium on Power Electronics for Distributed Generation Systems (PEDG), Florianópolis, Brazil, 17–20 April 2017; pp. 1–8.
5. Colak, I.; Kabalci, E.; Fulli, G.; Lazarou, S. A survey on the contributions of power electronics to smart grid systems. *Renew. Sustain. Energy Rev.* **2015**, *47*, 562–579. [CrossRef]
6. Teodorescu, R.; Liserre, M.; Rodriguez, P. *Grid Converters for Photovoltaic and Wind Power Systems*; John Wiley & Sons: Hoboken, NJ, USA, 2011.
7. Benysek, G.; Kazmierkowski, M.; Popczyk, J.; Strzelecki, R. Power electronic systems as a crucial part of Smart Grid infrastructure—A survey. *Bull. Pol. Acad. Sci. Tech. Sci.* **2011**, *59*, 455–473. [CrossRef]
8. Arbab-Zavar, B.; Palacios-Garcia, E.J.; Vasquez, J.C.; Guerrero, J.M. Smart Inverters for Microgrid Applications: A Review. *Energies* **2019**, *12*, 840. [CrossRef]
9. Jiang, H.; Cao, S.; Soh, C.B.; Wei, F. Unbalanced load modeling and control in microgrid with isolation transformer. In Proceedings of the IEEE International Conference on Electrical Drives & Power Electronics (EDPE), Dubrovnik, Croatia, 22–24 September 2021; pp. 129–135.
10. Özsoy, E.; Padmanaban, S.; Mihet-Popa, L.; Fedák, V.; Ahmad, F.; Rasool, A.; Şabanoviç, A. Control strategy for grid connected inverters under unbalanced network conditions-A DOb based decoupled current approach. *Energies* **2017**, *47*, 562–579.
11. Li-Jun, J.; Miao-Miao, J.; Guang-Yao, Y.; Yi-Fan, C.; Rong-Zheng, L.; Hai-Peng, Z.; Ke, Z. Unbalanced control of grid-side converter based on DSOGI-PLL. In Proceedings of the IEEE 10th Conference on Industrial Electronics and Applications (ICIEA), Auckland, New Zealand, 15–17 June 2015; pp. 1145–1149.
12. Suul, J.A. Control of Grid Integrated Voltage Source Converters under Unbalanced Conditions: Development of an on-Line Frequency-Adaptive Virtual Flux-Based Approach. Ph.D. Thesis, Norwegian University of Science and Technology, Trondheim, Norway, 2012.
13. Puranik, S.; Keyhani, A.; Chatterjee, A. Control of Three-Phase Inverters in Microgrid Systems. In *Smart Power Grids 2011*; Springer: Berlin/Heidelberg, Germany, 2012; pp. 103–176.
14. Brod, D.M.; Novotny, D.W. Current control of VSI-PWM inverters. *IEEE Trans. Ind. Appl.* **1985**, *3*, 562–570. [CrossRef]
15. Tenca, P.; Lipo, T.A. Synthesis of desired AC line currents in current-sourced DC-AC converters. In Proceedings of the IEEE Second International Conference on Power Electronics, Machines and Drives (PEMD), Edinburgh, UK, 31 March–2 April 2004; pp. 656–661.
16. Milosevic, M. Decoupling control of d and q current components in three-phase voltage source inverter. In Proceedings of the IEEE Power Systems Conference and Exposition (PSCE), Atlanta, GA, USA, 29 October–1 November 2006; pp. 34–44.
17. Sowmmiya, U.; Jamuna, V. Voltage control scheme for three phase SVM inverter fed induction motor drive systems. In Proceedings of the IEEE 1st International Conference on Electrical Energy Systems (ICEES), Chennai, India, 3–5 January 2011; pp. 207–211.
18. Sabanovic, A.; Ohnishi, K.; Sabanovic, N. Control of PWM three phase converters: A sliding mode approach. In Proceedings of the Conference Record of Power Conversion Conference, Yokohama, Japan, 19–21 April 1993; pp. 188–193.
19. Fiaz, A.; Rasool, A.; Ozsoy, E.E.; Sabanoviç, A.; Elitas, M. A robust cascaded controller for DC-DC Boost and Cuk converters. *World J. Eng.* **2017**, *14*, 459–466.
20. Vijay, A.S.; Doolla, S.; Chandorkar, M. Unbalance mitigation strategies in microgrids. *IET Power Electron.* **2020**, *13*, 1687–1710. [CrossRef]
21. Navas-Fonseca, A.; Burgos-Mellado, C.; Gómez, J.S.; Donoso, F.; Tarisciotti, L.; Saez, D.; Cardenas, R.; Sumner, M. Distributed predictive secondary control for imbalance sharing in AC microgrids. *IEEE Trans. Smart Grid* **2022**, *13*, 20–37. [CrossRef]
22. Utkin, V.; Guldner, J.; Shi, J. *Sliding Mode Control in Electro-Mechanical Systems*, 3rd ed.; Taylor & Francis, CRC Press: Boca Raton, FL, USA, 2017; pp. 333–374.

23. Wodyk, S.; Iwanski, G. Three-phase converter power control under grid imbalance with consideration of instantaneous power components limitation. *Int. Trans. Electr. Energy Syst.* **2020**, *30*, e12389. [CrossRef]
24. Rasool, A. Control of Three Phase Converters as Source for Microgrid. Ph.D. Thesis, Sabancı University, Tuzla, Istanbul, Turkey, 2017.

symmetry

MDPI

Article

Research on the Fault-Transient Characteristics of a DC Power System Considering the Cooperative Action of a Flexible Current-Limiting Device and a Circuit Breaker

Feng Zheng, Yaling Peng, Weidong Wang and Song Zheng *

College of Electrical Engineering and Automation, Fuzhou University, Fuzhou 350100, China
* Correspondence: s.zheng@fzu.edu.cn

Abstract: As an effective carrier of a new energy collection, the DC power grid has low inertia and weak damping characteristics, making it essential to limit fault current and isolate the DC system. To quickly and effectively suppress fault current, a flexible current-limiting device (FCLD) is proposed, which can realize transient fault self-recovery without circuit breaker action and permanent and quick isolation of a fault. It improves the operational ability of the DC system under an asymmetric condition. First, a rectifier provides a set-slope current to each cascade inductor, so the voltage of the inductor can be clamped. Second, a controlled current source (CCS) is applied to generate inverse flux to prevent the inductor from magnetic saturation. The protection action time of the DC circuit breaker is reformulated. Finally, by considering the synergistic action of the current-limiting device, the circuit breaker, and the transient characteristics of the DC grid fault, the protection scheme of the multi-terminal flexible DC system can be formulated. To verify the validity of the proposed flexible current-limiting device, a multi-terminal flexible DC simulation platform is established, and the faults of DC lines are simulated and analyzed.

Keywords: cascade inductor; flexible current limiting; controlled current source; multi-terminal flexible DC system

1. Introduction

A Flexible DC power grid has the advantages of large transmission capacity, low loss, and excellent safety performance. It can collect energy from the DC supply-and-demand equipment, such as the distributed energy resource, the DC load, and the energy storage device, attracting wide attention from scholars at home and abroad [1,2]. However, some of the active equipment will be off the network quickly and release energy rapidly through the small damping path when a fault occurs in the DC line. It will generate a large fault current and pose a serious threat to the safe operation of the power network [3–5].

To improve the toughness of the active DC power grid in response to line faults, current research focuses on joint cooperation among the DC circuit breaker, the converter, and the current-limiting device. Such cooperation can reduce the mechanical stress and the thermal stress of the disconnected functional equipment during the fault removal process and extend the protection action time of the disconnected equipment and the fault-removal capacity margin, thereby realizing the rapid isolation of the faulty line [6–12]. A previous study [6] integrated the resistance-inductive current-limiting device with the DC circuit breaker, which can limit the current to a certain extent and improve the dynamic recovery performance of the DC power grid. Another study [7] used modular multi-level converters to avoid the capacitors outputting DC voltage to the grid side, reducing the outlet voltage of the converter station to zero and effectively clearing the DC fault current. In [8], the flexible input of current-limiting components and self-bypass design were used to reduce the difficulty of breaking the circuit breaker. In [9,10], a hybrid DC circuit breaker with a current limiting function was used to reduce the fault-current rise rate and the impact damage of the fault current to the equipment. In [11,12], a pre-charged capacitor was used to access the fault circuit to increase the line-side voltage of the mechanical switch so that the mechanical switch could be turned off without arc and the fault could be cleared.

Although the abovementioned methods can play the roles of limiting current and isolating auxiliary fault to a certain extent, they are accompanied by a certain energy loss during the whole operation process, and they cannot quickly and completely eliminate the negative impact of the fault on the non-fault area. There has been no in-depth discussion on whether the circuit breaker can quickly perform the reclosing action again in response to a transient fault after the mechanical switch is broken without arc. To achieve quick reclosing, Refs. [13,14] proposed an adaptive reclosing scheme, which analyzed the characters of capacitive-coupling voltage; then, a fault identification criterion was constructed. However, reducing the converter output fault current by this method will inevitably weaken the DC dynamic recovery performance of the system. One study [15] presented an adaptive reclosing scheme based on pulse injection from a parallel energy-absorption module, but it applied only with a transmission line. Another study [16] proposed an adaptive reclosing scheme based on the phase characteristics, but it only worked in scenarios where the DC circuit breaker was not available.

Therefore, it is necessary to design a control method that is characterized by simple control, arc-free shutdown, and fast execution of the reclosing operation. The advantages of this method are that it can eliminate the negative impact of the fault area on the non-fault area in the DC grid, ensure the reliable operation of the DC circuit breaker in the non-fault area, minimize the impact of transient faults, and, ultimately, improve the resilience of the DC system to cope with different types of faults.

In this paper, a type of flexible DC current-limiting device that consists of a voltage-source converter and current-limiting inductors is proposed. This paper proposes a cascaded current-limiting inductor. The mechanism of magnetic saturation elimination by inverse flux is analyzed. The collaborative-action process of a current-limiting device and a circuit breaker is analyzed. The operational process of a current-limiting device under the permanent fault of a three-terminal DC system is analyzed by simulation.

2. New Type of Flexible DC Current Limiting

2.1. Topology

The proposed new FCLD topology, based on a series inductor with a clamp-voltage function, is shown in Figure 1. It can be seen from Figure 1 that the FCLD has a simple topology and is mainly composed of N inductors coupled in a series. Each series inductor is connected to the DC system after being connected in parallel with the corresponding rectifier.

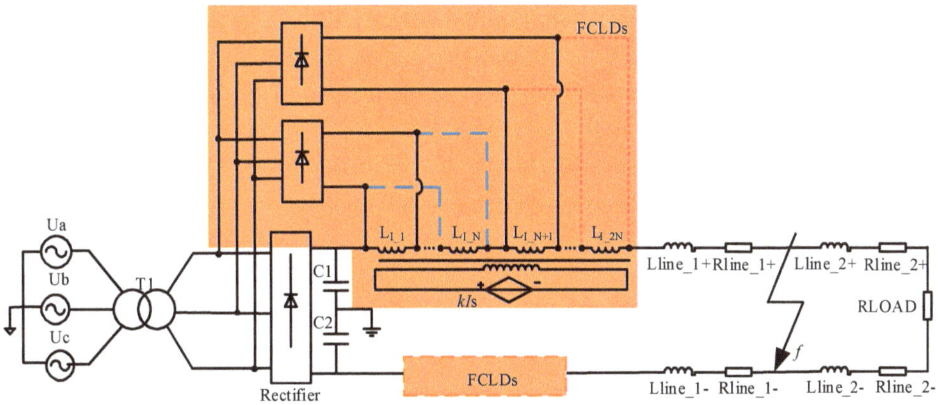

Figure 1. Spilt voltage flexible DC current limiting device.

In order to avoid the insufficient clamping voltage of the current-limiting inductor, the positive and negative lines at the outlet of each converter station are equipped with two current limiters that are powered by two converter stations. When one current-limiting inductor fails, the standby current-limiting device is put into service, which solves the problem of the current-limiting device being unable to provide sufficient clamp voltage due to the fault of the current-limiting inductor.

When the system is in normal operation, the FCLD realizes the no-voltage operation mode of the series inductor through its rectifier control. It can eliminate the voltage fluctuation of the current-limiting inductor that is caused by load fluctuation. When a fault occurs, the rectifier provides a set-slope linear current to the current-limiting inductor, which causes the current-limiting inductor to produce a stable DC voltage. However, due to the magnetic saturation of the current-limiting inductor, a controlled current source is added to the secondary side of the current-limiting inductor. It suppresses the magnetic flux saturation of the primary side by generating inverse magnetic flux to ensure the stable voltage of the primary current-limiting inductor.

2.2. Design of Application Layer Based on Hybrid Algorithm

According to the operational requirements of the FCLD, Figure 2 shows the block diagram of the control system of a single FCLD. Since the current-limiting device is connected in a series with the line, the voltage of the current-limiting inductor is controlled at 0 V when there is no fault, and the voltage of the current-limiting inductor is set to the default value when a fault occurs. Compared with the existing methods, this paper considers the combination of voltage differential and undervoltage detection methods to improve detection speed and accuracy [17].

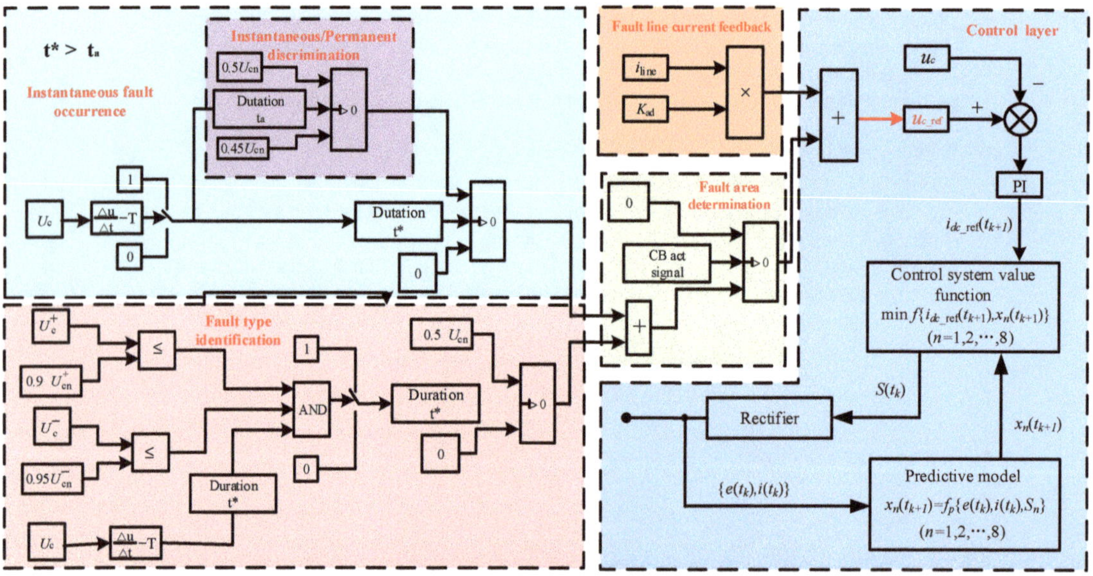

Figure 2. Control system block diagram of application layer.

According to Figure 2, when the DC system is operating normally, the external output voltage of the FCLD is 0 V. Therefore, the application layer output reference voltage u_{c_ref} is set to 0. When a fault occurs, the DC-side voltage changes instantaneously, and the voltage differential component presents a maximum value. According to the parameters of the DC system, different fault types, fault locations, transition resistances, and the influencing factors of the voltage differential component are determined. The minimum value T in the regular interval is used to set the control criterion [17,18]. If the voltage differential component is greater than the minimum value T, the trigger signal appears and it is kept for t^* seconds (t^* is greater than the whole process time of the circuit breaker operation). Fault type cannot be accurately identified according to an extreme value criterion, so it is necessary to carry out a secondary undervoltage judgment for the control input [17].

Because the fault circuit of the DC system can be equivalent to a second-order circuit, if the FCLD is not put into the converter station (as shown in Figure 1—f is the fault location), the discharge voltage of the outlet capacitor can be expressed as follows:

$$\begin{cases} u_c = e^{-\frac{R_{line\_1}}{2L_{line\_1}}t}\sqrt{\left(\frac{U_0 R_{line\_1}}{2\omega L_{line\_1}} - \frac{I_0}{\omega C}\right)^2 + U_0^2}\sin(\omega t + \varphi) \\ \varphi = \arctan U_0 / \left(\frac{U_0 R_{line\_1}}{2\omega L_{line\_1}} - \frac{I_0}{\omega C}\right) \\ \omega = \sqrt{1 - (CL_{line\_1}) - R_{line\_1}/2L_{line\_1}} \end{cases} \tag{1}$$

U_0 and I_0 are the initial values of the fault voltage and current at the output port of the DC converter station, and u_c is the discharge capacitor voltage at the output port of the converter station. $(R_{line\_1}/2L_{line\_1})$ is the voltage attenuation coefficient.

As can be seen from Formula (1), due to the large voltage attenuation coefficient of the general DC line, there is a delay of several milliseconds between the capacitor discharge and the judgment threshold in undervoltage protection, so judgment at this stage can be carried out simultaneously with an extreme value judgment. When the positive and negative voltages are both lower than 0.9 times their own voltage ratings (when the DC system voltage is as low as 90% of its rated value, it is considered that there is a voltage sag problem) and when $(\Delta u/\Delta t) > T$ is established at the same time, it is determined that an inter-electrode fault has occurred on the DC side. The positive and negative current-limiting application layer output reference voltage u_{c_ref} is set to u_{cn}. Then, the positive and negative capacitor voltage of the output of the DC side is raised to the rating value, so that the fault area is completely isolated. The fault current in the whole dynamic process is obviously suppressed. When either the positive or the negative output capacitor voltage is lower than 0.9 times its rated value, it is determined that a positive or negative ground fault has occurred on the DC side. It is necessary to adjust only the output reference voltage of the positive or negative current-limiting the application layer to $u_{c\_ref} = u_{cn}$. The output reference voltage of the non-fault current-limiting application layer maintains $u_{c\_ref} = 0$.

At the same time, in order to realize the coordination between the FCLD and the circuit breaker, we adopted adaptive adjustment to the output reference voltage u_{c_ref} of the application layer, and its specific control structure is shown in Figure 2. The current limiter adopts model predictive control [19], and its control method is shown in Figure 2.

The adaptive method of this paper takes the line current as the reference value, and outputs the regulated voltage Δu_{c_ref} through the adjustment coefficient K_{ad}. Because the adaptive adjustment control can correct the clamping voltage of the current-limiting device, the line current during the fault quickly drops to around 0 A. This avoids the tedious process of manual testing and realizes the automatic correction of the clamp voltage and the zero-crossing of the circuit-breaker current.

2.3. Core Saturation Suppression

In order to solve the magnetic saturation problem of the inductor, we eliminated the magnetic saturation of the primary-side inductor by generating reverse flux through the secondary-side controllable current source, based on the primary- and secondary-side magnetic linkage relationship of the closed-loop iron core [20].

Figure 3 shows the magnetic circuit diagram of the closed-loop iron core. It can be seen from Figure 3 that the primary-side flux linkage equation can be expressed as follows:

$$\begin{cases} \psi_1 = N_1\varphi_1 = N_1 L_1 I_1 \\ e_1 = \mathrm{d}\psi_1/\mathrm{d}t = N_1 \mathrm{d}(L_1 I_1)/\mathrm{d}t \end{cases} \tag{2}$$

where ψ_1 is the primary-side total magnetic flux, e_1 is the primary-side voltage, φ_1 is the single-turn coil magnetic flux, N_1 is the number of primary-side winding turns, and I_1 is the primary-side current. When the controlled current source ki_1 is added to the secondary side, the impedance value of the primary-side current-limiting inductor is as follows [20,21]:

$$L_1 = \frac{N_1^2 S_1}{l}\left(\frac{B}{H} - \frac{N_2 \lambda i_1 - N_1 i_1}{N_1}\frac{\mathrm{d}(\frac{B}{H})}{\mathrm{d}i_1}\right) \tag{3}$$

where N_2, S_1, l, B, and H are coil windings, cross-sectional area, magnetic circuit length, magnetic field density, and magnetic field strength, respectively. N_2, S_1, and l are the inherent properties of the closed-loop iron core inductor coil.

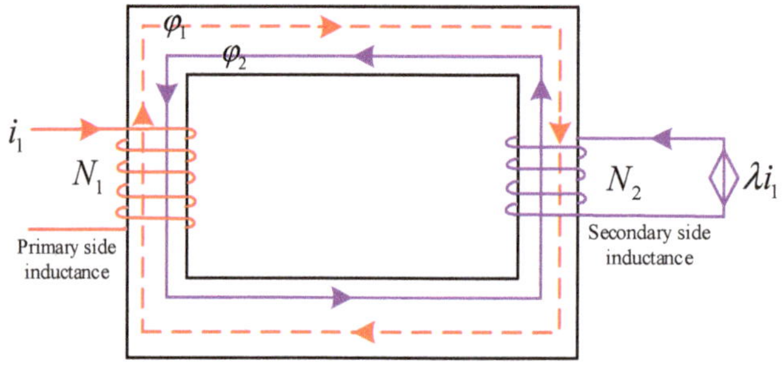

Figure 3. Magnetic circuit of iron core.

According to Equation (3), when the secondary winding is affected by a controlled voltage source, the ratio of B and H is changed to ensure that the primary winding is in an unsaturated state.

When the secondary side of the closed-loop iron core generates inverse magnetic flux before the magnetic flux is saturated, the total magnetic flux on the primary side is as follows:

$$\psi_1^* = \psi_1 - \psi_2 = B * S_1 \tag{4}$$

Figure 4 shows the change curves of B-H and μ-H. As can be seen from Figure 4, under the compensation effect of the secondary-side controlled current source, the saturation state of the primary winding is eliminated. The closed-loop iron core has the ability of self-adaptive adjustment. When the primary winding is about to reach the saturation state, the value of the controlled current source of the secondary winding will be increased, and the inverse magnetic flux will be generated to eliminate the saturation state of the primary winding. When the fault time is too long, the saturation state of the current-limiting inductor can be eliminated through multiple compensations, and the process of the compensations can be seen:

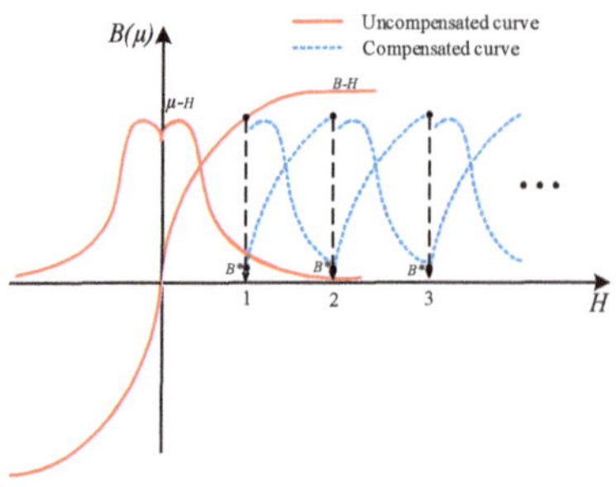

Figure 4. B/μ-H characteristic curve.

(1) If the inverse magnetic flux on the secondary side reduces B* to a small value, the magnetic flux saturation of the closed-loop iron core disappears, and the primary-side current continues to increase linearly. At this time, since the B-H curve (as shown in Figure 4) can be approximately regarded as a linear increase, L_1 is assumed to be constant. In this case, the inductance voltage can be obtained, as follows:

$$\begin{aligned} e_1 &= \frac{N_1^2 S_1}{l}\left(\frac{B}{H} - \frac{N_1 i_1 - N_2 \lambda i_1}{N_1}\frac{\mathrm{d}(\frac{B}{H})}{\mathrm{d}i_1}\right)\frac{di_1}{dt} \\ e_1 &= \frac{N_1^2 S_1}{l}\frac{B}{H}\frac{di_1}{dt} \end{aligned} \quad (5)$$

It can be seen from Formula (5) that only the linear change of i_1 can make the magnetic flux saturation disappear. Figure 5a,b show the current of the flexible current limiting inductor without and with the addition of the inverse flux at different clamp voltages. Figure 5c shows that when the clamp voltage increases, the change in current slope also increases.

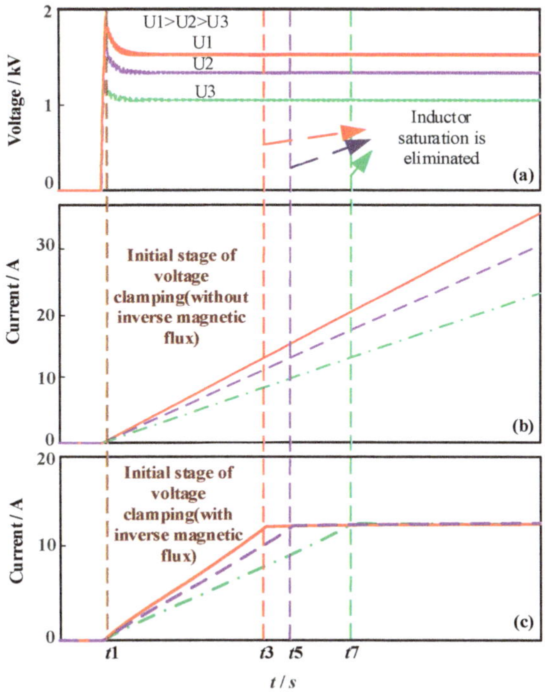

Figure 5. Inductance characteristics after compensation. (**a**,**b**) show the current of the flexible current limiting inductor without and with the addition of the inverse flux at different clamp voltages; (**c**) shows that when the clamp voltage increases, the change in current slope also increases.

(2) In order to prevent the primary- and secondary-side current from being too large within the set time, the secondary-side inverse magnetic flux eliminates the magnetic flux saturation in the early stage of the fault, and the magnetic flux saturation problem still exists in the later stage of the fault. Then, the primary current at the later stage of the fault remains at a stable value, as shown in Figure 5c. Therefore, at this time, it is only necessary to maintain the set value by adjusting the primary-side current-limiting inductance L_1. The primary-side voltage is as follows:

$$e_1 = -\frac{N_1 i_1 - N_2 \lambda i_1 B}{N_1}\frac{\mathrm{d}(\frac{1}{H})}{dt} \quad (6)$$

In the saturated state, B remains unchanged as H increases, so

$$e_1 = \frac{N_1^2 S_1 i_1 B}{l} \frac{d\frac{1}{H}}{dt} \tag{7}$$

According to Equation (7) and the definition formula of magnetic field strength $H = N_2 i_2 / l$, the λ of the secondary-side controlled current source can be obtained as follows:

$$\lambda = \frac{N_1^2 S_1 B}{e_1 N_2 t} \tag{8}$$

3. Synergistic Effect of Flexible Current Limiting and DC Circuit Breaker

3.1. Analysis on the Synergistic Process of Current-Limiting Inductor and DC Circuit Breaker

The studies in [5,22] analyzed the transient characteristics of the bipolar fault current of the DC power grid with a current-limiting inductor, and divided the whole operation process into six stages, as shown in Figure 6.

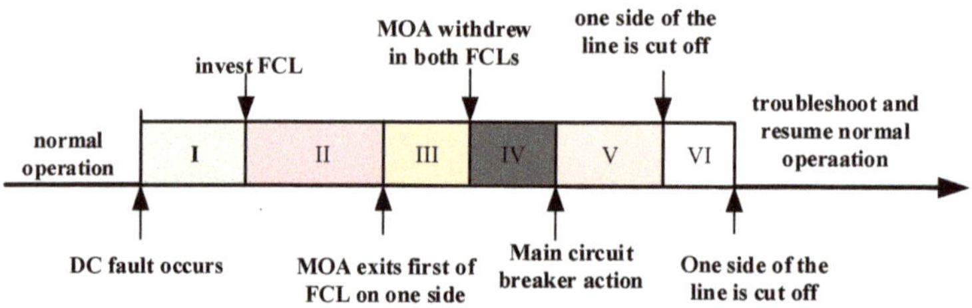

Figure 6. Cooperative action process of current limiting and DC circuit breaker.

If the synergistic effect of the current-limiting inductor and the circuit breaker is fully considered in this operation process, the transient current equation of a DC power grid system with N converter stations in the case of bipolar fault can be expressed as Formula (9) [22–25]:

$$\begin{cases} A\cdot u = \underbrace{\underbrace{\underbrace{R\cdot i_F + L_{line}\cdot i_F}_{\text{Level\_I}} + [u_{MF\_1}, u_{MF\_2}, 0, \cdots]^T}_{\text{Level\_II;III}} + [u_{MB\_1}, u_{MB\_2}, 0, \cdots]^T}_{\text{Level\_IV;V}} \\ u = C\cdot i_F \\ u_{MF\_1} = L_{F\_1} d(i_{F\_1} - i_{MOA\_1})/dt \\ u_{MF\_2} = L_{F\_2} d(i_{F\_2} - i_{MOA\_2})/dt \end{cases} \tag{9}$$

where u is the node capacitance voltage matrix; A is the association matrix of branches and nodes; i_F is the branch current matrix; R is resistance matrix; L is inductance matrix; C is capacitance matrix; i_{F_1} and i_{F_2} are the current on both sides of the fault line respectively. i_{MOA_1} and i_{MOA_2} are the metal oxide arrester (MOA) current of the current-limiting inductor on both sides, respectively. L_{F_1} and L_{F_2} are the inductances of the current-limiting inductor on both sides, respectively. U_{MF_1} and u_{MF_2} are the voltages of the current-limiting inductor. U_{MB_1} and u_{MB_2} are the voltages of the MOA in the DC circuit breaker. According to Formula (9) and Figure 6, if the above current-limiting and isolation methods are adopted when bipolar faults occur in the N-terminal DC system, the following problems exist in the six stages.

The first stage is from the occurrence of a DC fault to the input stage of the current-limiting inductor. The time for this stage is 3.5 ms [5,17]. In this stage, the fault current and the DC line loss increase.

The second stage is from the time when the current-limiting inductor begins to operate to the time when the MOA on one side of the fault line exits. The time is restricted by the *V-I* characteristic of the MOA. In this stage, although the MOA is triggered to absorb part of the energy, it will prevent the current-limiting inductor from generating overvoltage. The energy loss is obvious, and there is still a negative impact on the non-fault areas.

The third stage is from the time when the fault current-limiting MOA on the other side is activated until the MOA on both sides of the fault current-limiting inductor is withdrawn. The time is also restricted by the *V-I* characteristics of the MOA. The negative effects of the fault area and the energy loss still exist in this stage.

The fourth stage is when the current-limiting inductor is fully put into operation. Combined with the time of the second and third stages, the total time is about 0.5 ms to 1 ms. In this stage, the fault current is slightly reduced, but the energy loss and the negative influence of the fault area are similar to those in the second and third stages.

The fifth and sixth stages are when the DC circuit breakers on both sides start to cut off the fault current. The two stages include the MOA action of the circuit breakers on both sides. The time is about 2 ms [25]. In this stage, the fault current is cleared, but there is an MOA energy loss.

According to the above analysis, when the DC system with the traditional current-limiting inductor responds to the DC line two-pole fault, the process takes about 6 ms to 6.5 ms, during which there is an energy loss, and the negative impact of the fault area on the non-fault area cannot be eliminated [5]. If the time of the whole current-limiting operation process is taken as the reference, when dealing with a "transient fault" with a fault duration greater than 4.5 ms, the line protection is bound to act, increasing the number of switching actions and shortening its working life.

3.2. Analysis on the Synergistic Process of Flexible Current Limiting and DC Circuit Breaker

According to Figure 6, there is an obvious power loss in the whole process of cooperative protection between traditional current limiters and circuit breakers [17], and the system self-recovery without the circuit breaker cannot be realized through the auxiliary role of the traditional current-limiting device. The FCLD proposed in this paper uses differential undervoltage [26] and its own secondary undervoltage monitoring to enable it to operate instantaneously. By adjusting the u_{c_ref} control system, the DC output voltage of the converter station can be rapidly increased, the fault current can be suppressed, the fault type can be identified, the permanent fault can be quickly isolated, and the instantaneous fault can be eliminated without circuit breaker action.

After the flexible current limiter is added to the DC system, the transient current equation of a DC power grid system with N converter stations under the bipolar fault can be expressed as follows:

$$\begin{cases} A\cdot u = \underbrace{\underbrace{R\cdot i_{\mathrm{F}} + L_{line}\cdot i_{\mathrm{F}}}_{\text{level\_I}} + [u_{\mathrm{CLD\_1}}, u_{\mathrm{CLD\_2}}, 0, \cdots]^T}_{\text{level\_II, level\_III}} \\ u = C\cdot i_{\mathrm{F}} \\ u_{\mathrm{CLD\_1}} = L_{\mathrm{F\_1}}\frac{di_{\mathrm{F\_1}}}{dt} \\ u_{\mathrm{CLD\_2}} = L_{\mathrm{F\_2}}\frac{di_{\mathrm{F\_2}}}{dt} \end{cases} \quad (10)$$

U_{CLD_1} and u_{CLD_2} are the voltages of the FCLD. Problems existing in the six stages of traditional current limiting and isolation can be solved by the FCLD. The whole process of synergistic action between the proposed FCLD and the DC circuit breaker is shown in Figure 7.

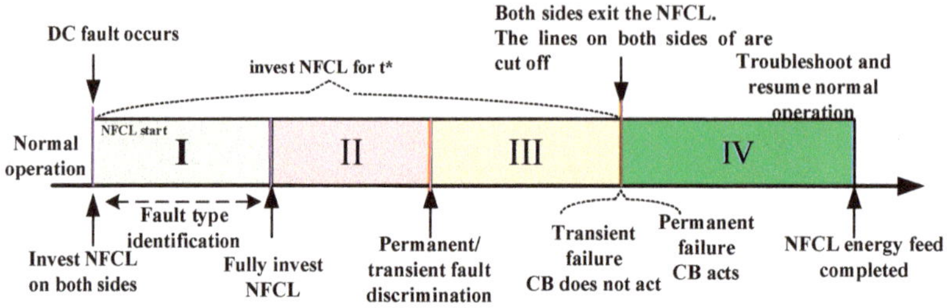

Figure 7. Cooperative action whole process of FCLD and DC circuit breaker.

The first stage is from the occurrence of the DC fault to the full operation of the current-limiting inductor. In this stage, the fault type is identified through undervoltage detection, the reference voltage of the application layer is adjusted, and the FCLD on both sides is operated. In the case of permanent fault, the time length of starting the FCLD is controlled by the attenuation coefficient of the capacitor voltage of the converter station. The FCLD has no energy loss and completely eliminates the negative impact of the fault area on the non-fault area. The time of the current limiting is shortened, compared with the traditional method.

It can be seen from Formula (10) that when the sum of the voltage of the current-limiting device and the impedance voltage of the line is greater than the voltage at the outlet of the converter station, the current of the DC system will continue to decrease until the current passes zero.

The second stage: By adjusting the control objectives of FCLD application layer, the permanent and transient faults of the system can be distinguished according to the output current variation characteristics of the DC circuit breaker. In this stage, there is no energy loss and no negative impact on the fault area. At this stage, the voltage of the circuit breaker is expressed as follows:

$$u_{\mathrm{DCCB}} = A \cdot u - \left\{ R \cdot i_{\mathrm{F}} + L_{line} \cdot i_{\mathrm{F}} + [u_{\mathrm{CLD\_1}}, u_{\mathrm{CLD\_2}}, 0, \cdots]^T \right\} \tag{11}$$

The third stage: According to the line current feedback process in Figure 2 and Formula (11), the sum of the voltage of the FCLD and the line impedance voltage is equal to the outlet voltage of the converter station, and the voltage of the DC circuit breaker is zero. Within t^* time, if the output current of the current-limiting device has the feature of fault disappearance, the normal working state is restored. Otherwise, the DC circuit breaker is switched off. After that, the FCLD exits the current-limiting state so that the fault line of the DC system can be switched off without arc under the condition of permanent fault.

This stage can improve the flexibility of the original traditional current-limiting device and the circuit breaker. The energy loss and negative impact of the fault area are the same as those in the second and third stages.

The fourth stage: The FCLD on both sides feeds the stored energy into the power grid through its rectifier. The input current of the rectifier is controlled by the energy feedback time. In this stage, the fault is cleared without energy loss.

According to the above four stages, the FCLD can significantly suppress the fault current and ensure that the output voltage of the converter station stays at the rated voltage and there is no voltage-sag problem. The negative impact of the fault area on the non-fault area is completely eliminated. Compared with the traditional six stages, the method proposed in this paper can identify instantaneous faults and greatly reduce the number of circuit breaker reclosing actions. Because the saturation problem of current-limiting inductance is eliminated by the inverse flux of the controllable power supply, the current-limiting operation time t^* can be set according to the actual protection requirements of the system.

4. Simulation Analysis

In order to verify whether the new FCLD can assist the DC system to achieve fault current suppression and fault area isolation, Matlab/Simulink simulation software was used to build a simulation platform of a DC $\pm$ 5 kV distribution network system with three terminals, as shown in Figure 8. The voltage source converter (VSC1) is the power end, and VSC2 and VSC3 are the load ends. Tables A1 and A2 in Appendix A provide the relevant simulation parameters of the DC system. According to the normal operation standard of the international DC distribution network system, the judgment criteria of whether the line fault area is completely eliminated are set: (1) During the fault, the system voltage must not be lower than 90% of the rated voltage, and the fault line current must not exceed the rated current of the circuit breaker. (2) The fault inrush current does not exceed 10 times the rated current of its circuit breaker. The fault stable current does not exceed 1 time the rated current of its circuit breaker. (3) The DC system runs rapidly and stably after the line fault, and the maximum voltage fluctuation of DC bus satisfies $\Delta U_N \leq 10\%\ U_N$ during the recovery process, and there is no distorted fluctuation current.

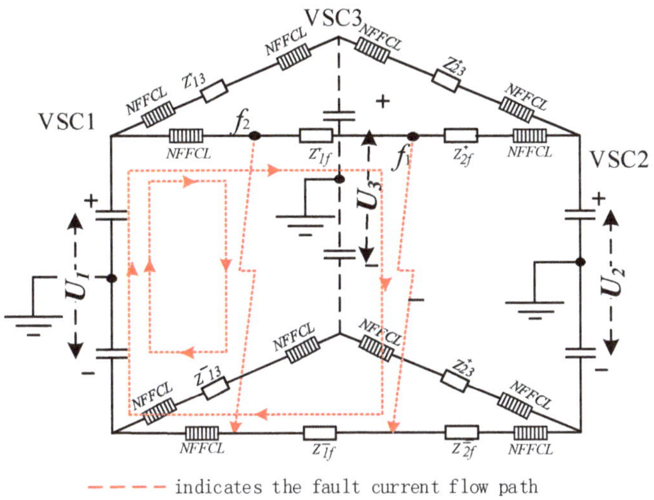

Figure 8. Three-terminal DC distribution network system.

4.1. Comparative Analysis of Current-Limiting Effect

In order to analyze the synergistic effect of the new FCLD and the circuit breaker, the current-limiting effects of the superconducting current-limiting device and the FCLD without the coordinated action of the circuit breaker are compared. Taking the interpole fault of the three-terminal DC system as an example, the simulation comparison diagram is shown in Figure 9.

It can be seen from Figure 8 that the peak value of the fault current can reach 550 A when an inter-pole fault occurs without a current-limiting inductor. The peak value of the fault current with the superconducting current-limiting inductor is up to 350.6 A, which is 36.25% lower than that without the superconducting current-limiting inductor. Table A3 compares the superconducting current limiter and the flexible current limiter in terms of economy and current-limiting effect.

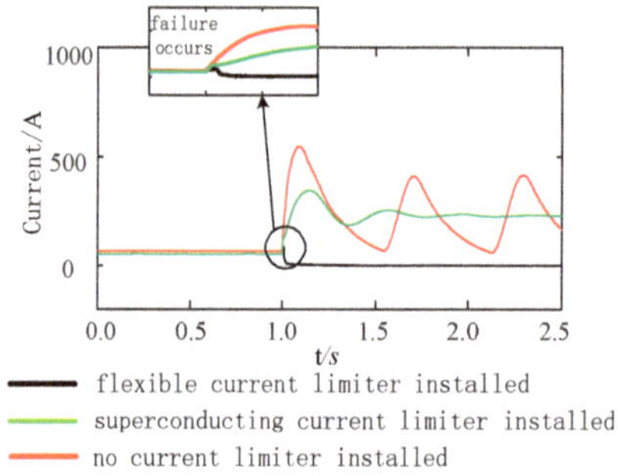

Figure 9. Comparison chart of current limiting.

When an FCLD is installed, the peak current of the fault line is suppressed to 85.7 A, which is 84.4% lower than the fault current without the current-limiting inductor and 75.6% lower than that with the superconducting current-limiting inductor. Because the clamping voltage of the FCLD can be flexibly adjusted according to the line current, it can be seen from Figure 9 that without the circuit breaker, the current in the faulty line is held steadily at zero. Compared with the superconducting current-limiting inductor, which only suppresses the faulty line current, the FCLD can provide a new breaking environment for the DC circuit breaker and realize zero-current and arc-free breaking.

4.2. Transient Analysis of Interpole Faults of Transmission Lines

(1) Permanent failure

When t = 0.5 s is set, a permanent fault occurs between the poles at position f_1 and position f_2 of line$_{1\text{-}2}$, respectively. As shown in Figure 8, t_a is set to 10 ms and t^* is set to 50 ms. The clamping voltage of the positive FCLD on the VSC1 output terminal can be seen in Figure 10.

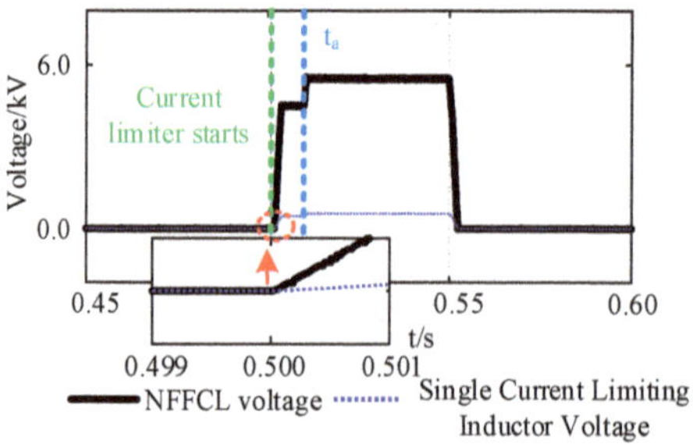

Figure 10. The clamping voltage of the positive FCLD on the VSC1 output terminal.

From the simulation results in Figures 10 and 11(a$_1$–a$_3$) and, it can be seen that under an inter-pole fault, the voltage differential component instantly activates the FCLD at each port of the converter station, so the FCLD clamping voltage on both sides of line$_{1\text{-}2}$ can take effect quickly. Figure 11(a$_1$–b$_3$) show the output voltage and the output current curve of the DC system, the VSC1 power supply terminal, and the VSC2 and VSC3 load terminal ports. Figure 11(c$_1$–c$_3$) provides the current variation curves of the FCLD on both sides of the fault position. The positive and negative output voltages at each port of the DC system are kept at ±5 kV within 0.506 s, and the voltage fluctuation of each port does not exceed 1 kV. That is, the voltage has no sag problem. Compared with the DC system without the current-limiting inductor, the output voltage of each port drops to 0 V when an inter-pole fault occurs. The installation of the current-limiting inductor effectively solves the problem of the voltage sag of the DC system when an inter-pole fault occurs.

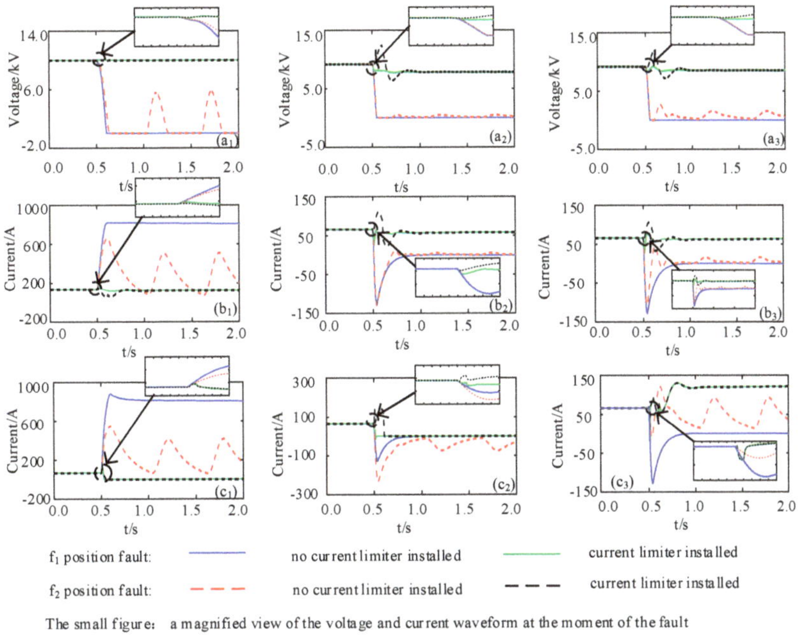

Figure 11. The voltage and current waveforms of each terminal of the DC system in permanent pole-to-pole fault. (**a$_1$–b$_3$**) show the output voltage and the output current curve of the DC system, the VSC1 power supply terminal, and the VSC2 and VSC3 load terminal ports; (**c$_1$–c$_3$**) provides the current variation curves of the FCLD on both sides of the fault position.

As can be seen from the comparison of the curve changes before and after the installation of the current-limiting inductor in Figure 11(b$_1$–b$_3$), the proposed current-limiting technique suppresses the inrush current caused by the line fault in fault position f_1 and position f_2. When a permanent fault occurs at position f_1 and position f_2, due to the action of the current-limiting inductor, the peak fault current of VSC1 is suppressed from 654.3 A and 816.6 A to 147.7 A and 173.2 A, respectively, the peak fault current of VSC2 is suppressed from -130.1 A and -130 A to 54.63 A and 27.58 A, respectively and the peak fault current of VSC3 is suppressed from -107.1 A and -128.1 A to 53.96 A and 26.34 A, respectively. The surge current during the fault is effectively suppressed, and the fault current is quickly suppressed to below 200 A.

As can be seen from the current change curve of the FCLD terminal in Figure 11(c$_1$–c$_3$), when the current-limiting inductor starts normally, the line current is made to cross zero at t = 0.533 s and t = 0.538 s at the positions f_1 and f_2, respectively, and the DC circuit breaker is switched off at the moment when the fault current passes zero so as to realize the arc-free shutdown of the DC circuit breaker and the rapid isolation of the fault line. When the power is supplied by the normal line, the current on this line increases.

(2) Instantaneous failure

When t = 0.5 s is set, a transient fault occurs between the poles at position f_1 and position f_2 of line$_{1-2}$, respectively. As shown in Figure 8, t^* and t_a are set to 50 ms. Figure 12 (a$_1$–b$_3$) show the output voltage and output-current curve of the DC system VSC1 power supply terminal and the VSC2 and VSC3 load terminal ports. Figure 11(c$_1$–c$_3$) provides the current variation curves of the FCLD on both sides of the fault position.

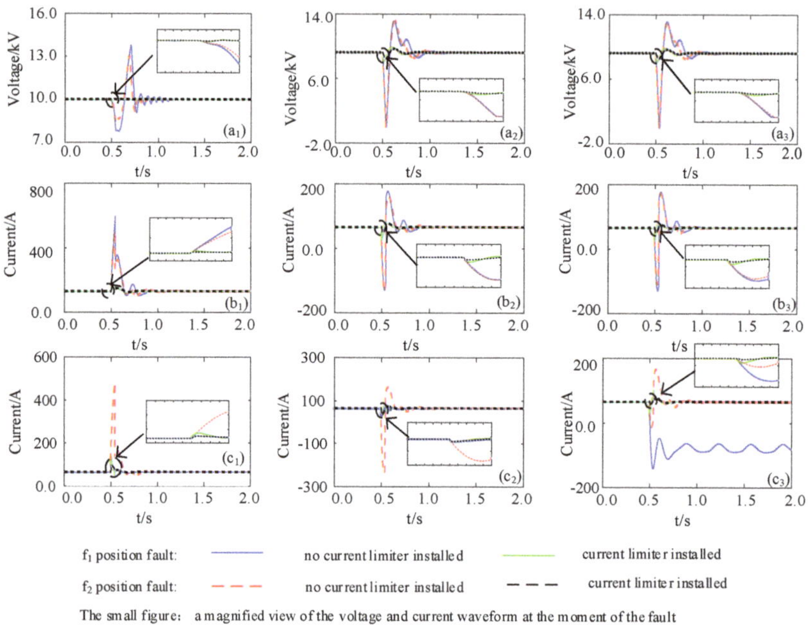

Figure 12. The voltage and current waveforms of each terminal of the DC system in transient pole-to-pole fault. (**a$_1$–b$_3$**) show the output voltage and output-current curve of the DC system VSC1 power supply terminal and the VSC2 and VSC3 load terminal ports; (**c$_1$–c$_3$**) provides the current variation curves of the FCLD on both sides of the fault position.

According to Figure 12(a$_1$–a$_3$), inter-pole faults occur at position f_1 and position f_2 without the current-limiting inductor, the peak-to-peak values of VSC1's output voltage are 4209 V and 5914 V, respectively, and the peak-to-peak values of VSC2's output voltage are 13,069.1 V and 12,939.1 V, respectively. The peak-to-peak output voltages of VSC3 are 12,679 V and 12,999 V, respectively. The positive and negative voltages of each output port of the DC system at the time of inter-pole failure are kept at ±5 kV at 0.506 s through the rapid rise of the clamp voltage of the FCLD, so that the voltage fluctuation of each port does not exceed 1 kV and the voltage-sag problem under the condition of failure is suppressed.

It can be seen from Figure 12(b$_1$–b$_3$) that when a unipolar fault occurs at position f_1 and position f_2 of the DC system line$_{1-2}$, and the DC system is not equipped with a current-limiting inductor, the peak fault current of the VSC1 output port within 50 ms reaches 506.1 A and 504.7 A, respectively, the peak-to-peak value of fault current of VSC2 output port reaches 283.8 A and 284.3 A, respectively, and the peak-to-peak value of fault current of VSC3 output port reaches 273.9 A and 273.9 A, respectively. The proposed current-limiting technology can rapidly increase the FCLD clamping voltage to suppress the inrush current and suppress the peak current of the VSC1 output port to 153.7 A and 177.2 A, respectively, when the f_1 and f_2 faults occur and suppress the peak current of the VSC2 output port to 29.7 A and 47.8 A, respectively, when the fault occurs. The peak-to-peak values of the VSC3 output port fault current are reduced to 28.3 A and 50.3 A, respectively.

From the current change curve of the FCLD terminal in Figure 12(c$_1$–c$_3$), it can be seen that when an inter-pole fault occurs, the current of the faulty line is quickly suppressed, and the current of the faulty line is kept at the normal system level after 34 ms. Therefore, the DC system can smoothly proceed through the inter-pole fault continuous process without the action of the circuit breaker and realize self-recovery.

Under normal operation of the three-terminal DC network, the two load sides are always operating symmetrically, due to the parameter setting of load–source–load. The current of the two source–load transmission lines is almost equal, and the current of the load–load transmission line is almost zero.

It is assumed that a short-circuit fault occurs in one of the source–load transmission lines in the three-terminal network, and the balance of the three-terminal DC network is destroyed and the voltage symmetry of the two load sides is broken. The fault line voltage is raised by a current-limiting device, so that the DC line current of the fault is quickly increased and the voltage on the load side is kept stable.

5. Conclusions

In view of the defects existing in the whole process of the synergistic action between the circuit breaker and the current limiting inductor in the existing DC system, this paper proposed a new FCLD suitable for the DC power grid, and we draw the following conclusions via theoretical analysis and simulation verification:

(1) By introducing reverse magnetic flux compensation of the controllable current source, the FCLD current-limiting reactance can obtain a stable clamping voltage. It can achieve obvious suppression on the fault current. It can eliminate the negative impact of the fault area on the non-fault area and ensure that distributed power sources in the network will not be disconnected due to voltage sags.

(2) The cooperation of the FCLD and the DC circuit breaker can realize the circuit-breaker close without arc. The device can eliminate the negative impact of the fault on the non-fault area and realize the arc-free shutdown of the permanent-fault DC circuit breaker. Through the identification of permanent fault and transient fault, the circuit breaker action can be prevented when an instantaneous fault occurs. It can reduce the times of the circuit breaker action, the instantaneous-fault DC circuit breaker action probability, and the user power-loss rate.

(3) According to the control scheme and the simulation results of the coordinated action of an FCLD and a DC circuit breaker, the FCLD proposed in this paper prolongs the protection action time of the circuit breaker and greatly reduces the interruption requirements of the DC circuit breaker. Furthermore, the validity of the proposed FCLD is verified, and a new protection idea is provided for the practical application of DC-system engineering.

(4) The FCLDs can adjust the internal voltage of the system to maintain symmetrical operation of the system in the event of asymmetrical operation of the DC system.

Author Contributions: Conceptualization and methodology, F.Z.; software, validation, and writing—original draft preparation, Y.P.; investigation, S.Z.; data curation, review, and editing, W.W. All authors have read and agreed to the published version of the manuscript.

Funding: This research was funded by National Natural Science Foundation of China (61903088).

Data Availability Statement: Not applicable.

Conflicts of Interest: The authors declare no conflict of interest.

Appendix A

Table A1. System parameters of three terminal DC distribution network.

	Converter Station 1	Converter Station 2	Converter Station 3
Rated Capacity of Converter Station (MV·A)	20	10	10
DC Voltage/kV	±5	±5	±5
DC Capacitance Value/mF	4	2	2
AC Reactance Value/mH	50	70	70
Smoothing Reactor/mH	20	20	20

Table A1. *Cont.*

	Converter Station 1	Converter Station 2	Converter Station 3
R	Constant Voltage	P/Q	P/Q
Parameter	Value	Parameter	Value
$R_{12}/R_{13}/R_{14}/R_{23}/R_{24}/R_{34}$	0.08/0.08/0.08/0.08/0.08	R_{1f}/R_{2f}	0.02/0.08
$L_{12}/L_{13}/L_{14}/L_{23}/L_{24}/L_{34}$	5 mH/5 mH/5 mH/5 mH/5 mH/5 mH	L_{1f}/L_{2f}	1 mH/4 mH

Table A2. FCLD system parameters.

Parameter	Value	Parameter	Value
k_{u_P}	1	$L^1{}_I$	0.1 mH
k_{u_I}	500	$L^2{}_I$	1 mH
R	0.15	n	5
L	1.5 mH	k	10
R_c	0.5	C	10
a	100 ms	t_b	100 ms
t^*	400 ms		

Table A3. Comparison of two kinds of current limiter.

	Superconducting Current-Limiting Inductor	Flexible Current-Limiting Inductor
Cost	USD 2,500,000	USD 250,000
Peak fault current	350.6 A	85.7 A
Time of the whole operation process	>200 ms	35 ms
Device applications	Interpole faults	Interpole faults/pole to ground fault
Arc	Circuit breaker break-off with arc	Arc free

References

1. Zhang, Y.; Meng, X.; Shotorbani, A.M.; Wang, L. Minimization of AC-DC Grid Transmission Loss and DC Voltage Deviation Using Adaptive Droop Control and Improved AC-DC Power Flow Algorithm. *IEEE Trans. Power Syst* **2021**, *36*, 744–756. [CrossRef]
2. Watson, J.D.; Lestas, I. Control of Interlinking Converters in Hybrid AC/DC Grids: Network Stability and Scalability. *IEEE Trans. Power Syst* **2021**, *36*, 769–780. [CrossRef]
3. Pan, Y.; Chen, L.X.; Yuan, Z.; Xu, M.; Liu, L. Research on Current Limiting and Energy Limiting Technology for DC Power Grid Fault. *Proc. CSEE* **2020**, *40*, 2006–2015.
4. Huang, H.; Xu, Z.; Lin, X. Improving Performance of Multi-Infeed HVDC Systems Using Grid Dynamic Segmentation Technique Based on Fault Current Limiters. *IEEE Trans. Power Syst.* **2012**, *27*, 1664–1672. [CrossRef]
5. Zhu, S.; Zhao, C.; Li, C.; Xu, J. The DC Fault Current Calculation of DC Fault Current Limiter Action Included in Bipolar MMC-HVDC Grid. *Proc. CSEE* **2019**, *39*, 469–478.
6. Wang, Y.; Yuan, Z.; Wen, W.; Ji, Y.; Fu, J.; Li, Y.; Zhao, Y. Generalized protection strategy for HB-MMC-MTDC systems with RL-FCL under DC faults. *IET Gener. Transm. Distrib.* **2018**, *12*, 1231–1239. [CrossRef]
7. Xu, J.; Zhu, S.; Li, C.; Zhao, C. The Enhanced DC Fault Current Calculation Method of MMC-HVDC Grid with FCLs. *IEEE J. Emerg. Sel. Top. Power Electron.* **2019**, *7*, 1758–1767. [CrossRef]
8. Gowaid, I.A. A Low-Loss Hybrid Bypass for DC Fault Protection of Modular Multilevel Converters. *IEEE Trans. Power Deliv.* **2017**, *32*, 599–608. [CrossRef]
9. Lyu, H.; He, J.; Li, B.; Li, Y.; Spier, D.W.; Prieto-Araujo, E.; Gomis-Bellmunt, O. An Improved Hybrid DC Circuit Breaker with Self-Adaptive Fault Current Limiting Capability. *IEEE Trans. Power Electron.* **2022**, *37*, 4730–4741. [CrossRef]
10. Sneath, J.; Rajapakse, A.D. Fault Detection and Interruption in an Earthed HVDC Grid Using ROCOV and Hybrid DC Breakers. *IEEE Trans. Power Deliv.* **2016**, *31*, 973–981. [CrossRef]
11. Ma, D.; Chen, W.; Ye, H.; Xue, C.; Pan, P.; Zhu, X. An Assembly High Voltage DC Circuit Breaker Based on Pre-charged Capacitors 2018. In Proceedings of the IEEE International Power Electronics and Application Conference and Exposition (PEAC), Shenzhen, China, 4–7 November 2018.

12. Ye, H.; Chen, W.; Pan, P.; Xue, C.; Azeem, S.W.; Zhu, X. A Novel Hybrid DC Circuit Breaker Based on Precharged Capacitors 2018. In Proceedings of the IEEE International Power Electronics and Application Conference and Exposition (PEAC), Shenzhen, China, 4–7 November 2018.
13. Yang, S.; Xiang, W.; Lu, X.; Zuo, W.; Wen, J. An Adaptive Reclosing Strategy for MMC-HVDC Systems with Hybrid DC Circuit Breakers. *IEEE Trans. Power Deliv.* **2020**, *35*, 1111–1123. [CrossRef]
14. Wei, X.; Zou, G.; Zhang, S.; Xu, C.; Sun, W. Adaptive Reclosing Scheme for flexible Multi-Terminal DC Distribution Grid 2021. In Proceedings of the 2021 6th Asia Conference on Power and Electrical Engineering (ACPEE), Chongqing, China, 8–11 April 2021; pp. 746–750.
15. Mei, J.; Ge, R.; Zhu, P.; Fan, G.; Wang, B.; Yan, L. An Adaptive Reclosing Scheme for MMC-HVDC Systems Based on Pulse Injection from Parallel Energy Absorption Module. *IEEE Trans. Power Deliv.* **2021**, *36*, 1809–1818. [CrossRef]
16. Zheng, T.; Lv, W.; Li, R. An Adaptive Reclosing Scheme Based on Phase Characteristics for MMC-HVDC Systems. *IEEE Trans. Power Deliv.* **2022**, *37*, 2986–2996. [CrossRef]
17. Liu, J.; Tai, N.; Fan, C. A hybrid current-limiting circuit for DC line fault in multi-terminal VSC-HVDC system. *IEEE Trans. Ind. Electron.* **2017**, *64*, 5595–5607. [CrossRef]
18. Zheng, F.; Zhang, J.; Lin, J.; Deng, C.; Huang, J. A Novel Flexible Fault Current Limiter for DC Distribution Applications. *IEEE Trans. Smart Grid* **2022**, *13*, 1049–1060. [CrossRef]
19. Qiu, W.; Guo, M.; Zheng, Z.Y. Flexible Arc-suppression Method Based on Single-DC-source Cascaded H-bridge Converter for Earth Fault in Distribution Networks. *Power Syst. Technol.* **2019**, *43*, 3848–3858.
20. Li, B.; Jing, F.; Jia, J.; Li, B. Research on Saturated Iron-Core Superconductive Fault Current Limiters Applied in VSC-HVDC Systems. *IEEE Trans. Appl. Supercond* **2016**, *26*, 5603805. [CrossRef]
21. Li, B.; Jing, F.; Li, B.; Chen, X.; Jia, J. Study of the Application of Active Saturated Iron-Core Superconductive Fault Current Limiters in the VSC-HVDC System. *IEEE Trans. Appl. Supercond.* **2018**, *28*, 1–6. [CrossRef]
22. Li, R.; Xu, L.; Holliday, D. Continuous operation of radial multiterminal HVDC systems under DC fault. *IEEE Trans. Power Deliv.* **2016**, *31*, 351–361. [CrossRef]
23. Li, C.; Zhao, C.; Xu, J. A pole-to-pole short-circuit fault current calculation method for DC grids. *IEEE Trans. Power Syst.* **2017**, *32*, 4943–4953. [CrossRef]
24. He, Z.; Wang, W.; Gu, H.; Li, G.; Xin, Y.; Wang, Z. Development Status and Prospect of Integrated Key Equipment of DC Power Grid with Fault Current Limiting and Breaking Functions. *Proc. CSEE* **2020**, *646*, 14–30.
25. Liu, G.; Xu, F.; Xu, Z. Assembly HVDC breaker for HVDC grids with modular multilevel converters. *IEEE Trans. Power Electron.* **2017**, *32*, 931–941. [CrossRef]
26. Qin, Y.; Wen, M.; Bai, Y. A Novel Distance Protection Scheme for HVDC Transmission Lines Based on Criterion Extremum Online Estimation. *Proc. CSEE* **2020**, *647*, 185–194.

symmetry

Article

Modern Active Voltage Control in Distribution Networks, including Distributed Generation, Using the Hardware-in-the-Loop Technique

Beshoy Nabil Fahmy Fanos [1], Mohammad H. Soliman [1], Hossam E. A. Talaat [2,*] and Mahmoud A. Attia [1,*]

[1] Electrical Power & Machines Department, Faculty of Engineering, Ain Shams University, Cairo 11517, Egypt
[2] Electrical Engineering Department, Future University in Egypt, Cairo 11517, Egypt
* Correspondence: hossam.eldeen@fue.edu.eg (H.E.A.T.); mahmoud.abdullah@eng.asu.edu.eg (M.A.A.)

Abstract: Voltage constraints usually place restrictions on how distributed generation (DG) can be connected to weak distribution networks. As DG capacity increases, active voltage control techniques are needed. Active approaches can greatly lower connection costs while boosting the capacity of connectable DG when used in place of the passive strategy. In this article, a modified active voltage control algorithm is used on an IEEE 33 bus system to test the robustness and reliability of the control algorithm under severe conditions. The simulations are carried out using the hardware-in-the-loop (HIL) method. Real-time simulations are used to test data transfer and the reliability of the control algorithm's execution. The analysis is based on a three-phase symmetric power system.

Keywords: active voltage regulation; coordinated control; distribution network; distributed generation (DG); hardware-in-the-loop; real-time simulations; integrated energy system

1. Introduction

Power systems are increasingly reliant on distributed generation (DG). External, non-electric factors such as appropriate geographic locations of wind and solar resources influence the location of DG units. As a result, whether planned or not, they are frequently attached to distribution networks at the closest location. Furthermore, all energy producers now have access to distribution networks as a result of energy market deregulation laws that include incentives for renewable energy production [1,2]. As a result, much research has been conducted to analyze issues concerning DG market access, such as locational fixed costs [3], the long-term implications of feed-in tariffs [4], and a carbon tax strategy for system cost-effectiveness [5]. Many strategies were proposed concerning the connection of DGs to the networks. A strategy proposed a fair and equitable electricity tariff for the calculation of the distribution network usage fee [6]. Another one suggested providing DGs with an incentive to locate themselves in the most needed areas of the network [7]. Optimization techniques were utilized for the optimization of the insertion of DGs into the network [8,9]. The voltage rise effect, particularly in weak distribution networks, is the most critical factor restricting the capacity of connected DGs. Currently, this is frequently handled by connecting DGs to specialized feeders or any other possible network upgrading, such that the power flow of the entire network keeps the extreme voltages (maximum generation/minimum loading) inside network boundaries, as they are traditionally connected in a "fit and forget" style [10]. Because the traditional distribution network is passive, this inefficient mode of operation is possible. Certainly, this significantly raises the cost of the DG connection because its control is solely dependent on the primary substation and ignores DG capabilities in network voltage regulation. In the passive configuration, if the increase in available low running cost DG units threatens network safety, the distribution network operator (DNO) must disconnect some DG parks, wasting some installed capacity. Instead, coordinated active control methods that make advantage of DG features can significantly increase the connected capacity [11].

Several coordinated control algorithms utilizing both the capabilities of DG units and the network were published [12,13]. It has been aimed to mitigate high ramp-rate variables in the network [14], while other articles implemented coordinated control strategies to minimize voltage deviations [15]. A hierarchical coordinated voltage correction scheme was proposed. This scheme consists of two layers of control: a regional layer based on the local measurement information and a feeder line coordination layer based on a communication network [16]. The charging and discharging of a battery-based storage system were controlled [17]. Control is divided into two layers: centralized for the entire network and decentralized in each DG [18]. An optimization problem is constructed and controlled by the DG output to make a balance between voltage regulation and reactive power transfer during islanding [19]. Moreover, these control strategies are accompanied by marginal incentives for DG owners to connect them at the optimized locations [20] and with incentives for more applications of customers' demand response [21]. An analysis of the latest studies on local, centralized, distributed, and decentralized voltage control algorithms and a comparison between them were introduced [22]. Moreover, the frequency was added to an adaptive coordinated control scheme [23]. Model predictive control techniques [24–27] and fuzzy control techniques [28] were utilized for time-saving during control. Optimization techniques were utilized to reach the optimal decisions for coordinated control [29,30]. Distributed control algorithms were applied with no need for communication infrastructure [31,32]. The symmetry of the voltage profile over the network is considered a major concern for electric power utilities. Thus, operators apply strict constraints to keep the voltage within acceptable boundaries.

This article follows the trend of a specific rule-based algorithm that selects the quantity to be controlled according to certain conditions and priorities [10,33].

In this article, the results of real-time simulations, including the hardware-in-the-loop (HIL) technique, of an optimal improvement of a control algorithm are demonstrated. The modified algorithm is applied to the IEEE 33 bus radial distribution network with arbitrary DG units. The goals of this application are to validate the coordinated control algorithm for more general networks, as well as to inspect new conditions and scenarios.

This article is organized as follows: Section 2 shows the tools, techniques, and coordinated control algorithm, Section 3 delves into real-time simulations of the algorithm being applied to the IEEE 33 bus radial distribution network in a variety of circumstances, and Section 4 wraps up the conclusions.

2. Coordinated Control Algorithm

The DNO is equipped with a number of tools to manage the network state. They are outlined beneath, along with the benefits and drawbacks of each. They are arranged according to increasing price:

1. On-load Tap Changer (OLTC): Utilizing the OLTC of the primary transformer is the simplest and cheapest way to regulate the network voltages, with no cost of high infrastructure for communication links or huge power losses through the network. However, it is inefficient in some cases, where the difference between the extreme (maximum and minimum) voltage of the network is close to the difference between the predefined limits. In these cases, when the maximum voltage exceeds the upper limit of the network and the OLTC decreases the whole network voltage, the minimum voltage of the network will decrease below the network's lower limit, causing the OLTC to be in a state of successive ups and downs. This problem is solved by deactivating the OLTC in case the other extreme voltage is too close to its limit [34]. Another adaptive OLTC voltage control focused only on the correction of the false image of the network load that has not taken the influence of DG into consideration [35]. In some networks, DNOs prefer to utilize other control tools before the OLTC for mechanical purposes [36].

2. Reactive Power Control Using DGs: DG units receive orders from the coordinated control algorithm to absorb or generate some reactive power within their limits. It is more efficient than the OLTC, although it requires a communication network among the DNO and DG units and highly increases network losses. However, it is still cost worthy, considering that normally the cost of losses is much less than the cost of curtailed energy of DGs. Voltage control loss factors were proposed as means of understanding the interactions between reactive power flows, losses, and curtailment [37].
3. Curtailing the Real Power of DGs: This tool, like the one before it, depends on a system of communication between DNO and DG units. It is regarded to be the most efficient method for maintaining the network voltages within limits (as proven in [11]). However, since it is the most expensive, it ought to be employed as the last choice after all other alternatives have been exhausted.

The Hardware-in-the-loop (HIL) simulation is a technique that substitutes the actual component of a machine or system with a simulation in order to design and test control systems that handle complicated machines and systems. This technique is used in this article to verify the reliability of the algorithm upon implementation in real networks, considering all technical issues such as delay periods, the convergence of calculations, data connection, safe testing, etc.; it is not just a theoretical algorithm.

The hardware component in this article is an Arduino Mega 2560 board utilizing an ATmega 2560 microprocessor. It is programmed with the control algorithm intended to govern the network's overall voltages. Applying MATLAB Simulink, the parameters of both the network and DGs' are modeled. Power flow analyses are carried out employing the forward/backward sweep method [38]. The forward/backward sweep method is selected in this article for its trustworthiness for radial networks in real-time tests since it converges to an output of reasonable accuracy in a relatively short number of iterations. While Simulink transmits the variables intending to evaluate the state of the network to Arduino, the latter returns the decisions to Simulink. This closed-loop system functions with a suitable sample time. Figure 1 summarizes the cycle of operations of the HIL technique. Figure 2 shows the experimental setup for the latter cycle. The image was taken while the configuration was displaying the live results of the simulation.

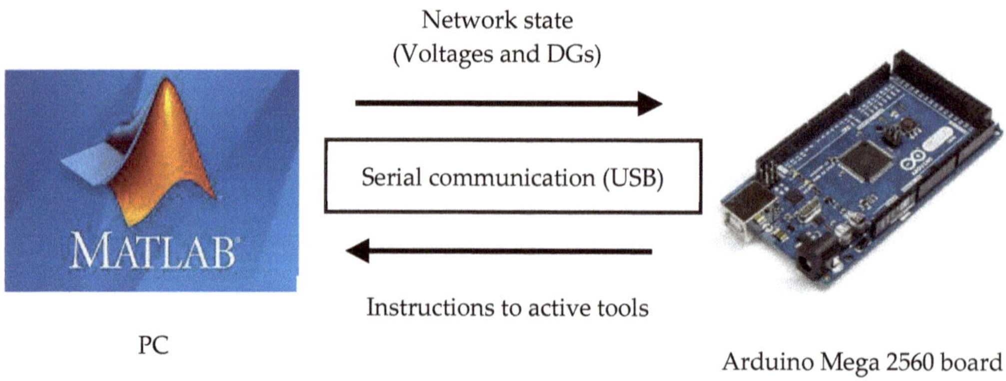

Figure 1. The cycle of operations of the HIL technique.

The control algorithm consists of two main portions: basic and restoring.

The basic part covers the measures necessary to keep network voltages within prescribed limits in the event that one of the extreme voltages exceeds the allowed limits. The restoring part tends to return both the power components of DGs, as well as the position of the OLTC to their optimal levels, as far as the network state permits. The restoring part functions only when all network voltages are within acceptable boundaries; consequently, the basic part is not functioning. Both parts employ an approximate law [39] to estimate the voltage sensitivities of buses of maximum and minimum voltages to the real and reactive powers of all DGs. It can also give a rough estimation of how much reactive power must be absorbed or how much real power must be curtailed of a particular unit in order to achieve a certain voltage change in a certain bus. It should be highlighted that this algorithm is applicable only to radial networks. The operational principles of the whole algorithm are illustrated in Figure 3.

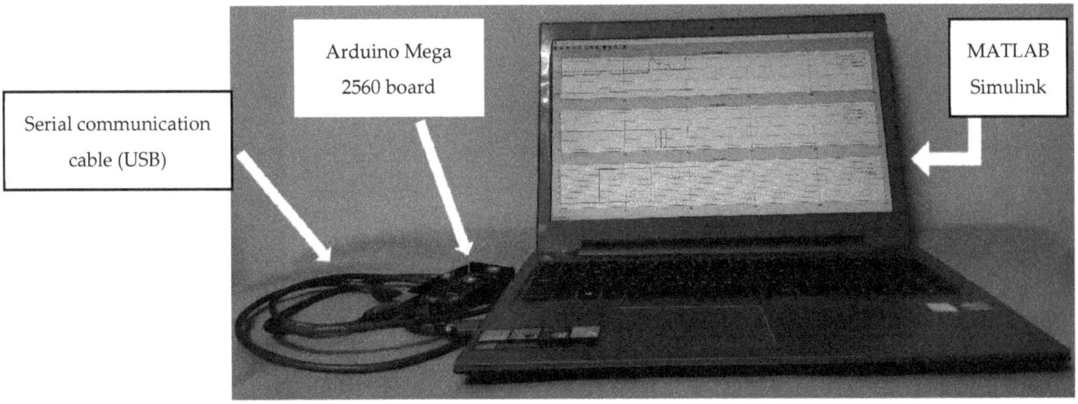

Figure 2. Experimental setup for the HIL technique.

The following active tools must be available for the actions to be applied:

1. OLTC: the OLTC is regarded as inaccessible for control if the other extreme network voltage is too near to the other boundary by less than a single tap step added to a reasonable margin. Otherwise, the desired number of taps is computed so that the overstepped voltage returns within boundaries. Consequently, it is actuated after a preset delay period of time to skip short voltage variations.
2. Absorption or production of reactive powers from DG units: The voltage sensitivities of the buses of extreme voltages with regard to all of the reactive power controllable units are estimated beforehand [39]. The highest sensitivity unit (i.e., lowest reactive power that is needed from it) to the node of the overstepped voltage is picked, provided that this unit has not reached its full capacity of reactive power and will not lead other voltages to cross the other boundary simultaneously. The quantity desired to be absorbed/produced by this unit is estimated, and the requisite quantity is subsequently implemented after a predefined delay period. In case the stated conditions are not fulfilled, then this tool is deemed unavailable for voltage regulation.
3. Curtailment of active power produced by DGs: It is similar to the preceding one except that it functions for the real powers of DGs, which only the DNO is capable of lowering. If the network cannot maintain all voltages within boundaries, even if no real power is injected into the network in any way, load shedding is the only alternative left in this situation.

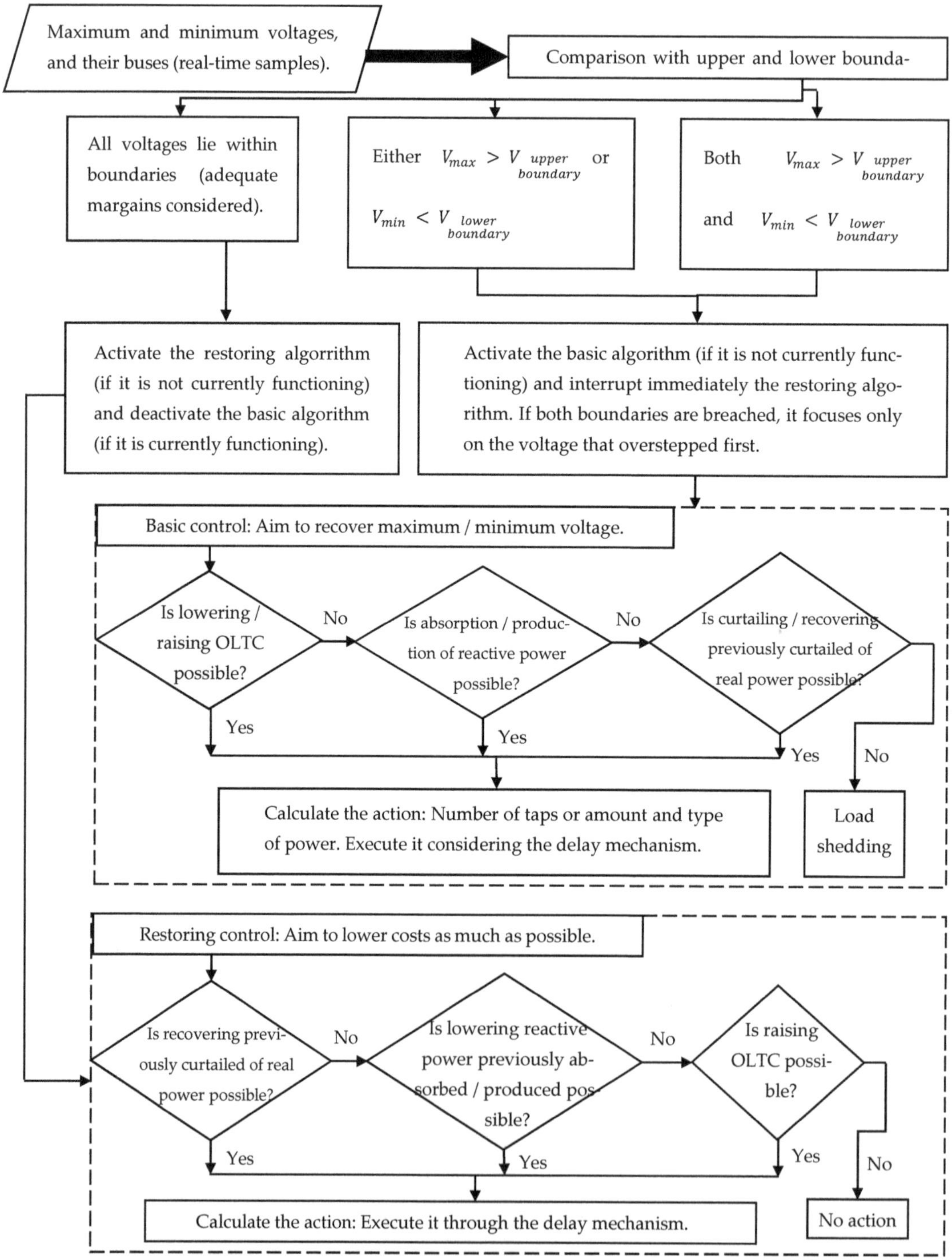

Figure 3. The operational principles of the coordinated control algorithm.

The basic control continues to employ the accessible active tools sorted ascendingly by their costs one after another until the achievement of its target. The basic control algorithm is improved such that if the two extreme voltages of the network violate their limits at a certain instant, the basic control tends to restore the voltage violating first (regardless of the implications of its actions on the other voltage), then tends to restore the other. The restoring control is modified to check if it is capable of recovering any decision conducted by the basic control, fully or partially, to reduce the running cost. That may be possible in case of the removal of a particular unit or an escalation in loading. It functions the opposite of how the basic part does. Real power is examined first to recover any prior curtailments since it causes the most expensive cost, followed by reactive power, and lastly boosting the OLTC as high as is feasible to reduce network losses. It ought to be regarded that restoring control picks the relevant resource with the lowest sensitivity coefficient in order to recover as much power as allowed.

The delay mechanism operates as follows:

- Both the basic and restoring control parts possess their own time delay. When activating a control part and deactivating the other one, it must wait for a preset time delay for the purpose of bypassing transient and rapid voltage surges.
- The DNO must wait after each decision for its special time delay. It ought to examine its implications in the network before deciding the next action. That time delay is prescribed based on the number of taps, mechanical activation time of the OLTC, time constants of generators, and the quantity of power upgrading.
- If a decision has been proven to be insufficient after its delay time, the system performs the next step immediately and does not apply the time delays of basic and restoring parts. The purpose is to prevent the risk of an extreme voltage violation for an extended period of time. The procedure is repeated until the regulation of all voltages for the basic part, or the optimal situation for restoring the part is achieved.
- In case a voltage boundary breach occurred during the application of a restoring control, the restoring control is immediately stopped, and the basic control restarts functioning.

3. Real-Time Simulations

The operation of the algorithm under investigation is tested in the IEEE 33 bus radial distribution network. Data of its loads are tabulated in Appendix A. The standard network is modified by adding three units of DGs, each one having a rated power of 6 MW. The units are located in buses 23, 19, and 22, respectively, as shown in Figure 4. The simulation is based on the assumption of a three-phase symmetric power system, as in practice, significant imbalances do not often take place in medium voltage networks. The tap step voltage of the main transformer is assumed to reach 1.67% of the nominal voltage and a 1 s mechanical activation time for each tap. The voltage of the substation is initially set to 2 taps above 1 p.u. with no intervention from DG units. The lower and upper limits of network voltages are set to be 0.95 and 1.05 p.u. The limits of DGs' reactive powers are $0.33 * P_{rated}$. The delay time is 4 s for the basic part and 6 s for the restoring part.

In the simulation test applied, the DGs are connected to the network (with their full capacity) at t = 30, 50, and 70 s, and then detached (modeling the case of non-availability of real power) at t = 110, 150, and 180 s, respectively. The simulation results are illustrated in Figure 5.

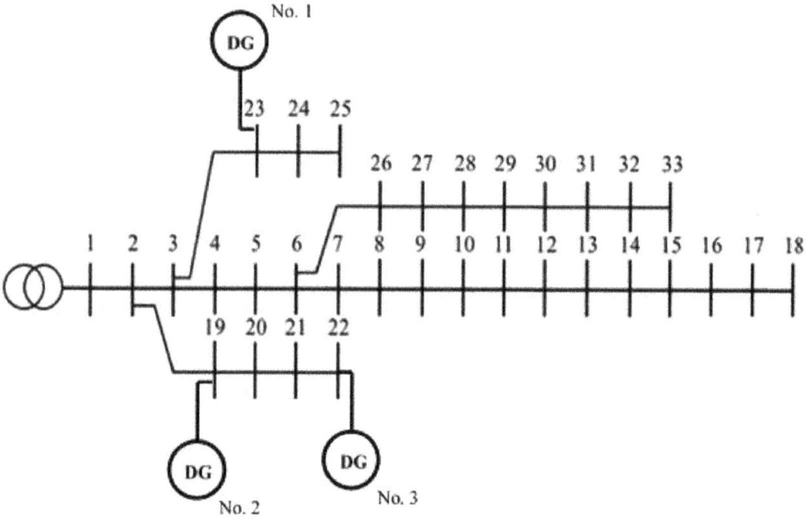

Figure 4. The network under inspection.

The network maximum voltage oversteps the upper boundary due to the connection of the first unit at a time of 30 s. The basic control drops the OLTC set point by a tap after the delay time has lapsed, and the tap changer functions after its mechanical activation time (1 s), returning the maximum voltage inside network boundaries.

None of the increasing network voltages has violated the voltage limits after the connection of the second unit at a time of 50 s. Therefore, no actions have been implemented by the coordinated control.

At a time of 70 s, the third unit is activated, and the network maximum voltage oversteps again. Initially, the basic control cannot set a further tap changer operation, as lowering the substation voltage further leads the minimum voltage to drop below the network voltage lower boundary. Thus, the OLTC option is inaccessible at this moment. Hence, reactive power control is selected. The entire reactive power control capacity of all units is employed one after another split by appropriate delays according to the time constants of the generators, such as the delay mechanism states. They are arranged in descending order by the coefficients of voltage sensitivity with respect to the bus of the overstepped voltage. The network maximum voltage continues to be beyond its boundary, even after the reactive power control, which has become currently inaccessible. As a result, real power control is employed. The third unit is picked for its highest sensitivity. Its real power is reduced by an appropriate quantity to maintain all voltages within acceptable limits.

When the first unit is removed at a time of 110 s, the network voltages decrease, causing some of them to violate the minimum voltage limits. The basic control algorithm has been reversed to boost the network voltage. No tap changer operations can be implemented, as raising the substation voltage would raise the network maximum voltage above the network voltage upper boundary. As a result, the availability of reactive power control is utilized. The reactive power control capability of the first two units is utilized not only to end the absorption of reactive power, but also to generate their full capacity of reactive power one after another as before. The minimum voltage is not yet restored within limits. However, the maximum voltage of the network also violated the upper boundary.

Some remarks should be noted:

- The control algorithm would not tend to recover the maximum voltage until the minimum voltage is restored within limits first (the first event to occur is handled first.)
- The option of the OLTC is not unavailable anymore, as the maximum voltage has overstepped the upper limit already. When such a scenario occurs (a violation of both

limits at the same time), the algorithm obviously cannot regulate both voltages at the same time without causing instability in the system. The voltage control algorithm is set to regulate the voltage violated limits, first neglecting its implications to the other extreme voltage, then regulating the latter voltage afterward.

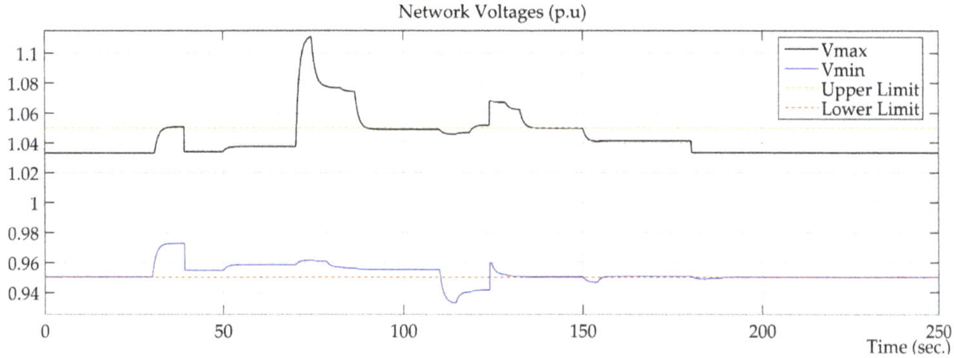

(**a**) The maximum & minimum voltages of the network with respect to the network limits.

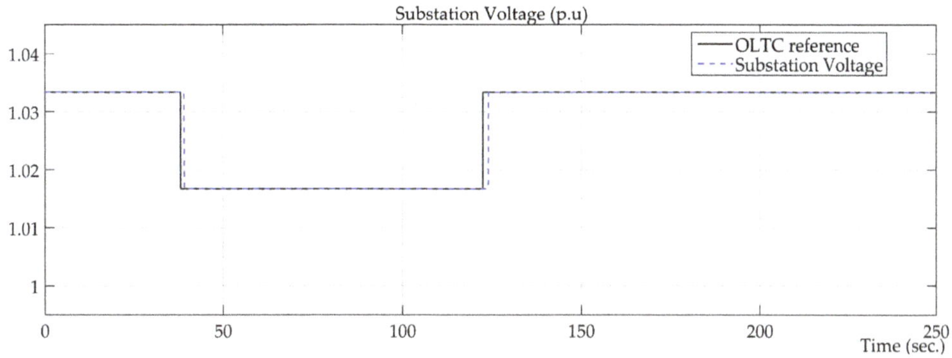

(**b**) The reference input to OLTC & the substation output voltage.

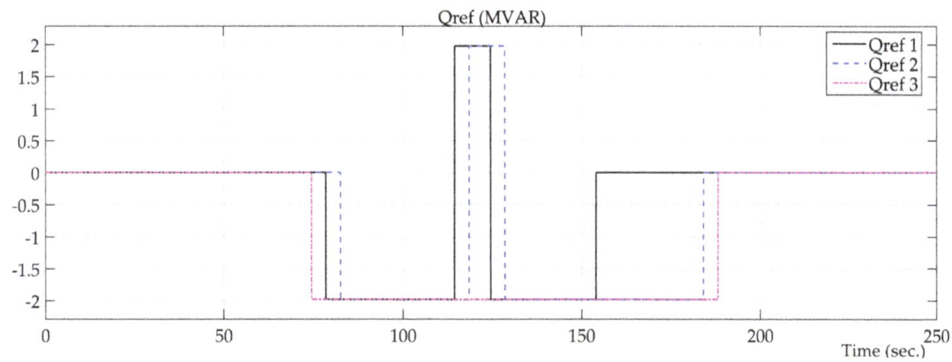

(**c**) The reference input signals of the reactive powers of the generators.

Figure 5. *Cont.*

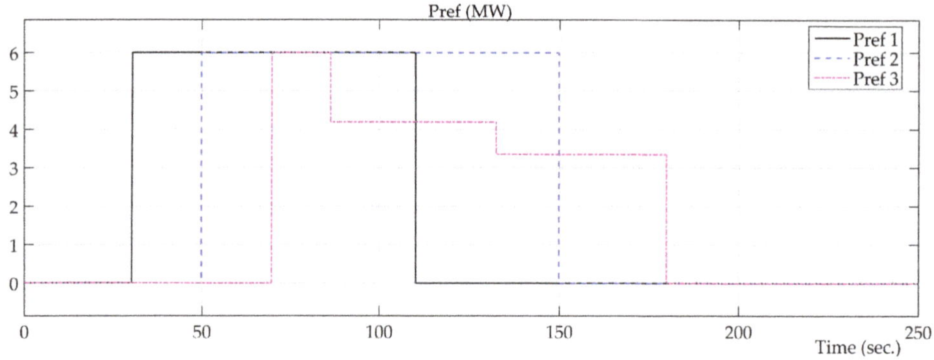

(**d**) The reference input signals of real powers of DGs.

Figure 5. The results of the execution of the control algorithm in real-time in four scopes. (**a**) The maximum and minimum voltages of the network with respect to the network limits. (**b**) The reference input to the OLTC and the substation output voltage. (**c**) The reference input signals of the reactive powers of the generators. (**d**) The reference input signals of real powers of DGs (the result of subtracting the value of curtailment powers set by the control algorithm from the available real powers).

The tap changer is raised by a single tap operation. Finally, the minimum voltage of the network is restored within limits. However, the maximum voltage is above limits, returning the network to a similar condition when all the units are connected. The reactive power of the first two units has been reversed to their full capacity of absorption. That is not adequate, leading to further curtailment of the real power of the third unit, restoring the network voltages within limits. The final steady state result is the utilization of the OLTC (which is not chosen first) and curtailing more active power, while keeping the reactive power control at its state. This result verifies the robustness of the modified algorithm for applying the optimized actions on the network under such scenarios.

The network minimum voltage decreases below the lower boundary again at time 150 s when the second unit is disconnected. The tap changer cannot be activated, as the maximum voltage is too close to the upper boundary. Consequently, the reactive power control is activated. It is adequate that the first unit gives up its reactive power absorption and operates at unity power factor to restore the network voltages within standard boundaries.

The same condition is repeated at a time of 180 s when the third unit is disconnected. The other units operate at a unity power factor as initial conditions.

Table 1 collects the data of special points in simulations to ease the follow-up process. It should be noted that:

- Time delays in both the algorithm and the connection of HIL components are apparent. For example, when the maximum voltage of the network exceeded its upper limit at t = 70.34 s., the first control action was taken at t = 74.43 s. after a period of time slightly larger than the 4 s. set for the delay of the basic control algorithm.
- Power losses may be relatively high in some instants. They increased from 186.27 kW in case of no contribution of DG units, to 801.72 kW (a steady state value at t = 89.83 s) in case of the maximum possible contribution of real powers from DG units and their full capacity of reactive power absorption. However, they are negligible compared to the real powers enabled by DG units (16.20 MW), which may be stored in energy storage systems or transported and sold to other MV networks, as the simulations emulate the case of maximum generation/minimum load (the worst case for voltage regulation algorithms).

Table 1. Data of special points in simulations.

Time (s)	V_{max} (p.u)	V_{min} (p.u)	OLTC Ref.	V_{ss} (p.u)	P_{ref} (MW)			Q_{ref} (MW)			P_{loss} (kW)
					DG_1	DG_2	DG_3	DG_1	DG_2	DG_3	
0.00	1.033400	0.950164	1.0334	1.0334	0.00	0.00	0.00	0.00	0.00	0.00	186.27
30.01	1.033400	0.950398	1.0334	1.0334	6.00	0.00	0.00	0.00	0.00	0.00	184.45
33.89	1.050005	0.972217	1.0334	1.0334	6.00	0.00	0.00	0.00	0.00	0.00	217.78
38.00	1.050722	0.972651	1.0167	1.0334	6.00	0.00	0.00	0.00	0.00	0.00	222.59
39.00	1.034271	0.954817	1.0167	1.0167	6.00	0.00	0.00	0.00	0.00	0.00	230.23
50.01	1.034308	0.954855	1.0167	1.0167	6.00	6.00	0.00	0.00	0.00	0.00	230.34
69.97	1.037571	0.958392	1.0167	1.0167	6.00	6.00	0.00	0.00	0.00	0.00	292.77
70.01	1.037603	0.958428	1.0167	1.0167	6.00	6.00	6.00	0.00	0.00	0.00	293.60
70.34	1.050374	0.959376	1.0167	1.0167	6.00	6.00	6.00	0.00	0.00	0.00	363.32
74.43	1.111057	0.961428	1.0167	1.0167	6.00	6.00	6.00	0.00	0.00	−1.98	857.86
78.75	1.077556	0.959709	1.0167	1.0167	6.00	6.00	6.00	−1.98	0.00	−1.98	976.50
83.22	1.075382	0.956471	1.0167	1.0167	6.00	6.00	6.00	−1.98	−1.98	−1.98	1048.58
87.51	1.057947	0.955625	1.0167	1.0167	6.00	6.00	4.20	−1.98	−1.98	−1.98	878.75
89.83	1.049988	0.955362	1.0167	1.0167	6.00	6.00	4.20	−1.98	−1.98	−1.98	801.72
110.01	1.049076	0.955115	1.0167	1.0167	0.00	6.00	4.20	−1.98	−1.98	−1.98	790.69
110.29	1.048322	0.949761	1.0167	1.0167	0.00	6.00	4.20	−1.98	−1.98	−1.98	729.14
114.47	1.045909	0.932724	1.0167	1.0167	0.00	6.00	4.20	1.98	−1.98	−1.98	697.21
118.70	1.048015	0.940633	1.0167	1.0167	0.00	6.00	4.20	1.98	1.98	−1.98	591.13
119.44	1.050030	0.941211	1.0167	1.0167	0.00	6.00	4.20	1.98	1.98	−1.98	575.12
123.00	1.051768	0.941708	1.0334	1.0167	0.00	6.00	4.20	1.98	1.98	−1.98	568.03
124.00	1.068187	0.959811	1.0334	1.0334	0.00	6.00	4.20	1.98	1.98	−1.98	550.40
124.52	1.068173	0.959664	1.0334	1.0334	0.00	6.00	4.20	−1.98	1.98	−1.98	549.99
128.55	1.066865	0.952039	1.0334	1.0334	0.00	6.00	4.20	−1.98	−1.98	−1.98	624.49
132.58	1.061454	0.950706	1.0334	1.0334	0.00	6.00	3.36	−1.98	−1.98	−1.98	667.21
136.44	1.049993	0.950303	1.0334	1.0334	0.00	6.00	3.36	−1.98	−1.98	−1.98	581.01
150.01	1.049718	0.950261	1.0334	1.0334	0.00	0.00	3.36	−1.98	−1.98	−1.98	578.51
150.11	1.048874	0.949928	1.0334	1.0334	0.00	0.00	3.36	−1.98	−1.98	−1.98	567.26
154.13	1.040971	0.947281	1.0334	1.0334	0.00	0.00	3.36	0.00	−1.98	−1.98	495.19
155.68	1.041271	0.950035	1.0334	1.0334	0.00	0.00	3.36	0.00	−1.98	−1.98	444.10
180.01	1.040835	0.950745	1.0334	1.0334	0.00	0.00	0.00	0.00	−1.98	−1.98	431.49
180.58	1.033400	0.949966	1.0334	1.0334	0.00	0.00	0.00	0.00	−1.98	−1.98	329.88
184.60	1.033400	0.949168	1.0334	1.0334	0.00	0.00	0.00	0.00	0.00	−1.98	282.40
188.62	1.033400	0.949810	1.0334	1.0334	0.00	0.00	0.00	0.00	0.00	0.00	213.11
189.43	1.033400	0.950011	1.0334	1.0334	0.00	0.00	0.00	0.00	0.00	0.00	193.08
194.73	1.033400	0.950164	1.0334	1.0334	0.00	0.00	0.00	0.00	0.00	0.00	186.29
250.00	1.033400	0.950164	1.0334	1.0334	0.00	0.00	0.00	0.00	0.00	0.00	186.27

4. Novelty/Contributions of the Article

The contributions of this article can be summarized as follows:

i. The basic part of the control algorithm was modified such that if the two extreme voltages of the network violate their limits simultaneously at a certain instant the basic control tends to restore the voltage violating first (regardless of the implications of its actions on the other voltage), and then it tends to restore the other one. That boosted the network's stability and prevented stalling of the system or even the occurrence of wrong actions that may cause unjustified losses. This is considered a major breakthrough since no publications, as far as the authors know, discussed such severe cases, as they often cause the traditional algorithms to get stuck and lead to instability of the network.

ii. The restoring control is modified to check if it is possible to restore any action performed by the basic control, totally or partially, to lower the running cost (due to a disconnection of a certain source or an increase in network loads). Thus, the proposed algorithm guaranteed that more curtailed active power and absorbed reactive power would be restored.

iii. Implementation of the HIL technique has proven the ability of the system to handle its actions at the right time, demonstrating its reliability and robustness.

5. Conclusions

In this article, a modified control strategy that is HIL-based is proposed to handle a severe situation. The modifications to the strategy proved the capability to maintain all network voltages within acceptable boundaries, even if both extreme voltages violated both limits simultaneously. Moreover, these modifications proved the reliability required to avoid instability or stalling of the system, or even wrong actions that may cause unjustified losses. This is considered a major breakthrough, as no publications, to the best of our knowledge, discussed that scenario. The modifications are performed upon restoring control within the proposed algorithm, guaranteeing more curtailed active power and consumed reactive power to be brought back, as the algorithm focuses on how much active and reactive power could be recovered, as long as the network voltages are maintained between the predefined boundaries without the need to set other stricter boundaries. Real-time domain simulations obtained from the implementation of the HIL technique verified the reliability and robustness of the modifications applied to the voltage control algorithm, as that algorithm applied its decisions at the optimal time.

Author Contributions: Conceptualization, B.N.F.F., M.H.S. and M.A.A.; methodology, B.N.F.F., M.H.S. and M.A.A.; software B.N.F.F. and M.H.S.; validation, B.N.F.F., M.H.S. and M.A.A.; investigation, B.N.F.F., M.H.S., M.A.A. and H.E.A.T.; writing—original draft preparation, B.N.F.F., M.H.S. and M.A.A.; writing—review and editing, M.H.S., M.A.A. and H.E.A.T.; supervision, M.H.S., M.A.A. and H.E.A.T.; funding acquisition, H.E.A.T. All authors have read and agreed to the published version of the manuscript.

Funding: This research was funded by Future University in Egypt (FUE), grant number 2000$ and The APC was funded by Future University in Egypt (FUE).

Institutional Review Board Statement: Not applicable.

Informed Consent Statement: Not applicable.

Data Availability Statement: Not applicable.

Conflicts of Interest: The authors declare no conflict of interest.

Nomenclature

DG	Distributed Generation
DNO	Distribution Network Operator
HIL	Hardware-In-the-Loop
MATLAB	Matrix Laboratory
OLTC	On-load Tap Changer

Appendix A

The IEEE 33 bus radial distribution system used in this article consists of 33 buses and 32 lines, and has a voltage of 12.66 kV. The data of its loads are tabulated in Table A1.

Table A1. Data of special points in simulations.

Node	P_{load} (kW)	Q_{load} (kVAR)
1	0	0
2	100	60
3	90	40
4	120	80
5	60	30
6	60	20
7	200	100
8	200	100
9	60	20
10	60	20
11	45	30
12	60	35
13	60	35
14	120	80
15	60	10
16	60	20
17	60	20
18	90	40
19	90	40
20	90	40
21	90	40
22	90	40
23	90	50
24	420	200
25	420	200
26	60	25
27	60	25
28	60	20
29	120	70
30	200	600
31	150	70
32	210	100
33	60	40

References

1. Jenkins, N.; Allan, R.; Crossley, P.; Kirschen, D.; Strbac, G. System studies. In *Embedded Generation*; The Institution of Electrical Engineers: London, UK, 2000.
2. Niemczyk, J.; Sus, A.; Borowski, K.; Jasiński, B.; Jasińska, K. The Dominant Motives of Mergers and Acquisitions in the Energy Sector in Western Europe from the Perspective of Green Economy. *Energies* **2022**, *15*, 1065–1081. [CrossRef]
3. Etchebehere, V.S.; Lima, J.W.M. Locational Tariff Structure for Radial Network Fixed Costs in a DER Context. *IEEE Access* **2022**, *10*, 597–607. [CrossRef]
4. Wong, S.; Bhattacharya, K.; Fuller, J.D. Long-Term Effects of Feed-In Tariffs and Carbon Taxes on Distribution Systems. *IEEE Trans. Power Syst.* **2010**, *25*, 1241–1253. [CrossRef]

5. Anzalchi, A.; Sarwat, A. Analysis of Carbon Tax as an Incentive Toward Building Sustainable Grid with Renewable Energy Utilization. In Proceedings of the 2015 Seventh Annual IEEE Green Technologies Conference, New Orleans, LA, USA, 15–17 April 2015.
6. Marcelo, J.A.; Rupolo, D.; Mantovani, J.R.S. A New Approach to Determine a Distribution Network Usage Fee for Distributed Generators. In Proceedings of the 2021 IEEE PES Innovative Smart Grid Technologies Europe (ISGT Europe), Espoo, Finland, 18–21 October 2021.
7. Trebolle, D.; Gómez, T. Reliability Options in Distribution Planning Using Distributed Generation. *IEEE Lat. Am. Trans.* **2010**, *8*, 557–564. [CrossRef]
8. Jain, N.; Singh, S.N.; Srivastava, S.C. A Generalized Approach for DG Planning and Viability Analysis Under Market Scenario. *IEEE Trans. Ind. Electron.* **2013**, *60*, 5075–5085. [CrossRef]
9. Santos, L.L.C.; Canha, L.N.; Bernardon, D.P.; Neto, N.K.; Pressi, R.A. Mapping of energetic potential in Southern Brazil to insertion of DG in distribution systems. In Proceedings of the 2015 IEEE PES Innovative Smart Grid Technologies Latin America (ISGT LATAM), Montevideo, Uruguay, 5–7 October 2015.
10. Kulmala, A.; Repo, S.; Järventausta, P. Coordinated voltage control in distribution networks including several distributed energy resources. *IEEE Trans. Smart Grid* **2014**, *5*, 2010–2020. [CrossRef]
11. Liew, S.N.; Strbac, G. Maximising penetration of wind generation in existing distribution networks. *IEE Proc.-Gener. Transm. Distrib.* **2002**, *149*, 256–262. [CrossRef]
12. Dutta, A.; Ganguly, S.; Kumar, C. Model Predictive Control based Coordinated Voltage Control in Active Distribution Networks utilizing OLTC and DSTATCOM. In Proceedings of the 2020 IEEE International Conference on Power Electronics, Drives and Energy Systems (PEDES), Jaipur, India, 16–19 December 2020.
13. Sun, X.; Qiu, J.; Yi, Y.; Tao, Y. Cost-Effective Coordinated Voltage Control in Active Distribution Networks With Photovoltaics and Mobile Energy Storage Systems. *IEEE Trans. Sustain. Energy* **2022**, *13*, 501–513. [CrossRef]
14. Chamana, M.; Chowdhury, B.; Jahanbakhsh, F. Distributed Control of Voltage Regulating Devices in the Presence of High PV Penetration to Mitigate Ramp-Rate Issues. *IEEE Trans. Smart Grid* **2018**, *9*, 1086–1095. [CrossRef]
15. Othman, M.M.; Ahmed, M.H.; Salama, M.M.A. A Coordinated Real-Time Voltage Control Approach for Increasing the Penetration of Distributed Generation. *IEEE Syst. J.* **2020**, *14*, 699–707. [CrossRef]
16. Zhang, Y.; Zhang, Q.; Wang, H.; Yu, D. Hierarchical Coordinated Voltage Correction Scheme for Active Distribution Network. In Proceedings of the 2020 IEEE 4th Conference on Energy Internet and Energy System Integration (EI2), Wuhan, China, 30 October–1 November 2020.
17. Tshivhase, N.; Hasan, A.N.; Shongwe, T. An Average Voltage Approach to Control Energy Storage Device and Tap Changing Transformers under High Distributed Generation. *IEEE Access* **2021**, *9*, 108731–108753. [CrossRef]
18. Jiao, W.; Chen, J.; Wu, Q.; Li, C.; Zhou, B.; Huang, S. Distributed Coordinated Voltage Control for Distribution Networks With DG and OLTC Based on MPC and Gradient Projection. *IEEE Trans. Power Syst.* **2022**, *37*, 680–690. [CrossRef]
19. Mohiuddin, S.M.; Qi, J. Optimal Distributed Control of AC Microgrids With Coordinated Voltage Regulation and Reactive Power Sharing. *IEEE Trans. Smart Grid* **2022**, *13*, 1789–1800. [CrossRef]
20. Nandasiri, N.; Pang, C.; Aravinthan, V. Marginal levelized cost of energy bases optimal operation of distribution system considering photovoltaics. In Proceedings of the 2017 North American Power Symposium (NAPS), Morgantown, WV, USA, 17–19 September 2017.
21. Shigenobu, R.; Yona, A.; Senjyu, T. Demand response considering participation rate in smart grid: Two level optimal management for DisCo and customers. In Proceedings of the 2017 17th International Conference on Control, Automation and Systems (ICCAS), Jeju, Republic of Korea, 18–21 October 2017.
22. Stanelytė, D.; Radziukynas, V. Analysis of Voltage and Reactive Power Algorithms in Low Voltage Networks. *Energies* **2022**, *15*, 1843–1868. [CrossRef]
23. Gu, M.; Meegahapola, L.; Wong, K.L. Coordinated Voltage and Frequency Control in Hybrid AC/MT-HVDC Power Grids for Stability Improvement. *IEEE Trans. Power Syst.* **2021**, *36*, 635–647. [CrossRef]
24. Han, H.; Li, Q.; Lv, Z. Multi-level Voltage Interaction Control in Active Distribution Network Based on MPC. In Proceedings of the 2019 Chinese Automation Congress (CAC), Hangzhou, China, 22–24 November 2019.
25. Guo, Y.; Wu, Q.; Gao, H.; Huang, S.; Zhou, B.; Li, C. Double-Time-Scale Coordinated Voltage Control in Active Distribution Networks Based on MPC. *IEEE Trans. Sustain. Energy* **2020**, *11*, 294–303. [CrossRef]
26. Kou, P.; Liang, D.; Gao, R.; Liu, Y.; Gao, L. Decentralized Model Predictive Control of Hybrid Distribution Transformers for Voltage Regulation in Active Distribution Networks. *IEEE Trans. Sustain. Energy* **2020**, *11*, 2189–2200. [CrossRef]
27. Maharjan, S.; Khambadkone, A.M.; Peng, J.C. Robust Constrained Model Predictive Voltage Control in Active Distribution Networks. *IEEE Trans. Sustain. Energy* **2021**, *12*, 400–411. [CrossRef]
28. Hou, Z.; Liu, Y.; Yan, J.; Liao, Y.; Miao, C.; Zhao, R. Voltage control strategy of AVC system based on Data Mining. In Proceedings of the 2022 14th International Conference on Measuring Technology and Mechatronics Automation (ICMTMA), Changsha, China, 15–16 January 2022.
29. Ping, B.; Zhang, X.; Song, Q.; Yu, Y.; Wu, N.; Ji, X. Voltage control strategy for integrated medium and low voltage distribution network based on active-reactive power coordination optimization. In Proceedings of the 2020 Chinese Automation Congress (CAC), Shanghai, China, 6–8 November 2020.

30. Sun, X.; Qiu, J.; Tao, Y.; Ma, Y.; Zhao, J. Coordinated Real-Time Voltage Control in Active Distribution Networks: An Incentive-Based Fairness Approach. *IEEE Trans. Smart Grid* **2022**, *13*, 2650–2663. [CrossRef]
31. Gerdroodbari, Y.Z.; Razzaghi, R.; Shahnia, F. Decentralized Control Strategy to Improve Fairness in Active Power Curtailment of PV Inverters in Low-Voltage Distribution Networks. *IEEE Trans. Sustain. Energy* **2021**, *12*, 2282–2292. [CrossRef]
32. Yu, P.; Wan, C.; Sun, M.; Zhou, Y.; Song, Y. Distributed Voltage Control of Active Distribution Networks With Global Sensitivity. *IEEE Trans. Power Syst.* **2022**, *37*, 4214–4228. [CrossRef]
33. Fahmy, B.N.; Soliman, M.H.; Talaat, H.E.A. Active Voltage Control in Distribution Networks including Distributed Generations using Hardware-in-the-Loop Technique. In Proceedings of the 2019 21st International Middle East Power Systems Conference (MEPCON), Cairo, Egypt, 17–19 December 2019.
34. Kulmala, A.; Repo, S.; Järventausta, P. Active voltage level management of distribution networks with distributed generation using on load tap changing transformers. In Proceedings of the Power Tech, 2007 IEEE Lausanne, Lausanne, Switzerland, 1–5 July 2007.
35. Švenda, G.; Simendić, Z. Adaptive on-load tap-changing voltage control for active distribution networks. *Electr. Eng.* **2022**, *104*, 1041–1056. [CrossRef]
36. Giacomuzzi, S.; Langwasser, M.; Carne, G.D.; Buja, G.; Liserre, M. Smart transformer-based medium voltage grid support by means of active power control. *CES Trans. Electr. Mach. Syst.* **2020**, *4*, 285–294. [CrossRef]
37. Deakin, M.; Morstyn, T.; Apostolopoulou, D.; McCulloch, M.D. Voltage control loss factors for quantifying DG reactive power control impacts on losses and curtailment. *IET Gener. Transm. Distrib.* **2022**, *16*, 2049–2062. [CrossRef]
38. Meerimatha, G.; Kesavarao, G.; Sreenivasulu, N. A Novel Distribution System Power Flow Algorithm using Forward Backward Matrix Method. *IOSR J. Electr. Electron. Eng.* **2015**, *10*, 46–51.
39. Conti, S.; Raiti, S.; Vagliasindi, G. Voltage sensitivity analysis in radial MV distribution networks using constant current models. In Proceedings of the 2010 IEEE International Symposium on Industrial Electronics, Bari, Italy, 4–7 July 2010.

symmetry

MDPI

Article

A State Machine-Based Droop Control Method Aided with Droop Coefficients Tuning through In-Feasible Range Detection for Improved Transient Performance of Microgrids

Mandarapu Srikanth and Yellapragada Venkata Pavan Kumar *

School of Electronics Engineering, VIT-AP University, Amaravati 522237, Andhra Pradesh, India
* Correspondence: pavankumar.yv@vitap.ac.in; Tel.: +91-7416285804

Abstract: The cascaded droop-voltage-current controller plays a key role in the effective operation of microgrids, where the controller performance is critically impacted by the design of the droop controller. Moreover, in critical loading (e.g.: connection/disconnection of large inductive load), the pre-set value of the droop coefficient brings asymmetry in transient performance leading to instability. Hence, to improve symmetry by reducing the trade-off between transient response and stability margin, this paper proposes a state machine-based droop control method (SMDCM) aided with droop coefficients' tuning through in-feasible range detection. Here, to realize the issues and the role of the droop controller's dynamics on the microgrid's stability, a small-signal stability analysis is conducted, thereby, an in-feasible range of droop values is identified. Accordingly, safe values for droop coefficients are implemented using the state machine concept. This proposed SMDCM is compared with the conventional constant droop control method (CDCM) and fuzzy logic-based droop control method (FLDCM) in terms of frequency/power/voltage characteristics subjected to different power factor (PF) loading conditions. From the results, it is seen that CDCM failed in many metrics under moderate and poor PF loadings. FLDCM is satisfactory under moderate PF loading, but, showed 54 Hz/48 Hz as maximum/minimum frequency values during poor PF loading. These violate the standard limit of ±2%, but SMDCM satisfactorily showed 50.02 Hz and 49.8 Hz, respectively. Besides, FLDCM levied an extra burden of 860 W on the system while it is 550 W with SMDCM. System recovery has taken 0.04 s with SMDCM, which completely failed with FLDCM. Similarly, voltage THD with FLDCM is 58.9% while with SMDCM is 3.08%. Peak voltage due to capacitive load switching is 340 V with FLDCM and 150 V with SMDCM. These findings confirm that the proposed SMDCM considerably improved the transient performance of microgrids.

Keywords: droop coefficients tuning; droop control; In-feasible range detection; microgrids; state machines; transient performance

1. Introduction

Engineers were driven to design an entirely new distribution system namely power electronic converters-based microgrids instead of traditional bulk power systems. These microgrids are low-to-medium voltage and power-rated generation systems normally constituted with renewable energy sources and are deployed locally at the load centers [1–3]. These systems normally suffer from stability issues due to the dependency on uncertain renewable energy sources and sensitive power converters. In microgrids, usually, a trade-off exists between the improvement of transient response and transient stability margin resulting in asymmetry between them. Enhancement of this symmetry depends on control system efficacy. The general structure of the controller consists of a cascaded interconnection of power, voltage, and current controllers as shown in Figure 1 [4].

There were some literature works discussed the effective design of voltage/current controller gain parameters, such as proportional-resonant controller-based voltage controller for better stability [5], internal model control-based voltage and current controllers for better transient response [6], pole-zero cancellation technique-based design for improved transient response [7], linear active disturbance rejection control based compensators for voltage and current controllers [8], and tuning of a modified resonant current controller considering time delays based on pole placement [9]. However, from the study of bandwidths, the dynamics of the power controller are found more significant than that of inner voltage-current controllers [10]. From this, it is understood that a modification to the power controller can offer better chances of getting improved transient stability. Since the performance of the traditional electrical machines-based power plants under droop management has been observed as desired, the droop control logic is considered to implement a power controller in the case of inverter-based microgrids [11]. As shown in Figure 1, the power/droop controller provides a reference value to the voltage controller, which in turn provides a reference value to the current controller. This indicates that the functionality of the droop controller is essential to the efficient operation of voltage and current controllers. Thus, the performance of the entire cascaded droop-voltage-current controller is majorly impacted by the design of the droop controller. This is the prime motivation for the proposed research work in this paper.

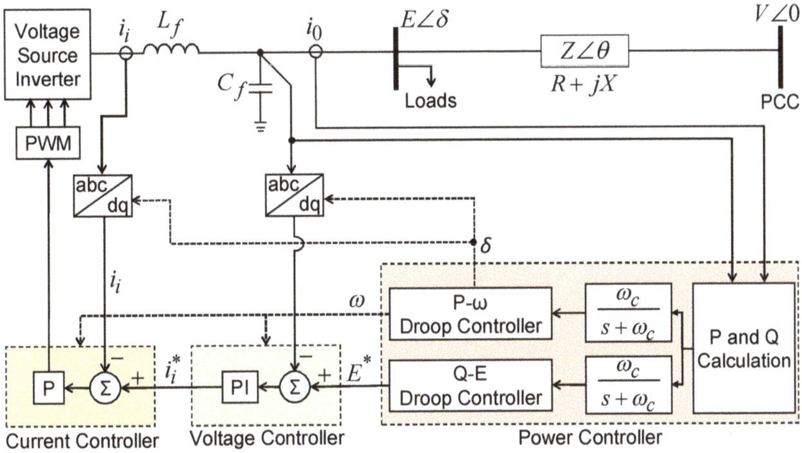

Figure 1. The layout of the microgrid includes cascaded droop, voltage, and current controllers.

Droop control is indeed a proportional controller, where the droop gain determines the allocation of steady-state power in the system. Corresponding to each unit, droop control subtracts proportionate values of active power from frequency (P-ω droop) and reactive power from voltage magnitude (Q-V droop). These will imitate the speed governor and automatic voltage regulator control loops respectively [12]. In this conventional droop control strategy, depending on the deviation from the scheduled value of active power, the inverter output frequency is controlled automatically [13]. This approach functions extremely well for inductive lines but poorly for resistive lines. Another notable disadvantage of droop control is the load-dependent frequency variation. As a result, there is a phase difference between the frequency of the inverter output voltage and the frequency of the utility main's input voltage [14]. As the droop value used in this conventional method is constant, it is referred to as the "constant droop control method (CDCM)" in this paper.

As a result of large changes in load and greater power needs, the low-frequency dominating modes of the microgrid move closer to the unstable zone. This causes the system to become more oscillatory and may potentially cause it to become unstable. The role of P-ω and Q-V droop control on the microgrid's transient stability was studied in [15,16]. It has been identified in [16] that the transient stability is positively correlated to the Q-V droop factor, whereas, the P-ω droop factor has a negative correlation with it. A sensitivity analysis was conducted, and the effects of the load characteristics on voltage and frequency transients are compared [17]. This sort of analysis has provided a deeper insight into CDCM. Significant efforts were made to enhance the droop control approach in overcoming the disadvantage of frequency change indicated with CDCM, particularly during transient conditions. Recent developments in this area have led to the design of variable droop coefficient adjustment techniques. Several offline/online-based variable droop coefficient design methods were attempted in the literature to address this issue.

In the offline tuning category, several works on droop gain adjustment based on the frequency and the rate of change of frequency to improve the transient response were discussed in the literature [18,19]. Further, the method based on the alternating principle discussed in [19] proved itself as an effective strategy for system stabilization. Further, it proved that the damping produced from the alternating principle is efficacious and produces similar outcomes under all conditions. This ensures the elimination of any transients before appearing. However, these offline techniques are very much dependent on the knowledge of the system's mathematical model.

Whereas, in the online tuning category, fuzzy logic techniques and artificial neural networks are state-of-the-art methods that were used effectively [20–26]. Here, the dependency on the mathematical model of the system is very minimum. Fuzzy logic has proven to be a successful method for enhancing power quality in microgrids [22,23]. Focusing on the application of fuzzy logic for droop control in microgrids for enhancement of transient response, [24] has presented a "fuzzy logic-based droop control method (FLDCM)" based on the frequency and the rate at which the frequency changes. Further, a novel dynamic fuzzy logic controller for P-V droop and Q-ω droop is presented in [25]. However, in this work, the knowledge of the mathematical model of the system is not included in fixing the fuzzy values. Without this knowledge, stability analysis cannot be conducted. In the absence of stability analysis, tuning of droop coefficient values simply based on experience can sometimes drive the system to instability. This issue is addressed in [26], where stability analysis was used to tune the fuzzy inference system so that the range of coefficients offered by fuzzy logic control ensures a large damping ratio to counteract the dynamics of dc voltage. However, the fundamental disadvantage of deploying fuzzy logic algorithms is they suffer from long computing times and computing burdens in real-time online applications.

Besides, a state machine-based design was discussed in [27,28] to coordinate numerous power sources, avoid transients, and minimize the effects of rapid changes in power demand. Experimentation has shown that state machines have a very minimal computing burden and are thus highly quick when compared to fuzzy logic and neural network approaches. Also, in the development of state machines, the given problem is addressed by decomposing into a finite number of states, where each of these states is associated with a specific outcome. Depending on the inputs it receives, the states within the state machine turn active alternatively. This condition is similar to the advantage presented by [19]. So, in view of these advantages, a state machine-based approach is adopted for the design of the droop controller in this paper.

With the help of a small-signal model, it is simple to carry out the stability analysis of the system with respect to changes in the parameters. For this purpose, an accurate small signal state-space model of the whole microgrid which includes the droop controller, network, and loads is utilized. From this analysis, the main control parameters of the inverter as well as their optimal bands, which have a substantial impact on system stability and dampening of oscillations associated with transient disturbance can be found [29–31]. Focusing on the role of droop coefficient value on microgrid stability, the authors in [32] have presented a method of control with two degrees of freedom that combines the traditional droop with a transient droop. In this paper, a complete small signal model of the studied microgrid is formulated and stability analysis is conducted to realize the role of droop coefficient value on the system stability when subjected to sudden connection/disconnection of large inductive loads. From this analysis, an infeasible range of droop coefficients leading to instability is identified and these values are purposefully avoided during the design stage itself of the proposed state machine-based droop controller. In the next stage, the selection of droop coefficient values which are designated as a specific outcome of a particular state of this state machine was carried out as similar to the fuzzy logic-based design.

In summary, to the best of the authors' knowledge, it is understood that a simple, fast, and robust droop control method for improving the transient performance of microgrids is not yet fully explored. With this motivation, this paper proposes a robust alternating "state machine-based droop control method (SMDCM)" whose droop coefficient tuning was carried out in offline mode based on the identification of an in-feasible range of values to enhance stability. Further, the proposed SMDCM is compared with conventional CDCM and FLDCM to prove its superiority in enhancing the said kind of symmetry.

The remaining sections of this paper are organized as follows. Section 2 describes the problem identification through a mathematical model and small signal stability analysis. Section 3 describes the identification of the in-feasible range of frequency droop coefficient values. Section 4 describes the implementation of the proposed SMDCM. Section 5 presents a case study to compare the conventional and proposed methods followed by a discussion and analysis. Finally, cumulative remarks and achievements of this paper are presented as the conclusion in Section 6.

2. Problem Formulation

In this section, the role of the droop coefficient value on the frequency stability of the microgrid followed by connect or disconnect of a large reactive load is investigated. Based on this, the setbacks associated with the conventional CDCM and the conventional FLDCM are highlighted. This investigation is carried out by conducting a small signal stability analysis on the mathematical model of the microgrid that is controlled by the cascaded droop-voltage-current-based controller.

2.1. Mathematical Model of Cascaded Droop-Voltage-Current Control

In a microgrid that is powered by inverters, the power stage is comprised of several sources of energy, along with DC-AC converters and optional energy storage. The inverters can provide a variety of capabilities, including voltage and frequency control as well as an improvement in power quality. In grid-connected mode, the local loads are fed jointly by the inverter and the grid, whereas in standalone mode, the local loads are fed only by the inverter. When compared to the bandwidth of the droop control, the bandwidth of the voltage-current control loop is significantly higher. Therefore, the dynamics of the droop controller are more significant than the inner voltage and current controllers. As a result, the references that are produced by the droop control technique are responsible for governing the magnitude and frequency of the inverter output voltage. The active power (P) and reactive power (Q) output of the inverter are expressed as Equations (1) and (2) respectively.

$$P = \frac{3E(ER - VR\cos\delta + VX\sin\delta)}{R^2 + X^2} \tag{1}$$

$$Q = -\frac{3E(VX\cos\delta - EX + VR\sin\delta)}{R^2 + X^2} \tag{2}$$

where E is the magnitude of the inverter output voltage, δ is the load angle or phase angle associated with E, V is the magnitude of the PCC voltage, R is the resistance of line impedance and X is the inductive reactance of the line impedance.

By introducing small disturbances around the equilibrium point, Equations (1) and (2) can be linearized as shown in Equations (3) and (4) respectively.

$$\Delta P = \left(\frac{\partial P}{\partial \delta}\right)\Delta\delta + \left(\frac{\partial P}{\partial E}\right)\Delta E \tag{3}$$

$$\Delta Q = \left(\frac{\partial Q}{\partial \delta}\right)\Delta\delta + \left(\frac{\partial Q}{\partial E}\right)\Delta E \tag{4}$$

The first-order partial derivatives of Equations (3) and (4) indicate the sensitiveness of P and Q concerning the changes in δ and E. These are shown as Equations (5) to (8).

$$\frac{\partial P}{\partial \delta} = \frac{3(EVX\cos\delta + ERV\sin\delta)}{R^2 + X^2} \tag{5}$$

$$\frac{\partial Q}{\partial \delta} = -\frac{3(ERV\cos\delta - EVX\sin\delta)}{R^2 + X^2} \tag{6}$$

$$\frac{\partial P}{\partial E} = \frac{3(2ER - VR\cos\delta - VX\sin\delta)}{R^2 + X^2} \tag{7}$$

$$\frac{\partial Q}{\partial E} = -\frac{3(VX\cos\delta - 2EX + VR\sin\delta)}{R^2 + X^2} \tag{8}$$

To simplify, consider a case where the nature of the line impedance is primarily inductive. In this case, the assumption can be safely considered. Applying these conditions to Equations (5) to (8), the simplified relations are shown as Equations (9) to (12).

$$\frac{\partial P}{\partial \delta} = \frac{3(EVX)}{R^2 + X^2} \tag{9}$$

$$\frac{\partial Q}{\partial \delta} = -\frac{3(ERV)}{R^2 + X^2} \tag{10}$$

$$\frac{\partial P}{\partial E} = \frac{3(2ER - VR)}{R^2 + X^2} \tag{11}$$

$$\frac{\partial Q}{\partial E} = -\frac{3(VX - 2EX)}{R^2 + X^2} \tag{12}$$

By observing Equations (9) to (12), the following comments can be made. The active power P is more sensitive to frequency fluctuations. This is because active power P is more sensitive to power angle and, as a result, frequency fluctuations. In comparison, reactive power Q is more sensitive to changes in output voltage magnitude. As a result of this, P-ω and Q-E droop control techniques are extremely popular in the field of power systems. Their expressions are shown as Equations (13) and (14).

$$\omega = \omega^* - kp(P - P^*) \tag{13}$$

$$E = E^* - kq(Q - Q^*) \tag{14}$$

where, ω is the frequency of the inverter output voltage, ω^* is the angular frequency reference, P^* is the active power reference and Q^* is the reactive power reference, kp is the droop gain of the P-ω droop controller, and kq is the droop gain of Q-E droop controller.

2.2. Small Signal Stability Analysis

Since it is simple to forecast how the system will react to changes in the parameters, the small-signal model is utilized extensively in estimating the transient stability of the inverter-controlled microgrid. Using this method to choose critical control parameters is advantageous. In addition, the small signal modeling and stability are affected by the configuration of the microgrid, the different operation modes, the locations of the load, and the connections made to the inverters. This method of analyzing small signals has a long history of application. Recent developments have allowed it to be applied in microgrid systems. In the research that has been done in [33–38], a comprehensive small-signal model microgrid is constructed. The results of this model are quite accurate when used to predict the stability and dynamics of the system.

2.2.1. Small Signal Model

In the small signal model, that was presented by [33], ω in X = ωL is believed to be static, rather than dynamic (where L is the inductance of the line impedance). This is the model's basic drawback. To circumvent this constraint, a model that is based on dynamic phasors is utilized [36–38]. Accordingly, the corresponding equations for P and Q in their dynamic form as suggested in [36] are given as Equations (15) and (16) respectively. Equations (17) to (20) are the linearized forms of Equations (13) to (16), respectively.

$$P = \frac{3E}{L^2} \cdot \left[\frac{(Ls + R)(E - V\cos\delta) + \omega LV \sin\delta}{s^2 + (2R/L)s + ((R^2/L^2) + \omega^2)} \right] \tag{15}$$

$$Q = \frac{3E}{L^2} \cdot \left[\frac{\omega L(E - V\cos\delta) + (Ls + R)V \sin\delta}{s^2 + (2R/L)s + ((R^2/L^2) + \omega^2)} \right] \tag{16}$$

$$\Delta\omega = \Delta\omega^* - kp(\Delta P - \Delta P^*) \tag{17}$$

$$\Delta E = \Delta E^* - kq(\Delta Q - \Delta Q^*) \tag{18}$$

$$\Delta P = k_{pd}\Delta\delta + k_{pe}\Delta E \tag{19}$$

$$\Delta Q = k_{qd}\Delta\delta + k_{qe}\Delta E \tag{20}$$

where k_{pd} is $\frac{\partial P}{\partial \delta}$, k_{pe} is $\frac{\partial P}{\partial E}$, k_{qd} is $\frac{\partial Q}{\partial \delta}$ and k_{qe} is $\frac{\partial Q}{\partial E}$, which are first-order partial derivatives. These are derived as given through Equations (21) to (24) respectively.

$$\therefore k_{pd} = \frac{\partial P}{\partial \delta} = \frac{3\omega E^2}{L} \cdot \frac{1}{s^2 + (2R/L)s + ((R^2/L^2) + \omega^2)} \tag{21}$$

$$\therefore k_{pe} = \frac{\partial P}{\partial E} = \frac{3E}{L^2} \cdot \frac{Ls + R}{s^2 + (2R/L)s + ((R^2/L^2) + \omega^2)} \tag{22}$$

$$\therefore k_{qd} = \frac{\partial Q}{\partial \delta} = -\frac{3E^2}{L^2} \cdot \frac{Ls + R}{s^2 + (2R/L)s + ((R^2/L^2) + \omega^2)} \tag{23}$$

$$\therefore k_{qe} = \frac{\partial Q}{\partial E} = \frac{3\omega E}{L} \cdot \frac{1}{s^2 + (2R/L)s + ((R^2/L^2) + \omega^2)} \tag{24}$$

A low-pass filter with a cut-off frequency (ω_c) is commonly employed in the measurement of the active and reactive power output of the inverter. Assuming ω^*, E^*, P^*, and Q^* to be constant, their deviation term can be neglected. Considering these assumptions, the dynamics of the frequency and voltage can be expressed as Equations (25) to (27).

$$\Delta\omega = -\frac{kp \cdot \omega_c}{s + \omega_c}\left(k_{pd}\Delta\delta + k_{pe}\Delta E\right) \tag{25}$$

$$\Delta E = -\frac{kq \cdot \omega_c}{s + \omega_c}\left(k_{qd}\Delta\delta + k_{qe}\Delta E\right) \tag{26}$$

$$\Delta\omega = s\Delta\delta \tag{27}$$

2.2.2. Influence of Reactive Load and Droop Coefficient on System Stability

Considering the interruption of large inductive kind of loads, connection or disconnection causes more serious stability issues. A sudden connection or disconnection of a large inductive load can be treated as turning an existing healthier system into a weaker system. This can be attributed to a large change in the grid inductance. This plays a major role in the stability of the microgrid. Hence, the stability issue associated with the connection/disconnection of large inductive loads is studied in this work. It is observed that frequency exhibits large overshoots and undershoots resulting in a stability crisis. Based on the regulations as allowed by the authorities, the maximum allowable frequency droop value is set to 0.0001. To understand the role of this droop coefficient on the change in line inductance, the small signal model derived in this section is utilized. The corresponding change of frequency ($\Delta\omega$) is shown in Figure 2.

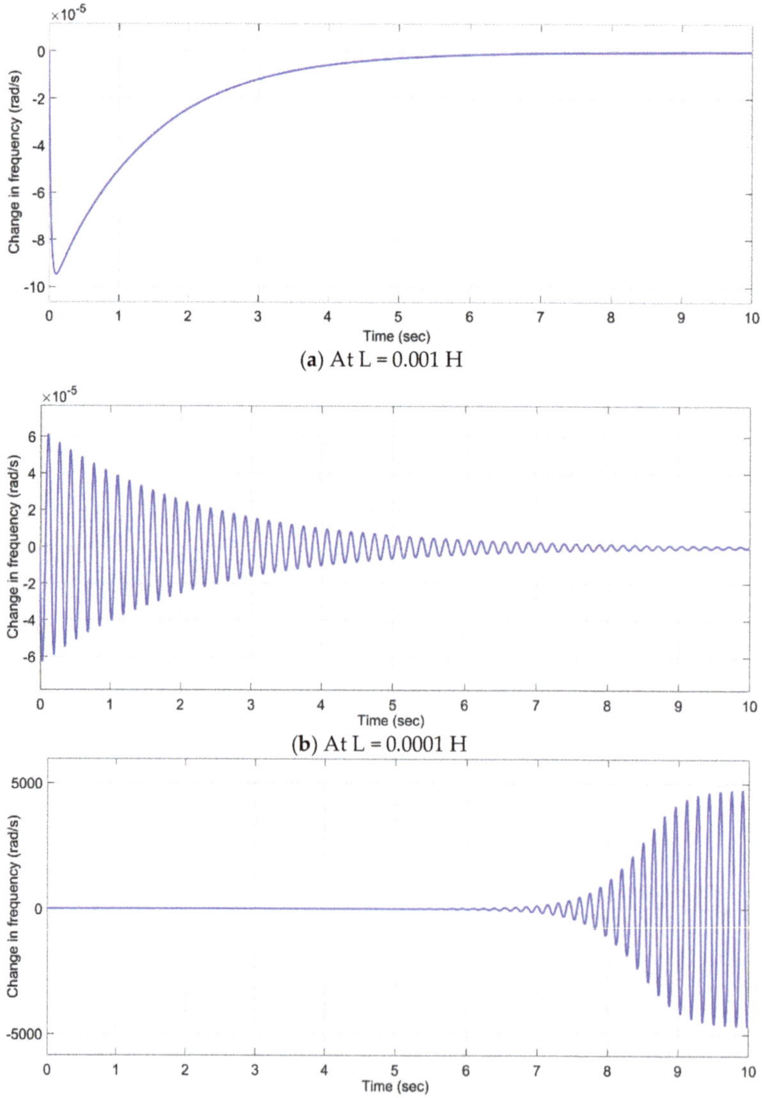

(**a**) At L = 0.001 H

(**b**) At L = 0.0001 H

(**c**) At L = 0.00001 H

Figure 2. Plot of change in frequency ($\Delta\omega$) with a step change in active power (ΔP) for various values of L at a fixed value of *kp* (=0.0001).

The frequency droop coefficient value (kp) is fixed at 0.0001 and with the normal inductance of 0.001 H, the frequency is settling to zero without any oscillations at 6 s. A little undershoot is noticed in the beginning. This can be noticed in Figure 2a. With L = 0.0001 H and kp at the same value, frequency is experiencing damped oscillations and finally settles around to zero at 10 s. This is shown in Figure 2b. However, it is substantial to notice from Figure 2c that with L = 0.00001 H and kp at the same value, the frequency is experiencing a large instability beginning from 7 s.

2.3. Summary of the Problem

A sudden connection/disconnection of a large inductive load makes the microgrid weak. This weakening can be attributed to a function of fall in the line inductance (L). Under this situation, even the normal value of kp that was set based on the regulatory authority guidelines can lead to serious transients in frequency leading to system instability as understood from Figure 2.

3. Identification of In-Feasible Range of Values for Frequency Droop Coefficient

Based on the small-signal model presented in Section 2, this section investigates the in-feasible range of kp values that can disturb the system stability after a sudden drop of a large inductive load. This in-feasible range identification was done by testing the system with different kp and L values. For this, the root locus plots given in Figure 3 are drawn.

Of all the poles, those which are nearer to the zero axis are the dominant poles of this system. The dynamic behavior of the system is determined by the location of these dominant poles. Hence these dominant poles shown as encircled with red color in Figure 3a–c are considered for the study. Their position in the s-plane determines the stability of the system. From Figure 3a, it can be understood that with L = 0.001 H and for all the values of $kp \leq 0.0001$, all the poles of the system are lying in the left half of the s-plane ensuring the stability of the system. Also, the poles are lying on the real axis presenting a completely damped response. From Figure 3b, it can be noticed that with L = 0.0001 H, the poles are much nearer to the right half of the s-plane when compared to the situation with L = 0.001 H. Moreover, the poles are exhibiting an oscillatory response. The situation comes more unstable with L = 0.0001 H, as shown in Figure 3c. with the values of kp = 0.00008, 0.00009, and 0.0001, the poles moved to the right half of the s-plane leading to instability. It has been verified that for these 3 values of kp, the system remains unstable with further lower values of L. Based on these observations, the present system is vulnerable to a zone of values $0.00008 \leq kp \leq 0.0001$ under disconnection of large reactive power loads. This infeasible region is further studied in light of CDCM where a constant droop value is used, whereas in FLDCM an adaptive droop value is deployed.

3.1. Limitations of Constant and Adaptive Droop Control

As a summary of the aforesaid analysis, the following reasons can be attributed to the system instability when constant (or adaptive)-droop control methods are employed.

- Conventional CDCM: In this scheme, a constant value is assigned to kp. A large value of kp implies a better transient response. The maximum value of kp which is however been limited by the regulatory bodies is adopted. This fixed value is acceptable to the system in terms of normal operating conditions. But in the case of transient conditions, especially when there is a change in line inductance, the same value of kp which proved to be good in normal conditions will now become detrimental.
- Adaptive Droop Control: This paper considers conventional FLDCM under this scheme for comparison with the proposed paper. When compared to CDCM, an adaptive droop scheme offers better transient performance. However, it is very much possible that kp values can fall in the in-feasible range leading to instability. Another major limitation is a requirement of an indispensable large computational effort.

To overcome these limitations, SMDCM is proposed in this work. In this proposed method, the droop values are set in such a way that, under transient conditions, the controller will not generate a value such that the resulting *kp* value will not fall in this in-feasible region.

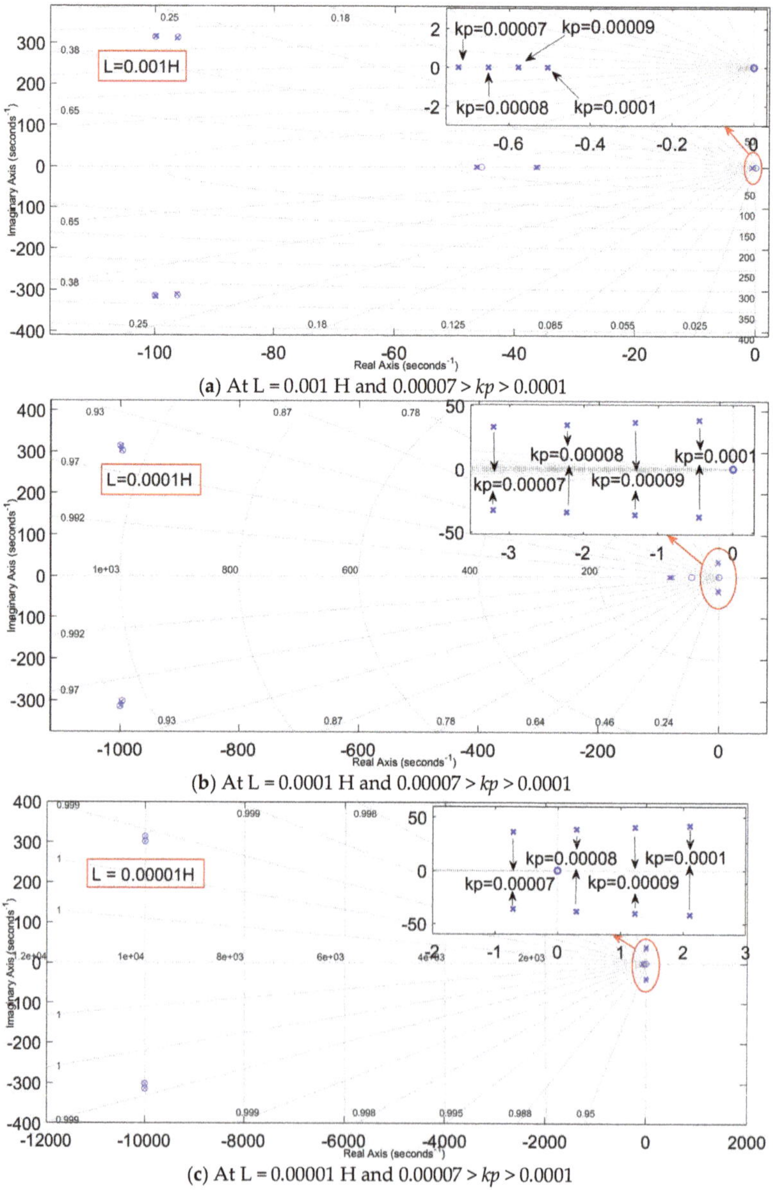

(**a**) At L = 0.001 H and 0.00007 > *kp* > 0.0001

(**b**) At L = 0.0001 H and 0.00007 > *kp* > 0.0001

(**c**) At L = 0.00001 H and 0.00007 > *kp* > 0.0001

Figure 3. Pole-Zero plot of the system for various values of *L* and *kp*.

4. Proposed State Machine-Based Droop Control Method (SMDCM)

As discussed in the literature, there were two approaches are being followed to design the droop coefficient value, namely, constant/fixed droop coefficient (named CDCM) [13] and adaptive droop coefficient tuning through fuzzy logic concept (named FLDCM) [24]. To overcome the limitations of these conventional methods that are discussed in Section 3.1, this section describes the proposed SMDCM, which is an alternating droop value adjustment method. This is implemented using state flow control concepts to resolve frequency transient performance issues in microgrids.

4.1. Frequency Droop Controller Coefficient and the Role of SMDCM in its Design

Instead of using a fixed frequency-droop controller coefficient that was defined by conventional CDCM, this paper adopts the design of the adaptive droop coefficient (*kp*) as a sum of the fixed value (*Fd*) and the variable value (*Md*), which was defined in [24] and given by Equation (28). In this, the *Dp* represents the maximum droop coefficient that is conventionally calculated by CDCM in terms of maximum and minimum frequencies (ω_{max} and ω_{min}) and *Md* is the variable parameter that is proposed to be designed by the SMDCM in this paper, which was conventionally done by fuzzy logic in [24].

$$\left.\begin{array}{l} kp = Fd + Md = \frac{Dp}{2} + Md \\ where, Fd = \frac{Dp}{2} \text{ and } Dp = \frac{(\omega_{max} - \omega_{min})}{P*} \end{array}\right\} \quad (28)$$

4.2. Description of the Proposed SMDCM

The structure of the proposed SMDCM for the design of the variable part (*Md*) of the frequency droop coefficient (*kp*) is given in Figure 4. It includes two parts, one is the peak detection unit and the second is the state machine, whose operations are described below.

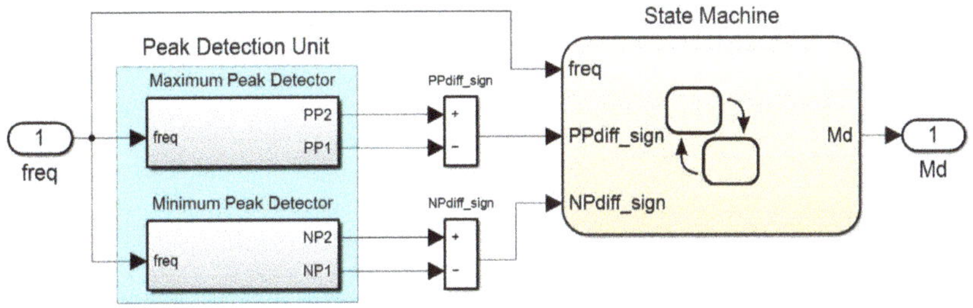

Figure 4. Schematic of the proposed SMDCM and its major parts.

4.2.1. Description of the Peak-Detection Mechanism

In this paper, a continuous type of three-phase phase-locked-loop is used for providing information about the frequency of the bus voltage to the proposed peak detection unit. The proposed peak detection unit involves both maximum and minimum peak detection mechanisms. To understand the operation of this unit, a typical frequency characteristic during transient conditions is considered as shown in Figure 5. Initially, the maximum peak detector detects the maximum local peak (PP1), and the minimum peak detector detects the minimum local peak (NP1) in the frequency waveform. This information is retained in memory as an earlier maximum peak and earlier minimum peak respectively. Further, the maximum peak detector detects the successive maximum value (PP2) and compares this value with the already recorded PP1 value, thus, calculating the sign of their difference (PPdiff_sign). Similarly, the minimum peak detector detects the successive minimum peak value (NP2) and compares this value with the already recorded NP1 value, thereby, calculating the sign of their difference (NPdiff_sign). In the next cycle, the earlier PP2/NP2 values are treated as PP1/NP1 and the newly acquired value will be updated for PP2/NP2, thereby, updating the signs of PPdiff_sign and NPdiff_sign. This process will continue, and these two signs are supplied as the control inputs to the proposed state machine implementation along with the information of the measured frequency (freq) as shown in Figure 4. Further, the "PPdiff_sign" is precisely indicated as "PPdiff_neg" (when PP1 > PP2) and "PPdiff_pos" (when PP1 < PP2). Similarly, the "NPdiff_sign" is precisely indicated as "NPdiff_neg" (when NP1 > NP2) and "NPdiff_pos" (when NP1 < NP2). It is worth noting that the accuracy of the peak detection unit depends on the sampling frequency chosen. However, the value of this sampling frequency brings in a trade-off between the accuracy of peak detection and the overall computational time. In view of this trade-off, a moderate and fixed value of 10 kHz is chosen as the sampling frequency for this work.

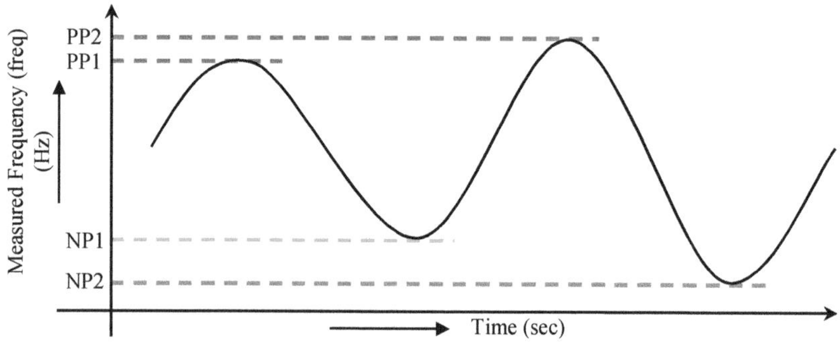

Figure 5. Typical frequency waveform under transient operating conditions.

4.2.2. Description of the State Machine for Generating Variable Droop Value

Based on the values of the three control inputs (freq, PPdiff_sign, NPdiff_sign), the state machine provides a variable value of *Md* as output. For the control input 'freq', the value of the nominal frequency (NF) is set to 50 Hz. This NF acts as pivot frequency based on which the tolerance values, namely, upper-frequency limit (UFL) and lower-frequency limit (LFL) are set. In this paper, the values for this UFL and LFL are set to 50.3 and 49.7 respectively to set a tight frequency tolerance. Based on these values, three frequency zones are defined in this work, namely, Zone-1 (freq_Normal), a normal-frequency zone whose range is defined as UFL $\geq$ freq $\geq$ LFL, Zone-2 (freq_BeyondNormal), a beyond-normal frequency zone whose range is defined as freq > UFL, and Zone-3 (freq_BelowNormal), a below-normal frequency zone whose range is defined as freq < LFL. The implementation of the proposed state machine is shown in Figure 6. Fifteen states are used in this state machine to implement the proposed logic under three different frequency zones that are defined above. For easy recognition, the numbering of various states and zones is shown as highlighted labels in Figure 6. The transitions between these states involve an alternating way of changing the *Md* values that are discussed in Section 4.3. It should be noted that six different values of *Md* (Md-I to Md-VI) are designed for the abovementioned fifteen states.

The working of the proposed state machine is explained through a series of flow charts shown in Figure 7, wherein all the fifteen states are classified under 3 levels based on the control input that regulates a particular level. In the first level, the control input "freq" determines the selection among the states I, II, and III. In the second level, the control input "PPdiff_sign" determines the selection among the states IV–IX. Similarly, in the third level, the control input "NPdiff_sign" determines the selection among the states X–XV. Here, the transition between one state to another state depends on the frequency transients that occur in the system, which further affects the condition of the three control inputs. Such conditions of the control inputs along with their corresponding fifteen states and six outcomes are summarized in Table 1.

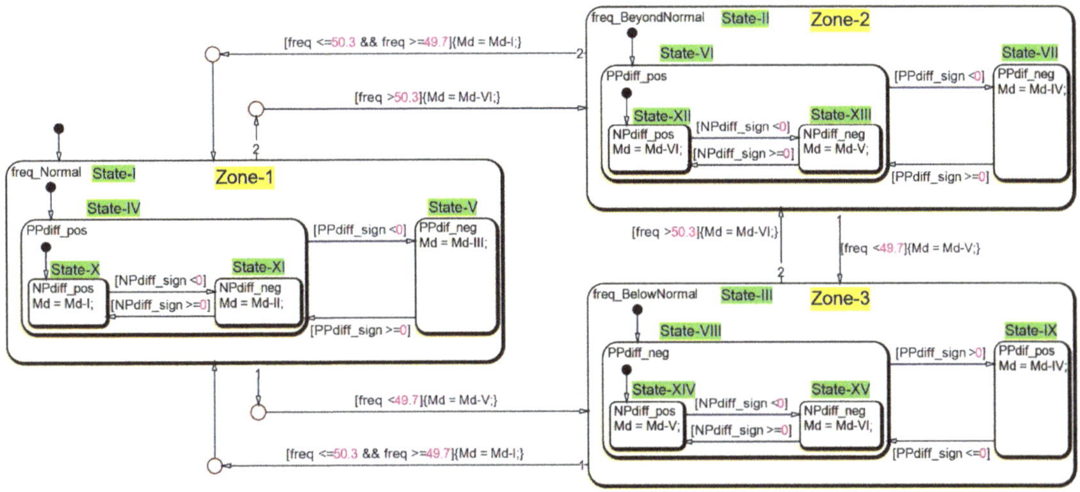

Figure 6. Implementation of the proposed state machine for SMDCM development.

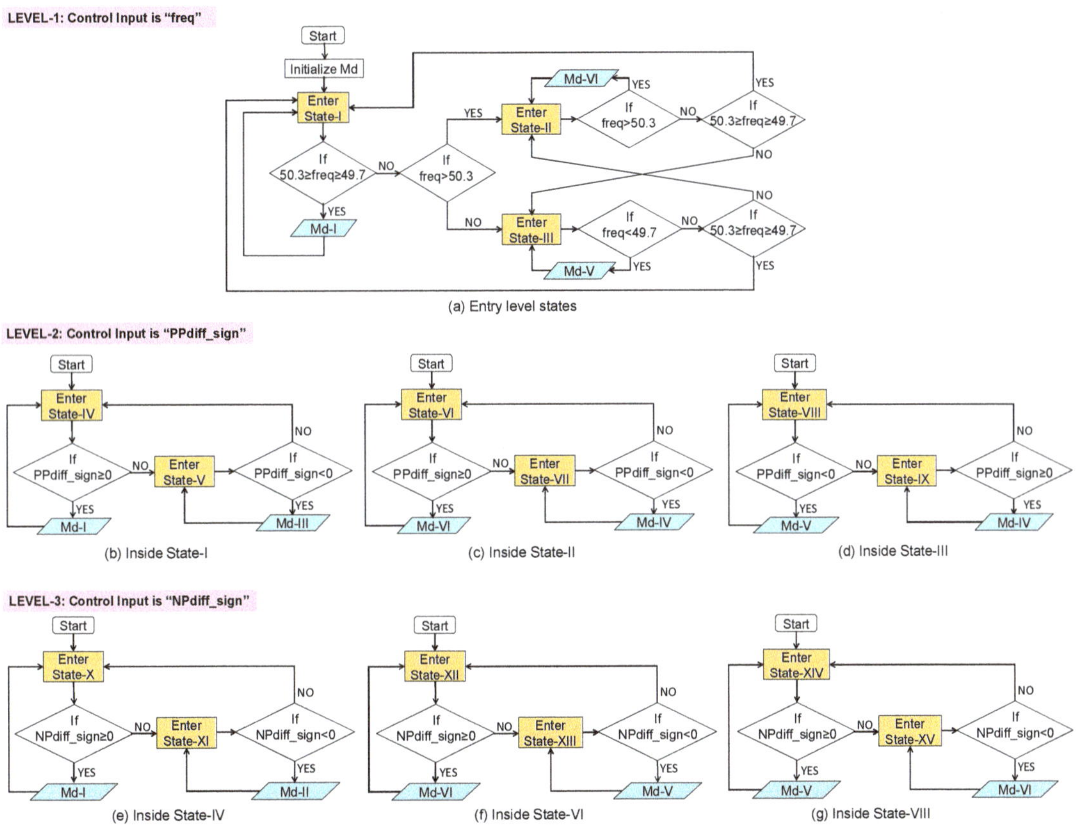

Figure 7. Flowcharts showing the sequence of transitions among various states of the state machine.

Table 1. States of the proposed state machine along with their entering conditions and outcomes.

Control Input	Control Input Condition	State	State Name	State Outcome
freq	50.3 ≥ freq ≥ 49.7	State-I	freq_Normal	Md-I
	freq > 50.3	State-II	freq_BeyondNormal	Md-VI
	freq < 49.7	State-III	freq_BelowNormal	Md-V
PPdiff_sign	50.3 ≥ freq ≥ 49.7 & PPdiff_sign ≥ 0	State-IV	PPdiff_pos	Md-I
	50.3 ≥ freq ≥ 49.7 & PPdiff_sign < 0	State-V	PPdiff_neg	Md-III
	freq > 50.3 & PPdiff_sign ≥ 0	State-VI	PPdiff_pos	Md-VI
	freq > 50.3 & PPdiff_sign < 0	State-VII	PPdiff_neg	Md-IV
	freq < 49.7 & PPdiff_sign < 0	State-VIII	PPdiff_neg	Md-V
	freq < 49.7 & PPdiff_sign ≥ 0	State-IX	PPdiff_pos	Md-IV
NPdiff_sign	50.3 ≥ freq ≥ 49.7 & PPdiff_sign ≥ 0 & NPdiff_sign ≥ 0	State-X	NPdiff_pos	Md-I
	50.3 ≥ freq ≥ 49.7 & PPdiff_sign ≥ 0 & NPdiff_sign < 0	State-XI	NPdiff_neg	Md-II
	freq > 50.3 & PPdiff_sign ≥ 0 & NPdiff_sign ≥ 0	State-XII	NPdiff_pos	Md-VI
	freq > 50.3 & PPdiff_sign ≥ 0 & NPdiff_sign < 0	State-XIII	NPdiff_neg	Md-V
	freq < 49.7 & PPdiff_sign < 0 & NPdiff_sign ≥ 0	State-XIV	NPdiff_pos	Md-V
	freq < 49.7 & PPdiff_sign < 0 & NPdiff_sign < 0	State-XV	NPdiff_neg	Md-VI

It is to be noted that all three control inputs appear continuously in parallel. But the priority of these control inputs in determining the state transition is different; where 'freq' has the highest priority, "PPdiff_sign" has the second priority, and "NPdiff_sign" has the last priority. To realize this, a sample operation with a test case is given below.

Test case: consider a case, where the control inputs "freq" is assumed as "freq > UFL", "PPdiff_sign" is assumed as "PPdiff_sign ≥ 0", and "NPdiff_sign" is assumed as "NPdiff_sign < 0". The state machine's operation, in this case, is described as follows.

Since "freq" has the highest priority, the state machine's operation starts from level 1. As shown in Figure 7a, initially, the loop enters the default State-I and proceeds to check the normal frequency control input condition (i.e., 50.3 ≥ freq ≥ 49.7), and finds it as false. So, in the next step, it verifies the transient control input conditions. Thus, it verifies the condition "freq > 50.3" and finds it as true, thereby, the loop enters State II. With the further confirmation of the condition "freq > 50.3", State-II turns active and temporarily generates the outcome as "Md-VI" before the loop enters into the inner state of State-II.

The inner state of State-II is determined by the next control input "PPdiff_sign", whose corresponding operation is shown in Figure 7c. Here, initially, the loop enters the default State-VI and proceeds to check the control input condition "PPdiff_sign ≥ 0". As this condition finds as true, State-VI turns active and temporarily generates an outcome as "Md-VI" before the loop enters into the inner state of State-VI.

The inner state of State-VI is determined by the third control input "NPdiff_sign", whose corresponding operation is shown in Figure 7f. Here, initially, the loop enters the default State-XII and proceeds to check the control input condition "NPdiff_sign ≥ 0". As this condition finds as false, the loop enters State-XIII. With the further confirmation of the condition "NPdiff_sign < 0" as true, State-XIII turns active and generates the outcome as "Md-V".

This "Md-V" is the final outcome of this test case, i.e., the desired variable droop coefficient (Md) generated by the state machine. The same procedure will be followed for any other conditions of the control inputs that are given in Table 1.

4.3. Design Philosophy of the Proposed SMDCM

The philosophy behind the design of the proposed SMDCM is derived by plotting the trajectories of maximum/minimum peaks of frequency characteristics under different transient conditions. The nature of these trajectories (whether rising/falling) can be ascertained based on the outputs of the peak detection unit viz., "PPdiff_sign" (PPdiff_neg, PPdiff_pos) and "NPdiff_sign" (NPdiff_neg, NPdiff_pos). Where "PPdiff_neg" leads to a "falling trajectory of positive peaks", "PPdiff_pos" leads to a "rising trajectory of positive peaks", "NPdiff_neg" leads to a "falling trajectory of negative peaks", and "NPdiff_pos" leads to a "rising trajectory of negative peaks".

For each frequency zone, 4 combinations of trajectories can be obtained, and in total 12 such combinations can be obtained for the 3 frequency zones, as depicted in Figure 8 and defined in Table 2. In Figure 8, the dotted lines in red color represents the trajectory of maximum peaks and that in blue color represents the trajectory of minimum peaks. These trajectories of the maximum and minimum peaks provide a qualitative inference of how much compensation is required for a particular situation. This can be like some generic terms viz., very large, large, medium, small, and no compensation, etc.

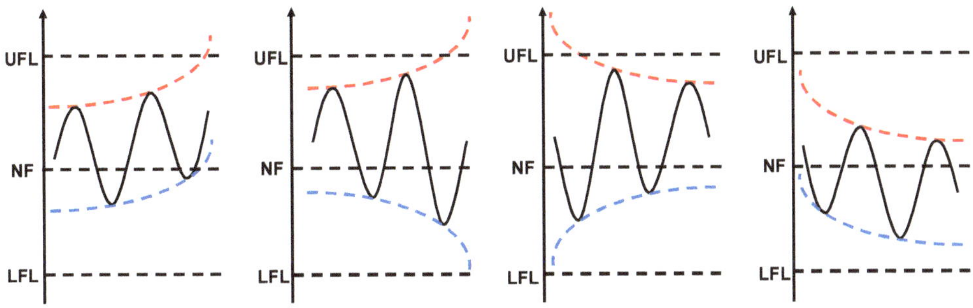

(A) PPdiff_pos, NPdiff_pos (B) PPdiff_pos, NPdiff_neg (C) PPdiff_neg, NPdiff_pos (D) PPdiff_neg, NPdiff_neg

(**a**) When the frequency is under Zone-1 (freq_Normal): UFL ≥ freq ≥ LFL.

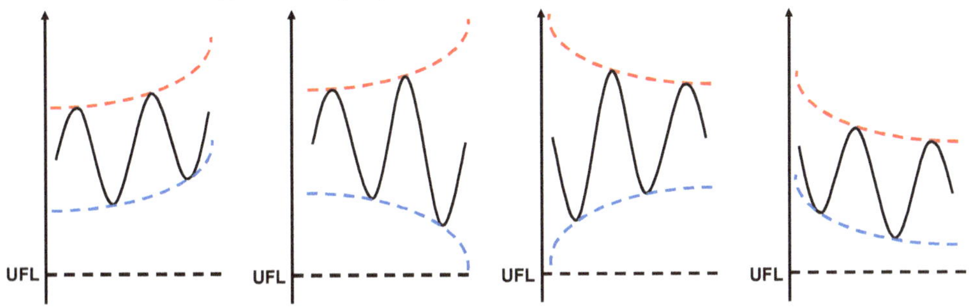

(E) PPdiff_pos, NPdiff_pos (F) PPdiff_pos, NPdiff_neg (G) PPdiff_neg, NPdiff_pos (H) PPdiff_neg, NPdiff_neg

(**b**)When the frequency is under Zone-2 (freq_BeyondNormal): freq > UFL.

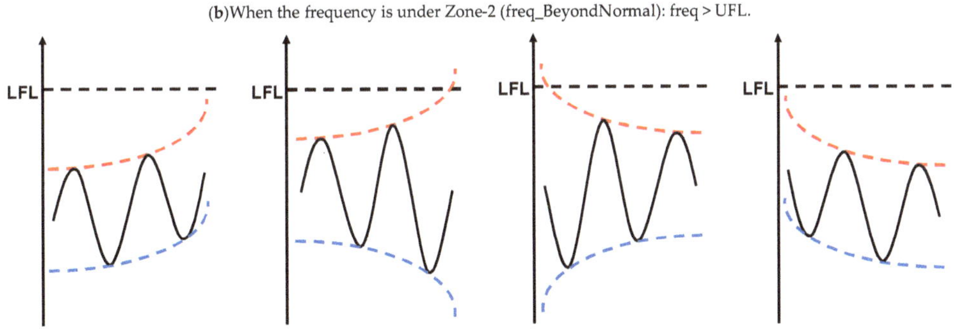

(I) PPdiff_pos, NPdiff_pos (J) PPdiff_pos, NPdiff_neg (K) PPdiff_neg, NPdiff_pos (L) PPdiff_neg, NPdiff_neg

(**c**) When the frequency is under Zone-3 (freq_BelowNormal): freq < LFL.

Figure 8. Various combinations of the trajectories obtained under 3 frequency zones.

Table 2. Summary of various conditions based on the trajectories of positive and negative peaks.

Inputs	**Different Combinations of the Control Inputs**											
freq	Zone-1: freq_Normal				Zone-2: freq_BeyondNormal				Zone-3: freq_BelowNormal			
PPdiff_sign	PPdiff_pos		PPdiff_neg		PPdiff_pos		PPdiff_neg		PPdiff_pos		PPdiff_neg	
NPdiff_sign	NPdiff_pos	NPdiff_neg	NPdiff_pos	NPdiff_neg	NPdiff_pos	NPdiff_neg	NPdiff_pos	NPdiff_neg	NPdiff_pos	NPdiff_neg	NPdiff_pos	NPdiff_neg

For example, consider Zone-2 in Figure 8b. From the comparison, it can be inferred that the average value of the frequency is steeply rising in Figure 8b(E) than in Figure 8b(F). Since the frequency is already beyond UFL, a large damping for compensation is required for Figure 8b(E) than for Figure 8b(F). Similarly, in the case of Figure 8b(H), since the inferred trajectory of average frequency is naturally restoring to normalcy, a little extra compensation is required. This kind of reasoning is the basis of how six different numeric values are designed for Md-I to Md-VI.

Thus, six different compensations are proposed to address various deviations of the frequency characteristic under different transient conditions that are given in Figure 8. These compensations are implemented through corresponding *Md* values as given in Table 3. The compensation and its corresponding *Md* value that is applied for each of the 12 conditions are mapped in Table 4. Moreover, in implementing this scheme, care is taken in fixing the values of Md-I to Md-VI such that the resultant *kp* value does not fall in the in-feasible region during unstable situations as discussed in Section 3. Therefore, the range of values described as zero compensation (ZC) is intentionally avoided as these values lead to *kp* falling to the in-feasible range.

Table 3. Various compensation levels and corresponding *Md* values.

Compensation	**Description**	**Md Value**	**Md Indication**
VPC	Very high positive compensation	0	Md-I
HPC	High positive compensation	1×10^{-5}	Md-II
NPC	Normal positive compensation	2×10^{-5}	Md-III
ZC	Zero compensation	$2 \times 10^{-5} < \mathrm{Md} < 5 \times 10^{-5}$	Avoided due to the in-feasible range of values
NNC	Normal negative compensation	-1.5×10^{-5}	Md-IV
HNC	High negative compensation	-2.5×10^{-5}	Md-V
VNC	Very high negative compensation	-5×10^{-5}	Md-VI

Table 4. Level of compensation and respective droop coefficient applied for various conditions.

Control Input		**Compensation Required**					
freq		Zone-1: UFL $\geq$ freq $\geq$ LFL		Zone-2: freq > UFL		Zone-3: freq < LFL	
PPdiff_sign		PPdiff_pos	PPdiff_neg	PPdiff_pos	PPdiff_neg	PPdiff_pos	PPdiff_neg
NPdiff_sign	NPdiff_pos	VPC (Md-I)	NPC (Md-III)	VNC (Md-VI)	NNC (Md-IV)	NNC (Md-IV)	HNC (Md-V)
	NPdiff_neg	HPC (Md-II)	NPC (Md-III)	HNC (Md-V)	NNC (Md-IV)	NNC (Md-IV)	VNC (Md-VI)

This proposed logic is implemented through the state machine. The corresponding compensation required for each of the conditions is realized by suitably activating various states of the state machine. The fifteen states of the state machine are classified under three control inputs as already shown in Table 1. It is reiterated that all three control inputs appear continuously in parallel which means that at any point in time, at least one state will be active under each of these control inputs. Because of this, a combination of any three states at any point in time corresponds to one of the twelve conditions given in Figure 8. The realization of these twelve conditions through a group of three states for each condition is shown in Figure 9.

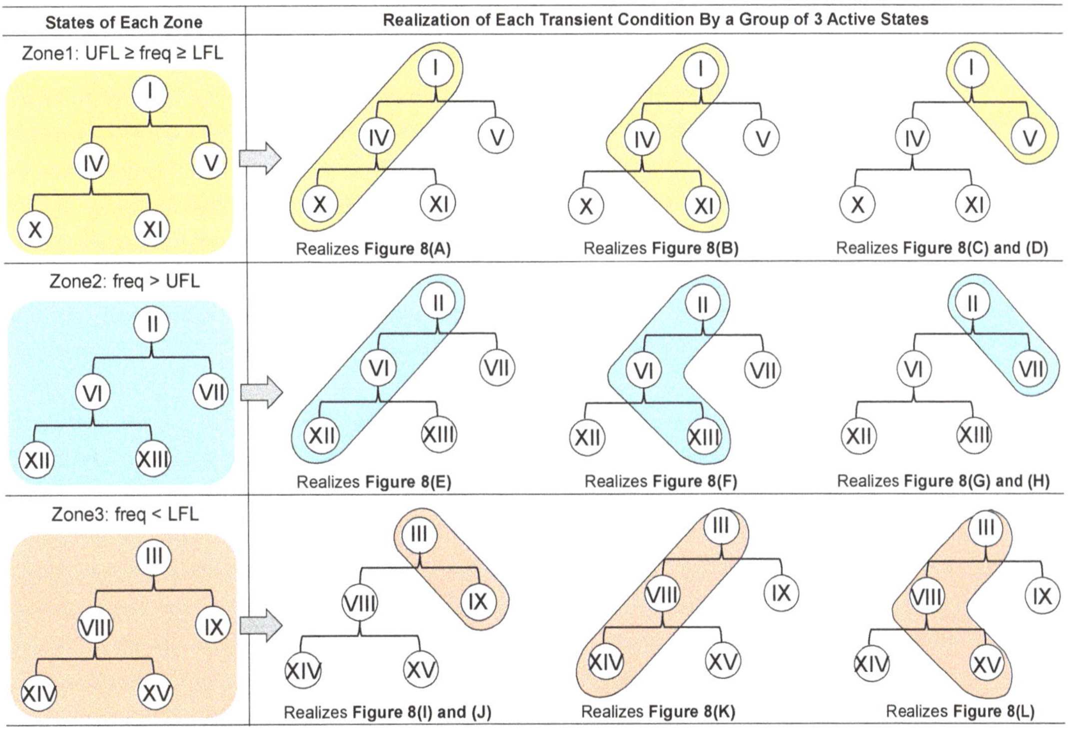

Figure 9. Realization of various transient conditions through the proposed combination of states.

From this Figure 9, the realization is done as follows.

- Activation of the states I, IV, and X realizes the transient condition shown in Figure 8A.
- Activation of the states I, IV, and XI realizes the transient condition shown in Figure 8B.
- Activation of the states I and V realizes the transient conditions shown in Figure 8C,D.
- Activation of the states II, VI, and XII realizes the transient condition shown in Figure 8E.
- Activation of the states II, VI, and XIII realizes the transient condition shown in Figure 8F.
- Activation of the states II and VII realizes the transient conditions shown in Figure 8G,H.
- Activation of the states III and IX realizes the transient conditions shown in Figure 8I,J.
- Activation of the states III, VIII, and XIV realizes the transient condition shown in Figure 8K.
- Activation of the states III, VIII, and XV realizes the transient condition shown in Figure 8L.

For a given real-time transient frequency characteristic, the proposed SMDCM follows the following procedure to activate the corresponding states and produce the required compensation.

- Step-1: The value of the measured frequency (freq) decides the zone out of the three zones. Thus, this reduces the choice of effective conditions from 12 to 4.
- Step-2: The "PPdiff_sign" decides between "PPdiff_pos" and "PPdiff_neg". This further reduces the choice of effective conditions from 4 to 2.
- Step-3: The "NPdiff_sign" decides between "NPdiff_pos" and "NPdiff_neg". With this, finally, 1 condition out of the available 2 becomes active. Thus, the resultant group of three states and the necessary compensation (any of Md-I to Md-VI) is realized.
- This value of *Md* will remain in a hold state till it gets updated to a new value based on the combination of the next three states.

5. Simulation Results and Comparative Analysis

The microgrid considered for this study includes a single DG unit and three three-phase loads. The DG unit is connected to its load bus through an LC filter followed by line inductance L. Load1, Load2 and Load3 are the three three-phase loads of the DG unit. Each three-phase load is represented by a series RLC branch at each phase. The detailed model of the system is implemented in MATLAB software environment and the electrical and control parameters of the system are shown in Table 5; where, *Kpv* and *Kiv* are the proportional and integral parameters of the voltage controller, and *Kpc* and *Kic* are the proportional and integral parameters of the current controller.

Table 5. Electrical and control parameters used in the simulation.

Parameter	Description	Rating
$Pr + jQr$	Rated power	25 kW + j25 kVar
Vdc	Input DC voltage of the inverter	540 V
Grid Line	Parameters of the line connecting to the grid	R = 1 Ω/km, L = 1 mH/km
Load Line	Parameters of the line connecting to load	R = 12.7 mΩ/km, L = 0.933 mH/km
Filter	Input LC filter parameters	Rf = 0.1 mΩ, Lf = 1 mH, Cf = 5 mF
Dp	Maximum frequency droop controller's coefficient	1×10^{-4}
Dq	Maximum voltage droop controller's coefficient	1.48×10^{-3}
Kpv	Proportional parameters of the voltage controller	5.65×10^{-4}
Kiv	Integral parameters of the voltage controller	0
Kpc	Proportional parameters of the current controller	0.12
Kic	Integral parameters of the current controller	6.7

Load1 (L1) is a continuous type of RL load whose value is fixed all the time. Load2 is a momentary RL load that gets connected to the system through a breaker with a normally open condition. Three different configurations of load2 (L21, L22, and L23) are designed to create three different loading effects on the system. Similarly, Load3 (L3) is also a momentary RC load that gets connected through a breaker with the normally open condition. The details of these test load configurations are shown in Table 6.

Table 6. Various test load configurations.

Load Component	Continuous Load	Momentary Load			
	Load1	Load2			Load3
	(L1)	(L21)	(L22)	(L23)	(L3)
R-load (W)	1200	300	300	300	300
L-Load (Var)	300	300	800	1200	0
C-load (Var)	0	0	0	0	300

In this study, three test scenarios with various loading conditions are considered viz., Case 1 with nominal power factor (P.F.) loading, Case 2 with moderate P.F. loading, and Case 3 with poor P.F. loading. A summary of all these test scenarios with the details of various test loads acting on the system with their respective timelines during which they occur is provided in Table 7. In this work, the following is the procedure adopted in fixing the values for different loads.

Table 7. Summary of test conditions.

Test Case	Applied Load During Various Instants of Time									
	Duration $(0 \leq t \leq 80)$s	P.F.	Duration $(80 < t \leq 90)$s	P.F.	Duration $(90 < t \leq 125)$s	P.F.	Duration $(125 < t \leq 135)$s	P.F.	Duration $(135 < t \leq 140)$s	P.F.
Case 1: (Nominal P.F.)	L1	0.97	L1+L21	0.928	L1	0.97	L1+L3	1	L1	0.97
Case 2: (Moderate P.F.)	L1	0.97	L1+L22	0.81	L1	0.97	L1+L3	1	L1	0.97
Case 3: (Poor P.F.)	L1	0.97	L1+L23	0.707	L1	0.97	L1	0.97	L1	0.97

During the process of selecting the loads, care has been taken in fixing the values more particularly for Load2 such that during 80–90 s, the total load seen by the system is Load1 + Load2. This value of Load2 is selected in such a way that it is the boundary limit beyond which one of the considered conventional methods will fail in maintaining stability. Three different values for Load2 namely L21, L22, and L23 are identified in a systematic order. In the first test case, the value fixed for Load2 is trivial and the total load seen during 80–90 s is L1 + L21 i.e., 1500 W + j600 Var. During this time, all three methods will work without losing system stability. In the second test case, Load2 is fixed to 300 W + j800 Var such that the total load will be L1 + L22 i.e., 1500 W + j1100 Var. It is from this loading onwards, conventional CDCM has lost its stability; while conventional FLDCM and proposed SMDCM continued to show stability. In the third test case, Load2 is fixed to 300 W + j1200 Var such that the total load will be L1+L23 i.e., 1500 W + j1500 Var. It is from this loading onwards, conventional FLDCM has also lost its stability; while the proposed SMDCM successfully continued to show stability. The sizing of Load3 and its effect on system response is discussed in the sections connected with their respective test cases. Various responses of the system such as frequency, voltage, and active power are plotted with conventional CDCM, conventional FLDCM, and the proposed SMDCM and are compared separately under each test case.

5.1. Case 1 (Nominal P.F. Loading)

In this case, the test scenario is such that, load1 is set to 1200 W + j300 Var, load2 is set to 300 W + j300 Var, and load3 is set to 300 W − j300 Var. The corresponding waveforms of frequency, output power, and voltage are shown in Figure 10, Figure 11, and Figure 12 respectively.

With the conventional CDCM, the frequency waveform settled to a normal frequency value of 50 Hz, and no deviations are found with the existing fixed load. During the connection and disconnection of Load2 also, no significant changes are noticed in the frequency. A little spike appeared at 125 s when load 3 is connected. However, the frequency value is still within the limit of 50 ± 1 Hz. These are identified in Figure 10a.

It is important to know whether the spike at 125 s in the frequency graph is due to capacitive load or due to an interaction of the capacitive loading effect and inductive load disconnection effect or solely due to the inductive load disconnection effect. In this case, the spike can be attributed to the capacitive loading effect since the inductive effect of Load2 is not existing at this time. The corresponding waveform of the output power and inverter voltage is shown in Figures 11a and 12a respectively. It can be noticed from Figure 11a that, the inverter momentarily failed to deliver power at 125 s indicating no more reserve margin.

In the case of the conventional FLDCM, as can be seen from Figure 10b, no deviations and surges are found in the frequency waveform. The resultant responses of the output power and inverter voltage are shown in Figures 11b and 12b respectively. Similarly, with the proposed SMDCM, no deviations and anomalies are noticed in the frequency waveform as seen in Figure 10c. The corresponding waveform of the output power and inverter voltage are shown in Figures 11c and 12c respectively indicating no anomalies.

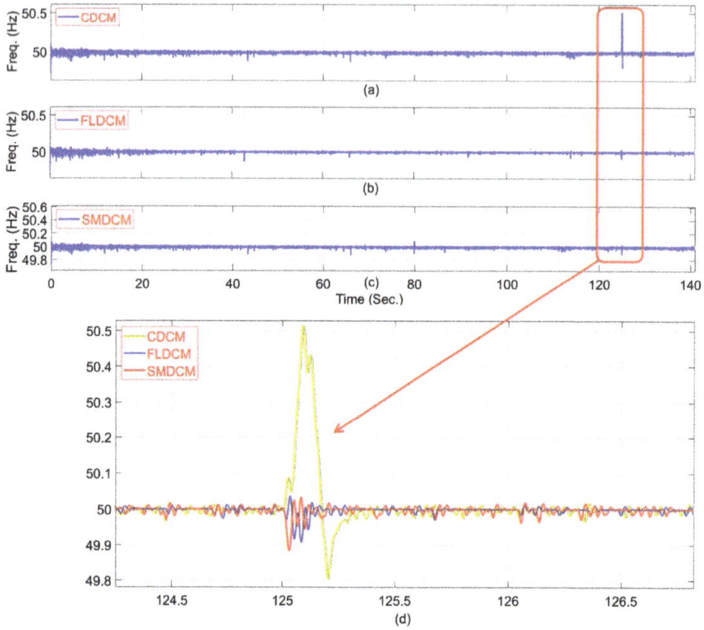

Figure 10. Frequency responses obtained with (**a**) CDCM, (**b**) FLDCM, and (**c**) SMDCM when subjected to Case 1. (**d**) zoom in of all frequency responses around 125 s.

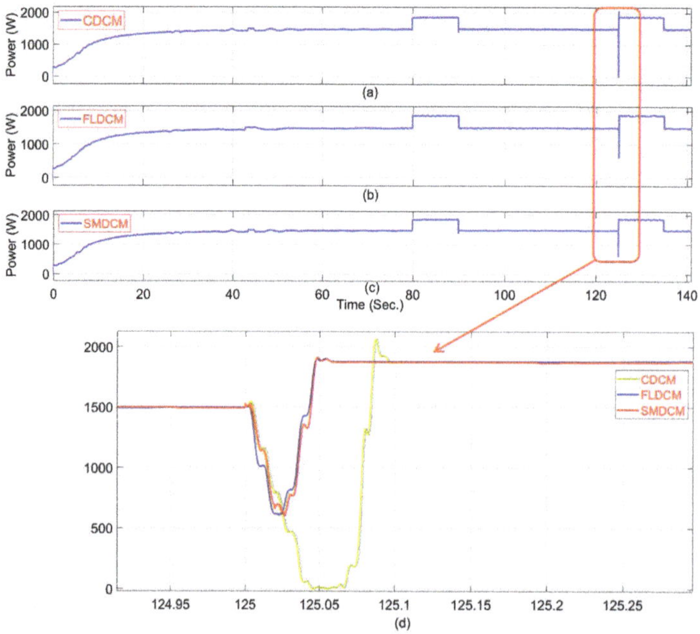

Figure 11. Output power responses obtained with (**a**) CDCM, (**b**) FLDCM, and (**c**) SMDCM when subjected to Case 1. (**d**) zoom in of all power responses around 125 s.

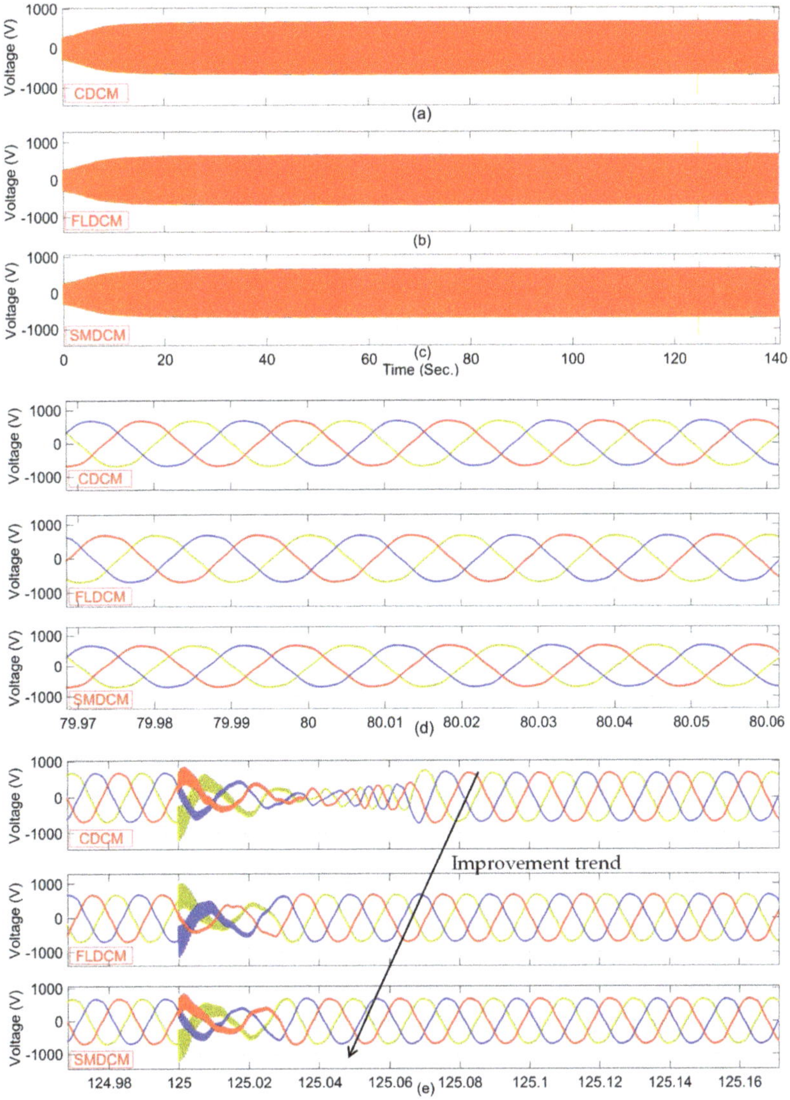

Figure 12. PCC voltage responses obtained with (**a**) CDCM, (**b**) FLDCM, and (**c**) SMDCM when subjected to Case 1. (**d**) zoom in of all voltage responses around 80 s and (**e**) zoom in of all voltage responses around 125 s.

The zoomed aspect of frequency, active power, and voltage waveforms at 125 s are shown in Figure 10d, Figure 11d, and Figure 12e respectively. The condition of voltage corresponding to switching at 80 s is shown in Figure 12d, where there is no deviation observed with all the methods.

From all these responses, it can be observed that the proposed SMDCM has shown superior performance in frequency, active power, and voltage characteristics when compared with the conventional CDCM and FLDCM.

5.2. Case 2 (Moderate P.F. Loading)

In this case, the test scenario is such that, load1 is set to 1200 W + j300 Var, load2 is set to 300 W + j800 Var, and load3 is set to 300 W − j300 Var. The corresponding waveforms for frequency, output power, and voltage are shown in Figure 13, Figure 14, and Figure 15 respectively. With the conventional CDCM, the frequency waveform settled to a normal frequency value of 50Hz, and no deviations are found with the existing fixed load. No major changes can be noticed in the frequency soon after the connection of Load2. However, from 86.6 s, the frequency waveform is seeing a noticeable distortion. These are identified in Figure 13a. The distortions are beyond the limit of 50 ± 1 Hz. Thus, it is justified from this that the larger the inductive load, the larger will be the instability. The corresponding output power and inverter voltage are shown in Figures 14a and 15a respectively. From Figure 14a, it is seen that the inverter failed to exhibit any reserve margin to maintain stability from 86.6 s onwards as indicated by repeated falling of power to zero. The voltage is completely distorted after 86.6 s as noticed in Figure 15a.

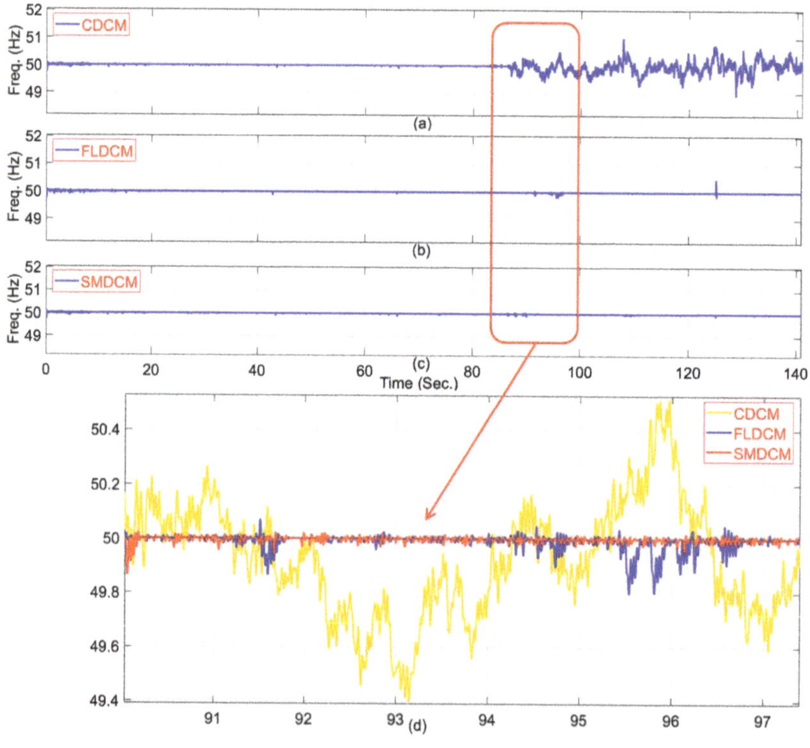

Figure 13. Frequency responses obtained with (**a**) CDCM, (**b**) FLDCM, and (**c**) SMDCM when subjected to Case 2. (**d**) zoom in of all frequency responses after 90 s.

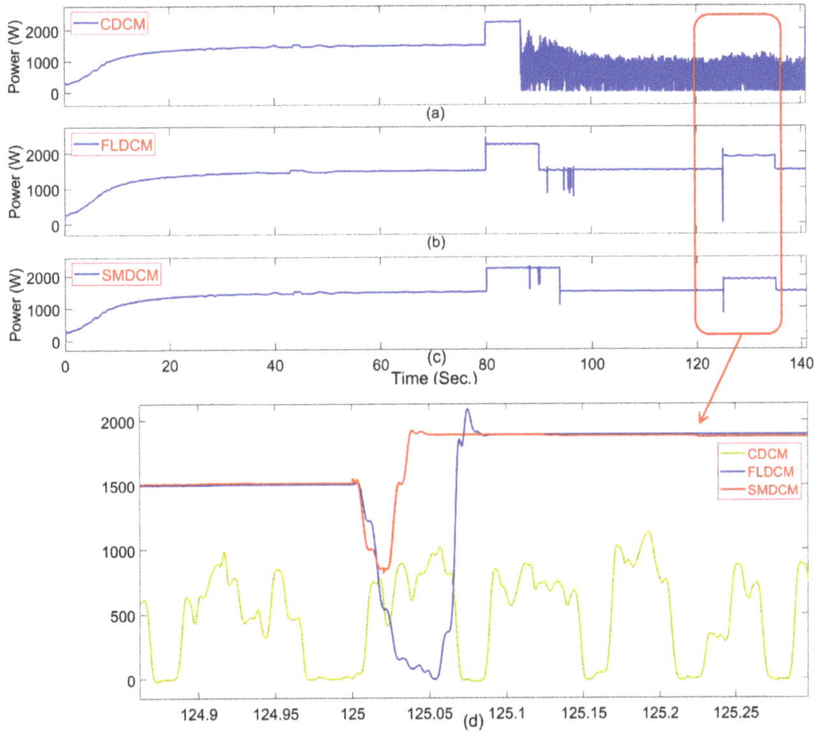

Figure 14. Output power responses obtained with (**a**) CDCM, (**b**) FLDCM, and (**c**) SMDCM when subjected to Case 2. (**d**) zoom in of all power responses around 125 s.

With the conventional FLDCM, as can be seen from Figure 13b, no deviations and surges are found in the frequency waveform. The corresponding waveform of the output power and inverter's voltage is shown in Figures 14a and 15a respectively. As seen from Figure 14b, the inverter momentarily failed to deliver power at 125 s indicating a similar kind of situation with the constant droop technique in case-1. It is worth noticing that, at 80 s, the output power shows a spike appeared reaching a value of 2500 W putting more burden on the energy source feeding the inverter. After 86.6 s, little deviations are seen in frequency and power waveforms. But, the proposed SMDCM did not allow any such deviations in frequency/power waveforms. This can be explained by Figure 14c that the SMDCM has provided an extra reserve margin even after 90 to 94 s. Also, no deviations/anomalies are noticed in frequency and voltage waveforms as seen in Figures 13c and 15c respectively. Voltage anomalies in the form of distortions and spikes appeared even after 90 s when FLDCM is employed; while no such issues exist with SMDCM. The occurrence of these deviations with FLDCM can be correlated with sudden droppings in power values occasionally during 90–100 s in Figure 14b. But, as shown in Figure 15c, no such deviations are noticed after 90 s, when SMDCM is employed. This confirms that SMDCM ensures a better transient response than FLDCM. The zoomed aspect of the frequency waveform after 90 s is shown in Figure 13d. The details of distortions in the waveform of active power can be found in Figure 14d. The condition of voltage corresponding to switching at 80 s and that after 86.6 s is shown in Figure 15d,e respectively. A typical example of an impact of a sudden dip in power values occasionally during 90–100 s on voltage is depicted in Figure 15f.

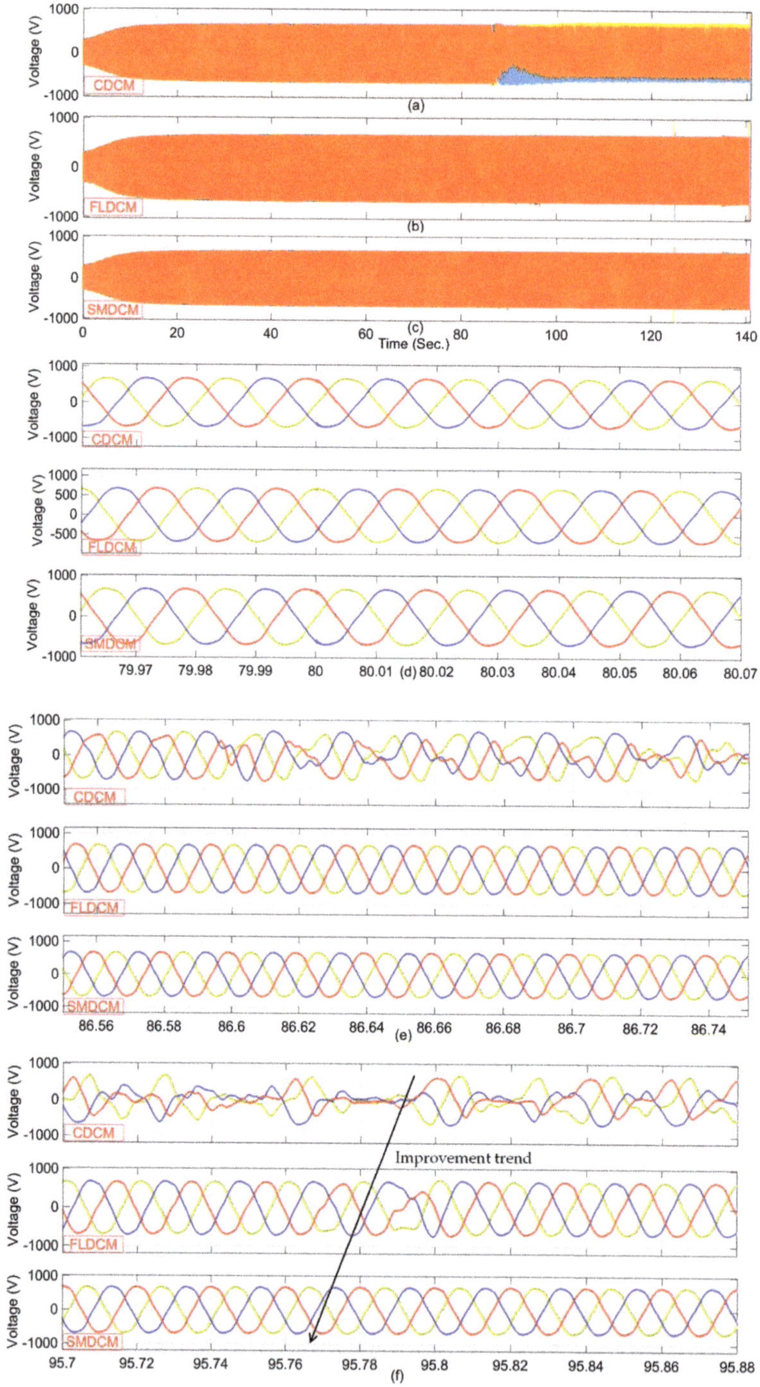

Figure 15. PCC voltage responses obtained with (**a**) CDCM, (**b**) FLDCM, and (**c**) SMDCM when subjected to Case 2. (**d**) zoom in of voltage responses around 80 s, (**e**) zoom in of voltage responses around 86 s and (**f**) zoom in of voltage responses around 95.8 s.

5.3. Case 3 (Poor P.F. Loading)

From Case 2, it is understood that connection/disconnection of large inductive loads brings larger instability situations. Thus, to focus on this aspect, Load3 is purposefully removed in this case and a large inductive load in the form of Load2 is introduced into the system. Thus, in the test scenario of this case, L1 is set to 1200 + j300, load2 is set to L23 i.e., 300 + j1200, and L3 is left unconnected. The corresponding responses for frequency, output power, and voltage are shown in Figure 16, Figure 17, and Figure 18 respectively.

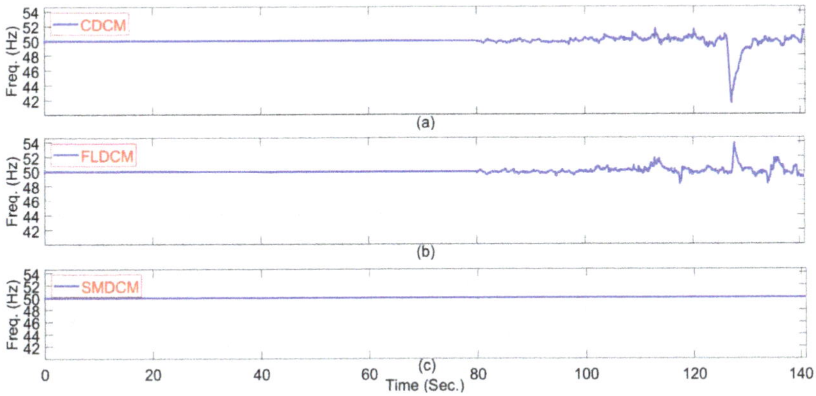

Figure 16. Frequency response obtained with (**a**) CDCM, (**b**) FLDCM, and (**c**) SMDCM when subjected to Case 3.

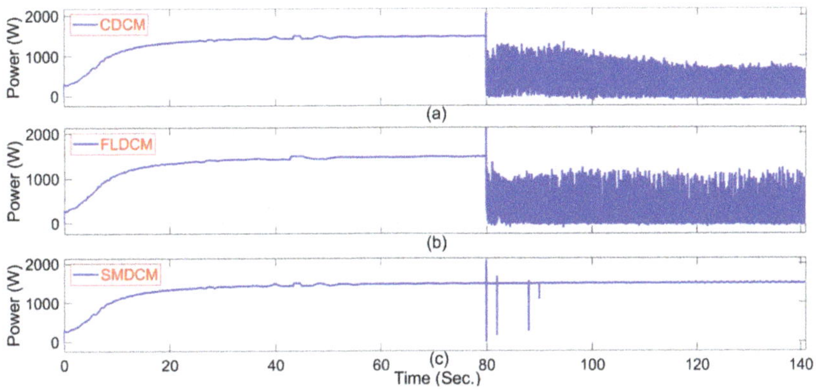

Figure 17. Output power responses obtained with (**a**) CDCM, (**b**) FLDCM, and (**c**) SMDCM when subjected to Case 3.

With the conventional CDCM, the frequency waveform settled to a normal frequency value of 50 Hz, and no deviations are found with the existing fixed load. No significant changes can be noticed in the frequency soon after with the connection of L23. However, a start in the growth of distortion in the frequency waveform is noticed from 80 s onwards. These are identified in Figure 16a. It can further be noticed from the same figure that the frequency sees a dip to 42 Hz at around 130 s leading to serious instability. It can be identified from the active power waveform as shown in Figure 17a that the inverter has failed to exhibit any reserve margin to maintain stability from 80 s onwards. Further, the voltage output of the inverter is completely distorted after 80 s as can be noticed in Figure 18a.

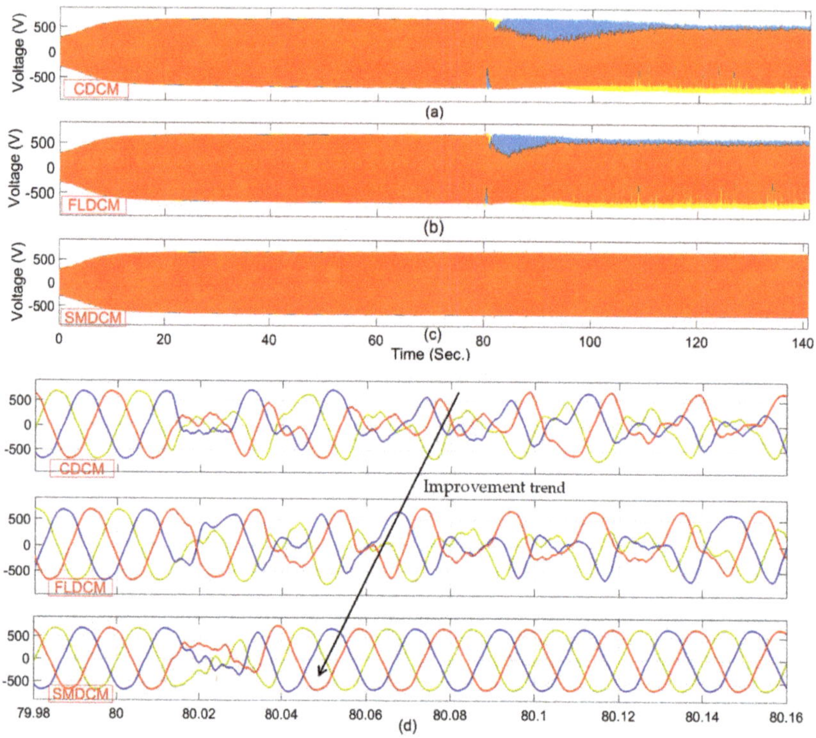

Figure 18. PCC voltage responses obtained with (**a**) CDCM, (**b**) FLDCM, and (**c**) SMDCM when subjected to Case 3. (**d**) zoom in of all voltage responses around 80 s.

Similarly, when the conventional FLDCM is employed, as can be seen from Figure 16b, the frequency waveform in this case sees a spike of 54 Hz around 130 s. This is a lesser deviation when compared with the constant droop approach. However, since this value is also beyond the limit of 50 ± 1 Hz, the system can be understood as the loss of its stability. The possible reason for loss of stability is because of possible chances of the *kp* value falling in the in-feasible zone. The situation with power output and inverter output voltage as shown by Figures 17b and 18b further confirms, that FLDCM has failed to maintain the stability of the system in this case.

With the caution of not allowing the *kp* value to fall into an in-feasible zone while implementing the state machine, the proposed SMDCM successfully mitigated any possible deviations in the frequency, output power, and voltage as can be noticed from Figure 16c, Figure 17c, and Figure 18c respectively. This proves the superiority of the proposed technique in upholding the stability of the system when subjected to connection/ disconnection of large inductive loads in the system. The zoomed aspect of distortion in voltage waveforms corresponding to switching at 80 s is shown in Figure 18d. From this figure, it is clear that there is an increasing trend from conventional to proposed methods with respect to the waveform shape, which justifies the importance of the proposed SMDCM in this paper.

As this test Case 3 leads to severe voltage distortions compared to the previous two test cases, the total harmonic distortion (THD) analysis is performed in this case. The voltage THD values with conventional and proposed methods are depicted in Figure 19. As shown in Figure 19a,b, the THD value with conventional CDCM and conventional FLDCM are obtained as 74.10% and 58.99% respectively. These values largely violated the standard 5% tolerance that is defined by IEEE 1547 standard. But, the voltage THD value shown in

Figure 19c, which is obtained with the proposed SMDCM is computed as 3.08%, which adheres to the standard 5% tolerance.

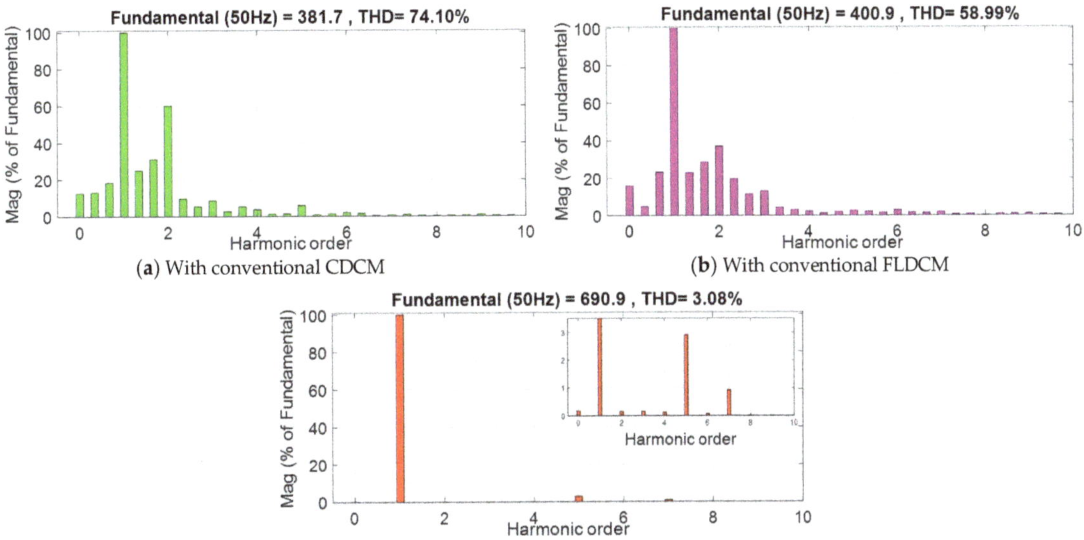

(**a**) With conventional CDCM (**b**) With conventional FLDCM (**c**) With proposed SMDCM

Figure 19. Voltage THDs obtained with CDCM, FLDCM, and SMDCM subjected to Case 3.

The abovementioned comparative performance analysis of the conventional CDCM, conventional FLDCM, and the proposed SMDCM are summarized in Table 8 to understand the superiority of the proposed SMDCM.

Table 8. Comparison of conventional and proposed methods in various performance aspects.

Performance Parameter		Test Cases		Conventional CDCM [13]	Conventional FLDCM [24]	Proposed SMDCM	Superior Method
Frequency characteristics Standard Limit: ▪ ±2% i.e., 49 to 51 Hz for a nominal frequency of 50 Hz (IEEE 1547)	Max value (Hz)	Case-1	Inductive load	50.1	50.1	50.1	All
			Capacitive load	50.5	50.1	50.1	All
		Case-2	Inductive load	50.1	50.1	50.1	All
			Capacitive load	51	50.47	50.03	FLDCM/SMDCM
		Case-3	Inductive load	53 (violated)	54 (violated)	50.02	SMDCM
	Min value (Hz)	Case-1	Inductive load	49.9	49.9	49.9	All
			Capacitive load	49.8	49.9	49.9	All
		Case-2	Inductive load	49.9	49.9	49.9	All
			Capacitive load	48.9 (violated)	49.8	49.8	FLDCM/SMDCM
		Case-3	Inductive load	42 (violated)	48 (violated)	49.8	SMDCM
Power characteristics Desired: ▪ Extra burden–Low ▪ Reserve margin–High	Extra burden (watts)	Case-1	Inductive load	0	0	0	All
			Capacitive load	170	30	30	FLDCM/SMDCM
		Case-2	Inductive load	0	190	0	CDCM/SMDCM
			Capacitive load	Failed	200	0	SMDCM
		Case-3	Inductive load	Failed	860	550	SMDCM
	Reserve margin (watts)	Case-1	Inductive load	1500	1500	1500	All
			Capacitive load	0 (recovered)	600	600	FLDCM/SMDCM
		Case-2	Inductive load	Failed	1500	2250	SMDCM
			Capacitive load	Failed	0 (recovered)	820	SMDCM
		Case-3	Inductive load	Failed	Failed	0 (recovered)	SMDCM
Voltage characteristics Standard Limits: ▪ Peak change: 200 V max for 600 V rated peak value (UL 1449, IEEE C62.41.2) ▪ THD: 5% (IEEE 519, IEEE 1547)	Peak change (volts)	Case-1	Inductive load	0	0	0	All
			Capacitive load	168	318 (violated)	158	SMDCM
		Case-2	Inductive load	Failed	58	48	SMDCM
			Capacitive load	Failed	340 (violated)	150	SMDCM
		Case-3	Inductive load	Failed	Failed	64	SMDCM
	Disturbance period (s)	Case-1	Inductive load	0	0	0	All
			Capacitive load	0.07	0.03	0.03	FLDCM/SMDCM
		Case-2	Inductive load	Failed	16.6	10	SMDCM
			Capacitive load	Failed	0.06	0.02	SMDCM
		Case-3	Inductive load	Failed	Failed	0.04	SMDCM
	THD (%)	Case-3	Inductive load	74.1 (violated)	58.9 (violated)	3.08	SMDCM

6. Conclusions

In this paper, to improve the symmetry in the enhancement of both transient response and stability of the microgrid when subjected to connection/disconnection of large inductive loads, a simple, fast, and robust technique based on the identification of an in-feasible range of droop coefficients is proposed. This proposed "state machine-based droop control method (SMDCM)" is implemented through the development of a state machine. The superiority of the proposed method is proved by comparing it with conventional CDCM and FLDCM. The salient achievements of the proposed work are given as follows.

- Based on the time domain analysis and small signal model, a strong correlation is identified between connection/disconnection of large inductive loads and fall in equivalent line inductance. Further, the role of the droop coefficient value on stability based on the change in line inductance is studied.
- The identification of an in-feasible range of the frequency droop coefficient values based on the eigenvalue analysis allows to avoid only those range of dangerous values; thereby, providing the user with a widened chance to pick the desired droop values.
- Since the proposed method of fixing the droop coefficients is an offline approach, the computation time and burden are very much reduced.
- The proposed SMDCM outperformed the conventional CDCM and FLDCM. It also provides robust control of both voltage and frequency in terms of transient stability and response. The same is proved through the comparative metrics given in Table 8. The following are the salient points that can be summarized from this table.
  - In terms of the frequency with Case 3 loading, both CDCM and FLDCM violated the standard limit of $\pm$ 2% limit while the proposed SMDCM scheme exhibited 50.02 Hz and 49.8 Hz as maximum and minimum values respectively ensuring stability.
  - In terms of extra burden during Case 3, CDCM has failed the system. While, FLDCM initially had presented an extra burden of 860 W on the system, while SMDCM had presented a lesser burden of 550 W on the system. With respect to the reserve margin, also, both CDCM and FLDCM have made the system fail. While SMDCM even though it has temporarily seen zero reserve power momentarily, recovered immediately ensuring the stability of the system.
  - With respect to voltage characteristics, with capacitive load switching in FLDCM, the peak change in voltage is 318 V and 340 V during Case 1 and Case 2 respectively, whereas, 200 V is the maximum allowable peak change. While with SMDCM, the peak changes recorded are 158 V and 150 V respectively. With inductive load switching in Case 3, both CDCM and FLDCM failed, while with SMDCM it is just 64 V.
  - In terms of disturbance period, in Case 2, CDCM has failed, with FLDCM it is 16.6 s, while the same with SMDCM is 10 s only. Further, in Case 3, both CDCM and FLDCM failed, while SMDCM recovered the response in 0.04 s.
  - Similarly, during Case 3, the voltage THD of CDCM and FLDCM is 74.1% and 58.9% respectively. The same with SMDCM is only 3.08% which is less than the standard limit of 5%.

Thus, the above-mentioned summary indicates the superiority of the proposed SMDCM for improving the transient performance of microgrids over conventional CDCM and conventional FLDCM.

Limitation and Future Scope

This research work is particularly designed to address the drawbacks of the fuzzy logic-based droop controller coefficient adjustment method. This is the limitation of the proposed work in this paper. Further, with respect to the contribution deposited in this paper, the following extensions can be considered as future scope.

- Design of droop coefficient with other artificial intelligence or machine learning-based methods. Their effectiveness can be tested with respect to the proposed SMDCM in this paper. Further, experimental verification of all these methods can be considered for future work.

- Retaining the proposed state machine-based droop controller, the existing conventional voltage and current controllers can be replaced with modern compensators to further enhance the transient performance of the microgrid.

Author Contributions: Conceptualization, M.S.; Formal analysis, M.S.; Funding acquisition, Y.V.P.K.; Investigation, M.S.; Methodology, Y.V.P.K.; Resources, Y.V.P.K.; Supervision, Y.V.P.K.; Validation, Y.V.P.K.; Visualization, M.S.; Writing—original draft, M.S.; Writing—review & editing, Y.V.P.K. All authors have read and agreed to the published version of the manuscript.

Funding: The APC was funded by VIT-AP University.

Institutional Review Board Statement: Not applicable.

Informed Consent Statement: Not applicable.

Data Availability Statement: Not applicable.

Acknowledgments: The authors would like to thank the Start-up Research Grant (SRG) scheme of the Science and Engineering Research Board (SERB), a statutory body under the Department of Science and Technology (DST), Government of India, for supporting this research work under project grant number SRG/2019/000648 and VIT-AP University, Amaravati, Andhra Pradesh, India for funding the open access publication fee for this research work.

Conflicts of Interest: The authors declare no conflict of interest.

References

1. Green, T.C.; Prodanović, M. Control of Inverter-Based Micro-Grids. *Electr. Power Syst. Res.* **2007**, *77*, 1204–1213. [CrossRef]
2. Norouzi, M.; Aghaei, J.; Pirouzi, S.; Niknam, T.; Fotuhi-Firuzabad, M.; Shafie-khah, M. Hybrid Stochastic/Robust Flexible and Reliable Scheduling of Secure Networked Microgrids with Electric Springs and Electric Vehicles. *Appl. Energy* **2021**, *300*, 117395. [CrossRef]
3. Sabzalian, M.H.; Pirouzi, S.; Aredes, M.; Wanderley Franca, B.; Carolina Cunha, A. Two-Layer Coordinated Energy Management Method in the Smart Distribution Network Including Multi-Microgrid Based on the Hybrid Flexible and Securable Operation Strategy. *Int. Trans. Electr. Energy Syst.* **2022**, *2022*, 3378538. [CrossRef]
4. Pavan Kumar, Y.V.; Bhimasingu, R. Electrical Machines Based DC/AC Energy Conversion Schemes for the Improvement of Power Quality and Resiliency in Renewable Energy Microgrids. *Int. J. Electr. Power Energy Syst.* **2017**, *90*, 10–26. [CrossRef]
5. Hosseinpour, M.; Dejamkhooy, A. Control and Power Sharing among Parallel Three-Phase Three-Wire and Three-Phase Four-Wire Inverters in the Presence of Unbalanced and Harmonic Loads. *IEEJ Trans. Electr. Electron. Eng.* **2018**, *13*, 1027–1033. [CrossRef]
6. Leitner, S.; Yazdanian, M.; Mehrizi-Sani, A.; Muetze, A. Small-Signal Stability Analysis of an Inverter-Based Microgrid with Internal Model-Based Controllers. *IEEE Trans. Smart Grid* **2018**, *9*, 5393–5402. [CrossRef]
7. Pavan Kumar, Y.V.; Bhimasingu, R. Design of Voltage and Current Controller Parameters Using Small Signal Model-Based Pole-Zero Cancellation Method for Improved Transient Response in Microgrids. *SN Appl. Sci.* **2021**, *3*, 836. [CrossRef]
8. Liu, S.; You, H.; Li, J.; Kai, S.; Yang, L. Active Disturbance Rejection Control Based Distributed Secondary Control for a Low-Voltage DC Microgrid. *Sustain. Energy Grids Netw.* **2021**, *27*, 100515. [CrossRef]
9. Stojić, Đ.M.; Šekara, T.B. A New Digital Resonant Current Controller for AC Power Converters Based on the Advanced Z-Transform. *ISA Trans.* **2022**, *139*, 535–545. [CrossRef]
10. Iyer, S.V.; Belur, M.N.; Chandorkar, M.C. A Generalized Computational Method to Determine Stability of a Multi-Inverter Microgrid. *IEEE Trans. Power Electron.* **2010**, *25*, 2420–2432. [CrossRef]
11. Schiffer, J.; Ortega, R.; Astolfi, A.; Raisch, J.; Sezi, T. Conditions for Stability of Droop-Controlled Inverter-Based Microgrids. *Automatica* **2014**, *50*, 2457–2469. [CrossRef]
12. De Brabandere, K.; Bolsens, B.; Van den Keybus, J.; Woyte, A.; Driesen, J.; Belmans, R. A Voltage and Frequency Droop Control Method for Parallel Inverters. *IEEE Trans. Power Electron.* **2007**, *22*, 1107–1115. [CrossRef]
13. Guan, M.; Pan, W.; Zhang, J.; Hao, Q.; Cheng, J.; Zheng, X. Synchronous Generator Emulation Control Strategy for Voltage Source Converter (VSC) Stations. *IEEE Trans. Power Syst.* **2015**, *30*, 3093–3101. [CrossRef]
14. Guerrero, J.M.; Vasquez, J.C.; Matas, J.; de Vicuna, L.G.; Castilla, M. Hierarchical Control of Droop-Controlled AC and DC Microgrids—A General Approach Toward Standardization. *IEEE Trans. Ind. Electron.* **2011**, *58*, 158–172. [CrossRef]
15. Xiao, Z.; Fang, H. Impacts of P-f & Q-V Droop Control on MicroGrids Transient Stability. *Phys. Procedia* **2012**, *24*, 276–282. [CrossRef]
16. Yu, M.; Huang, W.; Tai, N.; Zheng, X.; Wu, P.; Chen, W. Transient Stability Mechanism of Grid-Connected Inverter-Interfaced Distributed Generators Using Droop Control Strategy. *Appl. Energy* **2018**, *210*, 737–747. [CrossRef]

17. Rios Penaloza, J.D.; Adu, J.A.; Borghetti, A.; Napolitano, F.; Tossani, F.; Nucci, C.A. Influence of Load Dynamic Response on the Stability of Microgrids during Islanding Transition. *Electr. Power Syst. Res.* **2021**, *190*, 106607. [CrossRef]
18. Dheer, D.K.; Soni, N.; Doolla, S. Improvement of Small Signal Stability Margin and Transient Response in Inverter-Dominated Microgrids. *Sustain. Energy Grids Netw.* **2016**, *5*, 135–147. [CrossRef]
19. Alipoor, J.; Miura, Y.; Ise, T. Stability Assessment and Optimization Methods for Microgrid with Multiple VSG Units. *IEEE Trans. Smart Grid* **2018**, *9*, 1462–1471. [CrossRef]
20. Habibi, F.; Shafiee, Q.; Bevrani, H. Online Generalized Droop-Based Demand Response for Frequency Control in Islanded Microgrids. *Electr. Eng.* **2019**, *101*, 409–420. [CrossRef]
21. Kaushal, J.; Basak, P. Power Quality Control Based on Voltage Sag/Swell, Unbalancing, Frequency, THD and Power Factor Using Artificial Neural Network in PV Integrated AC Microgrid. *Sustain. Energy Grids Netw.* **2020**, *23*, 100365. [CrossRef]
22. Rao, S.N.V.B.; Kumar, Y.V.P.; Pradeep, D.J.; Reddy, C.P.; Flah, A.; Kraiem, H.; Al-Asad, J.F. Power Quality Improvement in Renewable-Energy-Based Microgrid Clusters Using Fuzzy Space Vector PWM Controlled Inverter. *Sustainability* **2022**, *14*, 4663. [CrossRef]
23. Kumar, Y.V.P.; Rao, S.N.V.B.; Padma, K.; Reddy, C.P.; Pradeep, D.J.; Flah, A.; Kraiem, H.; Jasiński, M.; Nikolovski, S. Fuzzy Hysteresis Current Controller for Power Quality Enhancement in Renewable Energy Integrated Clusters. *Sustainability* **2022**, *14*, 4851. [CrossRef]
24. Kumar, Y.V.P.; Bhimasingu, R. Fuzzy Logic Based Adaptive Virtual Inertia in Droop Control Operation of the Microgrid for Improved Transient Response. In Proceedings of the 2017 IEEE PES Asia-Pacific Power and Energy Engineering Conference (APPEEC), Bangalore, India, 8–10 November 2017; pp. 1–6. [CrossRef]
25. Choudhury, S.; Bhowmik, P.; Rout, P.K. Robust Dynamic Fuzzy-Based Enhanced VPD/FQB Controller for Load Sharing in Microgrid with Distributed Generators. *Electr. Eng.* **2018**, *100*, 2457–2472. [CrossRef]
26. Chen, X.; Wang, L.; Sun, H.; Chen, Y. Fuzzy Logic Based Adaptive Droop Control in Multiterminal HVDC for Wind Power Integration. *IEEE Trans. Energy Convers.* **2017**, *32*, 1200–1208. [CrossRef]
27. Li, Q.; Yang, H.; Han, Y.; Li, M.; Chen, W. A State Machine Strategy Based on Droop Control for an Energy Management System of PEMFC-Battery-Supercapacitor Hybrid Tramway. *Int. J. Hydrogen Energy* **2016**, *41*, 16148–16159. [CrossRef]
28. Kulkarni, O.V.; Doolla, S.; Fernandes, B.G. Mode Transition Control Strategy for Multiple Inverter-Based Distributed Generators Operating in Grid-Connected and Standalone Mode. *IEEE Trans. Ind. Appl.* **2017**, *53*, 5927–5939. [CrossRef]
29. Yu, K.; Ai, Q.; Wang, S.; Ni, J.; Lv, T. Analysis and Optimization of Droop Controller for Microgrid System Based on Small-Signal Dynamic Model. *IEEE Trans. Smart Grid* **2015**, *7*, 695–705. [CrossRef]
30. Wu, X.; Shen, C.; Iravani, R. Feasible Range and Optimal Value of the Virtual Impedance for Droop-Based Control of Microgrids. *IEEE Trans. Smart Grid* **2017**, *8*, 1242–1251. [CrossRef]
31. Ferreira, R.V.; Silva, S.M.; Antunes, H.M.A.; Venkataramanan, G. Dynamic Analysis of Grid-Connected Droop-Controlled Converters and Synchronverters. *J. Control Autom. Electr. Syst.* **2019**, *30*, 741–753. [CrossRef]
32. Dehkordi, N.M.; Sadati, N.; Hamzeh, M. Robust Tuning of Transient Droop Gains Based on Kharitonov's Stability Theorem in Droop-controlled Microgrids. *IET Gener. Transm. Distrib.* **2018**, *12*, 3495–3501. [CrossRef]
33. Coelho, E.A.A.; Cortizo, P.C.; Garcia, P.F.D. Small Signal Stability for Single Phase Inverter Connected to Stiff AC System. In Proceedings of the Conference Record of the 1999 IEEE Industry Applications Conference, Thirty-Forth IAS Annual Meeting (Cat. No.99CH36370), Phoenix, AZ, USA, 3–7 October 1999; Volume 4, pp. 2180–2187. [CrossRef]
34. Pogaku, N.; Prodanovic, M.; Green, T.C. Modeling, Analysis and Testing of Autonomous Operation of an Inverter-Based Microgrid. *IEEE Trans. Power Electron.* **2007**, *22*, 613–625. [CrossRef]
35. Barklund, E.; Pogaku, N.; Prodanovic, M.; Hernandez-Aramburo, C.; Green, T.C. Energy Management in Autonomous Microgrid Using Stability-Constrained Droop Control of Inverters. *IEEE Trans. Power Electron.* **2008**, *23*, 2346–2352. [CrossRef]
36. Guo, X.; Lu, Z.; Wang, B.; Sun, X.; Wang, L.; Guerrero, J.M. Dynamic Phasors-Based Modeling and Stability Analysis of Droop-Controlled Inverters for Microgrid Applications. *IEEE Trans. Smart Grid* **2014**, *5*, 2980–2987. [CrossRef]
37. Wang, R.; Sun, Q.; Ma, D.; Liu, Z. The Small-Signal Stability Analysis of the Droop-Controlled Converter in Electromagnetic Timescale. *IEEE Trans. Sustain. Energy* **2019**, *10*, 1459–1469. [CrossRef]
38. Rui, W.; Qiuye, S.; Pinjia, Z.; Yonghao, G.; Dehao, Q.; Peng, W. Reduced-Order Transfer Function Model of the Droop-Controlled Inverter via Jordan Continued-Fraction Expansion. *IEEE Trans. Energy Convers.* **2020**, *35*, 1585–1595. [CrossRef]

symmetry

MDPI

Article

Research on a Non-PLL Control Strategy for a Flexible Fault Current Limiter and Its Application in Improving the FRT Capability of Microgrids

Feng Zheng [1], Guojing Wu [1], Bo Qu [2], Xingyao Xiang [2] and Song Zheng [1,*]

1 College of Electrical Engineering and Automation, Fuzhou University, Fuzhou 350100, China
2 Shiyan Power Supply Company, State Grid Hubei Electric Power Company, Shiyan 442000, China
* Correspondence: s.zheng@fzu.edu.cn

Abstract: Aiming to achieve fault ride-through (FRT) in a microgrid under the condition of an asymmetric power grid, this paper proposes a scheme that applies the cascaded H-bridge flexible fault current limiter to realize both voltage regulation and current limiting at the point of common coupling (PCC). In order to overcome the instability of phase-locked loop (PLL) and its other problems, in the double-synchronous rotating coordinate system, the *dq*-axis components of each sequence electric quantity can be extracted without PLL by setting the rotational angular frequency of the coordinate axis to a given value. To eliminate the deviation between the set frequency and the actual frequency of the power grid, the calculation formula for the frequency correction amount is deduced, and, based on this, an iterative correction method for the set frequency of the rotating coordinate system is proposed. In this way, the set frequency can adaptively follow the actual frequency of the power grid, which ensures the effectiveness of the non-PLL scheme. Based on this, a sequence decomposition and compensation control strategy for the CHB-FFCL is presented. Finally, through the MATLAB/Simulink platform simulation, the effectiveness of the proposed strategy is verified and thereby shown to improve the FRT capability of microgrids under complex grid conditions.

Keywords: flexible fault current limiter; fault ride-through; no phase-locked loop; frequency correction; sequence decomposition and compensation control

1. Introduction

In recent years, with the development of renewable energy generation technology, microgrids using new energy sources have become effective ways to solve the conflict between energy demand and environmental protection. When a microgrid is connected to a grid, the protection device will disconnect the microgrid from the external grid if a fault occurs in the external power grid. However, when grid-connected power is high, the unplanned removal of a microgrid may lead to grid power imbalance and affect the safe and stable operation of the system; therefore, the microgrid system should have FRT capability to reduce the impact of off-grid events caused by external faults [1].

The fault current limiter (FCL) is an effective measure to limit short-circuit current in power systems. With the improvement of its power quality-control function, its application scenarios have been expanded. For microgrid systems, a FCL can limit fault current, reduce bus voltage sag and improve FRT capability.

At present, a considerable number of articles have reported research on superconducting FCLs to improve the FRT capability of microgrids [2–4]. The results show that FCLs of this sort can limit short-circuit current and improve the FRT of systems, but these FCLs have some problems, such as high dependence on cooling systems and long recovery times after operation. Relying on power electronic devices, the solid-state FCL has better controllability and fast action speed. However, limited by the device's low withstand voltage level and current withstand level, this type of FCL cannot be widely used at present [5–7].

The current work on the abovementioned FCLs to enhance the FRT capability of microgrids mainly focuses on fault current limiting, and the voltage regulation capability of FCLs has not been studied in depth. In Ref. [8], a dynamic voltage restorer (DVR) with current limiting function was proposed which limits short-circuit current by inputting current limiting impedances in the grid during faults and switches to DVR working mode after the fault is cleared. However, this scheme still has some problems that need to be solved, such as high heat dissipation requirements and the insufficient withstand voltage of devices. In ref. [9], the cascaded H-bridge flexible fault current limiter (CHB-FFCL) was proposed, which is connected to the grid through capacitive coupling instead of a transformer, thereby reducing the device size of the FCL. In addition, the cascaded multi-level structure adopted has the advantages of large capacity, good output characteristics, flexible controllability, etc., and reduces the withstand voltage and current of a single power electronic device. Therefore, applying the CHB-FFCL to improve the FRT of a microgrid can overcome many practical application problems that beset traditional FCLs.

Since the negative sequence components of unbalanced grid voltage will have a serious impact on the quality of PCC voltage, it is of great importance to study the recovery control strategy for PCC voltage under the condition of unbalance [10]. In ref. [11], a PCC voltage control scheme based on a sliding mode structure was proposed which can reduce the impact of grid voltage disturbance on PCC voltage. However, this scheme did not consider the influence of grid unbalanced voltage.

Considering the three-phase independent structure of the CHB-FFCL and the flexible output characteristics of the multi-level inverter structure, the CHB-FFCL has advantages in dealing with unbalanced grid faults. Usually, the realization of its control requires the obtainment of amplitude and phase angle information on the electric quantities through PLL. However, the feedback control-based PLL technique suffers from weak anti-interference capability and poor dynamic performance [12], especially in the case of an unbalanced power grid. For this reason, many scholars have proposed improved schemes for PLL. In ref. [13], the first-order low-pass filter (LPF) was used in the control structure of the traditional PLL to eliminate the impact of negative sequence or harmonic voltage components, but the delay and phase offset of the LPF affected the identification speed of the PLL. Therefore, in ref. [14], a notch filter was used to replace the first-order LPF to improve the stability of PLL. However, if the grid voltage contains multiple disturbance components, it is necessary to connect multiple notch filters with different frequencies in series, which increases the difficulty of PLL implementation and computational burden. In ref. [15], a positive-sequence fundamental frequency voltage component extraction method based on a differential equation was proposed which ensures that PLL is free from the influence of non-ideal voltage components. However, due to the differential calculation involved, the anti-interference performance of the system is reduced. The abovementioned improvement scheme increases the order of PLL and the poles of the original closed-loop transfer function, which may reduce the stability margin for PLL and cause instability in PLL. In addition, impedance modeling analysis shows that the introduction of PLL increases the coupling between system parameters in a grid-connected inverter system. Moreover, the increase in proportional or integral parameters in PLL will lead to large amplitude and phase fluctuations in the fundamental frequencies of the inverter impedance characteristics, which will lead to potential system stability problems [16,17]. Therefore, it is necessary to study the inverter control scheme without PLL to overcome the abovementioned problems caused by PLL. In refs. [18,19], the control scheme for an inverter without PLL based on preset rotational frequency was studied which improved the operating performance of a DFIG in an unbalanced grid. However, this strategy cannot adapt to grid frequency offset.

To address these problems, this paper analyzes the principle of achieving FRT with the CHB-FFCL as the research object. The *dq*-axis components are extracted without PLL in a double synchronous rotating coordinate system by setting the rotation angular frequency of the coordinate axis. Aiming to solve the problem of the set frequency deviating from the frequency of the grid, a calculation formula for frequency correction is derived. Through iterative correction, the set frequency can be corrected without feedback control. A sequence decomposition and compensation control strategy for the CHB-FFCL is presented. Finally, through simulation analysis, the effectiveness of the above strategy is verified.

2. The Topological Structure and Working Principle of the CHB-FFCL

The topology of the CHB-FFCL is shown in Figure 1, where U_S is the power supply on the distribution grid side and U_{MG} is the power source on the microgrid side. It adopts a three-phase independent inverter structure with independent control of each phase, aiming to reduce the coupling between the three phases and more flexibly cope with three-phase unbalanced faults. In the figure, $C_{ai/bi/ci}$ and $L_{ai/bi/ci}$ (i = 1...m) are the filter partial-voltage capacitors and filter inductors of CHB-FFCL, n and m represent the number of cascades and the number of partial-voltage capacitors (also called the number of multiples) of each phase of the CHB-FFCL, and U_{dc} is the DC side voltage of the H-bridge inverter, which is supplied by the distributed generation (DG). By connecting m partial-voltage capacitors in series and connecting n H-bridge units in cascade, the voltage drop added to an individual power electronic device is greatly reduced, and it is also convenient to expand the capacity of the CHB-FFCL.

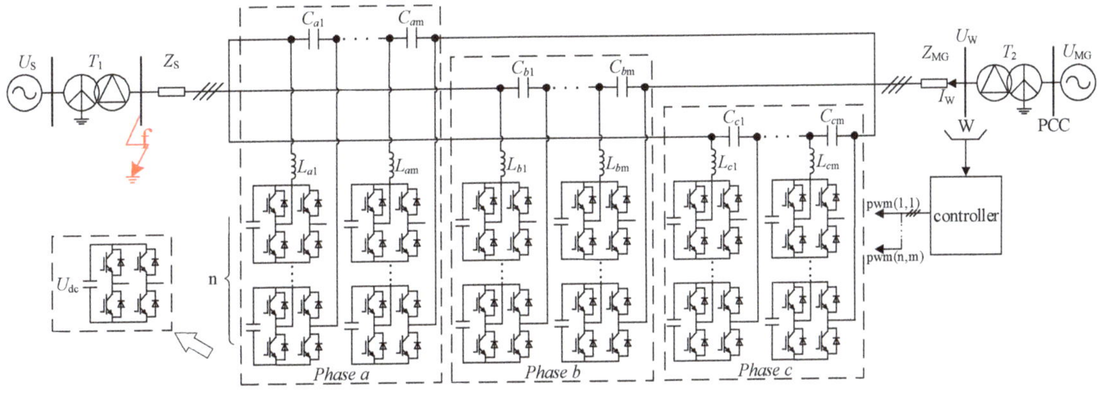

Figure 1. Topological structure of the CHB-FFCL.

Based on this topology, the operating principle of the CHB-FFCL for achieving both fault voltage recovery and fault current limitation is discussed.

Firstly, the regulation effect of the CHB-FFCL on PCC voltage when a fault occurs outside the microgrid is discussed.

As shown in Figure 2a, when a fault occurs outside the microgrid, the CHB-FFCL can be equivalent to a voltage-controlled voltage source (VCVS), which is shown in Figure 2b. The output voltage of the VCVS is U_C, and for the fault point f the direction of U_C is opposite to the direction of the microgrid supply voltage U_{MG}. The voltage of the bus W can be expressed as:

$$\dot{U}_W = \dot{U}_f + \dot{U}_C \quad (1)$$

where $\dot{U}_f$ is the voltage of bus W in case of a fault without the CHB-FFCL, and its value is equal to the voltage of short-circuit point f when the line impedance between the short-circuit point f and bus W is not considered. It can be seen from Equation (1) that by controlling the output voltage $\dot{U}_C$ of the CHB-FFCL, the bus W voltage can be compensated, thereby reducing the voltage sag of the PCC.

Figure 2. Schematic diagram of the CHB-FFCL voltage regulation principle (single-phase). (**a**) Grid connection diagram of the CHB-FFCL. (**b**) Equivalent circuit diagram of when a fault occurs.

Next, this paper analyzes the current limiting principle of the CHB-FFCL when a fault occurs. The system in Figure 1 can be equivalent to a complex power system shown in Figure 3, where G1 is the distribution network power endpoint and G2 is the microgrid power endpoint.

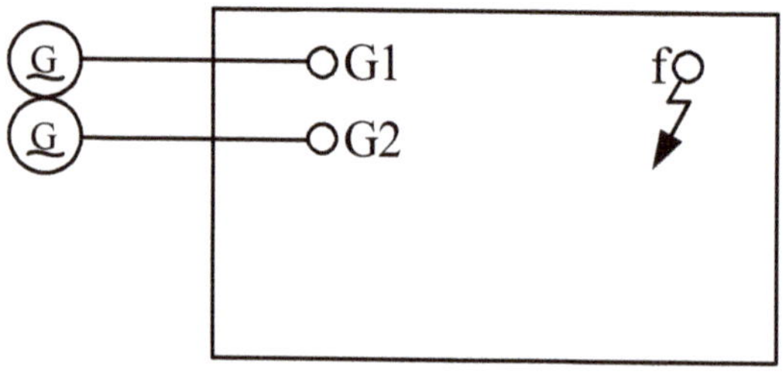

Figure 3. Complex power system diagram.

According to the symmetrical component method, the short-circuit fault current of this power system is analyzed using the general compound sequence network. Taking the two-phase (phase b and phase c) short-circuit fault at point f as an example and considering the boundary conditions, the short-circuit current expression of fault phase without the CHB-FFCL is obtained as follows:

$$\dot{I}_{fb} = -\dot{I}_{fc} = a^2\dot{I}_{f(1)} + a\dot{I}_{f(2)} = (a^2 - a)\frac{\dot{U}_{f|0|}}{z_{\Sigma(1)} + z_{\Sigma(2)}} = -j\sqrt{3}\frac{\dot{U}_{f|0|}}{z_{\Sigma(1)} + z_{\Sigma(2)}} \quad (2)$$

where

$$\begin{cases} a = e^{j120^\circ} = -\frac{1}{2} + j\frac{\sqrt{3}}{2} \\ a^2 = e^{j240^\circ} = -\frac{1}{2} - j\frac{\sqrt{3}}{2} \end{cases} \quad (3)$$

and $\dot{I}_{f(1)}$ and $\dot{I}_{f(2)}$ represent the positive and negative sequence currents flowing into the fault point, respectively; $\dot{U}_{f|0|}$ is the open-circuit voltage at point f when no fault occurs; and $z_{\Sigma(1)}$ and $z_{\Sigma(2)}$ are the positive sequence and negative sequence equivalent impedances viewed from fault point f, respectively.

The CHB-FFCL is switched on when a fault occurs, which is equivalent to connecting a voltage source whose value is k ($0 \leq k \leq 1$) $U_{f|0|}$ and whose direction is opposite to the power supply voltage in series, as shown in Figure 4. In this case, the open-circuit voltage at fault point f is reduced to 1-k times the original value, and the short-circuit current of the corresponding fault phase is reduced to:

$$\dot{I}_{fb} = -\dot{I}_{fc} = a^2\dot{I}_{f(1)} + a\dot{I}_{f(2)} = (a^2 - a)\frac{(1-k)\dot{U}_{f|0|}}{z_{\Sigma(1)} + z_{\Sigma(2)}} = -j\sqrt{3}\frac{(1-k)\dot{U}_{f|0|}}{z_{\Sigma(1)} + z_{\Sigma(2)}} \quad (4)$$

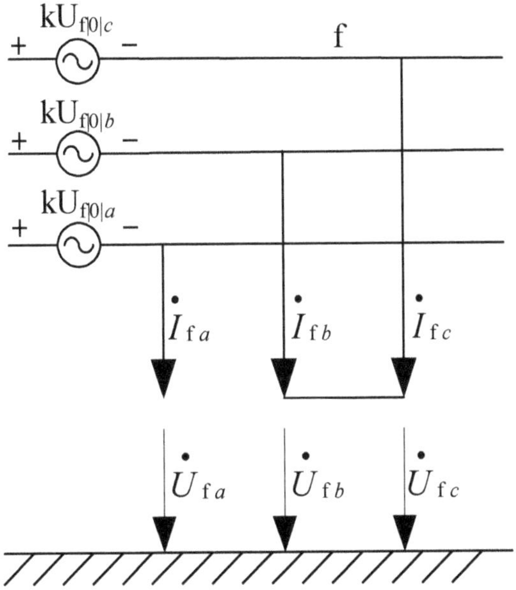

Figure 4. Currents and voltages at the fault point in the case of a two-phase short-circuit fault with the CHB-FFCL.

Similarly, when other types of short-circuit fault occur on the distribution network side, combined with ref. [20], the corresponding fault current expression after the CHB-FFCL input can be obtained according to the above analysis method. It can be seen from Figure 2 that the fault current is the current flowing through the PCC during the fault.

Comprehensive analysis shows that the structure of the CHB-FFCL ensures the control of output voltage, which helps to realize the dual function of voltage regulation and current limitation in case of a fault, thus improving the FRT capability of the microgrid.

3. The Non-PLL Control Principle of the CHB-FFCL in a Decoupled Double Synchronous Rotating Coordinate System

When designing the controller, compared with the scheme for controlling the AC signal in three-phase and two-phase stationary coordinates, converting the AC signal to the DC quantity in dq rotating coordinates for PI control can better realize the zero steady-state error characteristics of the control [21,22]. Ignoring coupling disturbance, when the load is only powered by CHB-FFCL, its d-axis voltage and current double closed loop control block diagram is shown in Figure 5. In the figure, G_v(s) and G_i(s) are PI regulators of voltage loop and current loop, respectively; the electric quantities with the subscript "refd" represent the corresponding d-axis reference values; the subscript "od" indicates the value of the load; and K_{pwm} is the inverter gain.

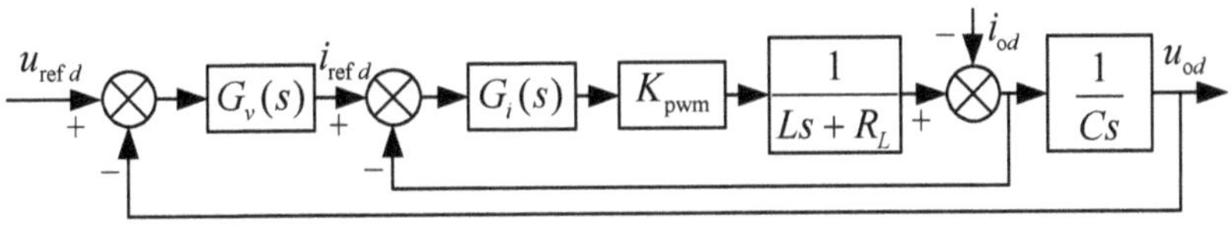

Figure 5. Block diagram of double closed-loop control for d-axis voltage and current.

The q-axis electrical quantities have the same control structure and each order's harmonic quantities can be controlled through the control loop, and the controlling vectors are synthesized and outputted. Therefore, obtaining the dq-axis components of the electric quantity for each order is a prerequisite for realizing the above CHB-FFCL control strategy.

In this paper, the current is taken as the object of analysis, and the voltage component is subjected to the same analytical process.

Referring to the principle of the decoupled double synchronous reference frame phase-locked loop (DDSRF-PLL), positive sequence and negative sequence rotation coordinate systems are added to the static abc three-phase coordinate system, as shown in Figure 6. When the synchronization condition is reached, Equation (5) is established. After that, Equation (6) is transformed into Equation (7), so the required positive and negative sequence dq-axis DC electrical components can be extracted through the decoupling network and low-pass filter [23]. In addition, the decoupling network of the DDSRF-PLL cancels out the double frequency oscillations at 2ω in $i^*_{q^{+1}\_DC}$ (feedback control estimates the current of the DDSRF-PLL); therefore, there is no need to reduce the control loop bandwidth, and the real amplitudes of the positive sequence and negative sequence components can be exactly detected. The amplitudes of positive sequence and negative sequence components can be obtained from Equation (8).

$$\omega' = \omega i \tag{5}$$

$$i_{abc} = \begin{bmatrix} i_a \\ i_b \\ i_c \end{bmatrix} = i_{abc}^{+1} + i_{abc}^{-1} = I^{+1}\begin{bmatrix} \cos(\omega t + \phi^{+1}) \\ \cos(\omega t - \frac{2\pi}{3} + \phi^{+1}) \\ \cos(\omega t + \frac{2\pi}{3} + \phi^{+1}) \end{bmatrix} + I^{-1}\begin{bmatrix} \cos(-\omega t + \phi^{-1}) \\ \cos(-\omega t - \frac{2\pi}{3} + \phi^{-1}) \\ \cos(-\omega t + \frac{2\pi}{3} + \phi^{-1}) \end{bmatrix} \tag{6}$$

$$i_{dq^x} = \begin{bmatrix} i_{d^x} \\ i_{q^x} \end{bmatrix} = \underbrace{I^x \begin{bmatrix} \cos(\phi^x) \\ \sin(\phi^x) \end{bmatrix}}_{\text{DC}} + \underbrace{I^{-x}\begin{bmatrix} \cos(-2x\omega t + \phi^{-x}) \\ \sin(-2x\omega t + \phi^{-x}) \end{bmatrix}}_{\text{AC}}, \quad (x = \pm 1) \tag{7}$$

$$I^x = \sqrt{(I^x\cos(\phi^x))^2 + (I^x\sin(\phi^x))^2} = \sqrt{(i_{d^x\_DC})^2 + (i_{q^x\_DC})^2}, \quad (x = \pm 1) \tag{8}$$

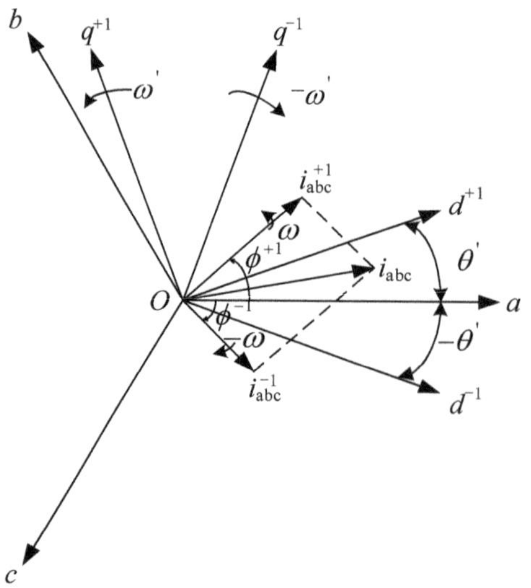

Figure 6. Schematic diagram of asymmetric current vector decomposition and the double synchronous reference coordinate system.

DDSRF-PLL achieves the above synchronization conditions through PLL, but this method has problems, such as frequency oscillation and poor anti-interference. According to the national standard, the power frequency of the power grid is a stable quantity.

Therefore, the synchronization condition can be guaranteed by directly setting the rotation angular frequency of the rotating coordinate system to be equal to the angular frequency corresponding to the power frequency, so that the *dq*-axis DC electrical components can be extracted without PLL.

4. Frequency Adaptive Correction Strategy of The Proposed Non-PLL Method

The realization of the abovementioned non-PLL method and the corresponding control strategy presupposes that the given angular frequency ω' of the rotating coordinate system is the same as the angular frequency ω of the fundamental wave of the power grid, which requires the power frequency of the power grid to remain constant. However, in national regulations, the actual frequency of the power grid is not constant but fluctuates within a certain range. Taking the Chinese standard as an example, the nominal frequency of the power grid is 50 Hz, the frequency offset of high-capacity power grids cannot exceed ± 0.2 Hz and the frequency offset of low-capacity power grids can be relaxed to ± 0.5 Hz. Therefore, there is an error between the given ω' and the fundamental angular frequency of the power grid, and an adaptive correction of the frequency is required.

It is assumed that there is a deviation $\Delta\omega$ between the rotation angular frequency of the given positive sequence fundamental wave rotating coordinate system and the actual power grid fundamental wave rotation angular frequency ω, as shown in Equation (9):

$$\omega' = \omega - \Delta\omega \tag{9}$$

then

$$\theta' = (\omega - \Delta\omega)t \tag{10}$$

Taking the positive sequence fundamental current into consideration, Equation (7) is converted into Equation (11), and, similarly, the same equation structure can be deduced when considering the negative sequence fundamental current.

$$i_{dq^{+1}} = \begin{bmatrix} i_{d^{+1}} \\ i_{q^{+1}} \end{bmatrix} = \underbrace{I^{+1}\begin{bmatrix} \cos(\Delta\omega t + \phi^{+1}) \\ \sin(\Delta\omega t + \phi^{+1}) \end{bmatrix}}_{AC1} + \underbrace{I^{-1}\begin{bmatrix} \cos(-2\omega t + \Delta\omega t + \phi^{-1}) \\ \sin(-2\omega t + \Delta\omega t + \phi^{-1}) \end{bmatrix}}_{AC2} \tag{11}$$

It can be seen from Equation (11) that in the case of deviation the *dq*-axis component is composed of two AC oscillation quantities. According to the Chinese national standard, the value of $\Delta\omega$ is within $[-\pi, \pi]$. Obviously, ω is much larger than $\Delta\omega$, and therefore the first AC quantity "AC1" in Equation (11) can be retained after passing through the low-pass filter. As a result, the decoupling network [23] regards the first AC quantity "AC1" in Equation (11) as the DC quantity obtained after filtering.

$$\begin{cases} i^*_{d^{+1}\_DC} = i_{d^{+1}} - i_{d^{-1}\_DC} \cdot \cos(2\theta') - i_{q^{-1}\_DC} \cdot \sin(2\theta') \\ i^*_{q^{+1}\_DC} = i_{q^{+1}} - i_{q^{-1}\_DC} \cdot \cos(2\theta') + i_{d^{-1}\_DC} \cdot \sin(2\theta') \end{cases} \tag{12}$$

Through the decoupling network, the decoupling expression of the positive sequence current is obtained, as shown in Equation (12). Inserting Equation (7), Equation (10) and the first AC quantity of Equation (11) into Equation (12), the estimated value of the *dq*-axis DC quantity can be obtained, as shown in Equation (13).

$$\begin{cases} i^*_{d^{+1}\_DC} = I^{+1}\cos(\Delta\omega t + \phi^{+1}) \\ i^*_{q^{+1}\_DC} = I^{+1}\sin(\Delta\omega t + \phi^{+1}) \end{cases} \tag{13}$$

From Equation (13), it is not difficult to see that, after decoupling, the obtained *dq*-axis DC estimator is the first AC component of Equation (11), which indicates that the decoupling network is still applicable in the case of frequency deviation. However, the *dq*-axis components obtained in this case are AC components, and it is difficult to obtain the target electric quantity in the control of CHB-FFCL. Therefore, it is necessary to correct the given frequency.

According to the analysis of Equation (13), in the case of frequency deviation, the frequency deviation amount can be calculated by the *dq*-axis DC estimated value obtained without PLL, and the calculation formula is shown in Equation (14).

$$\Delta f = \frac{\Delta\omega}{2\pi} = \frac{1}{2\pi}\frac{d(\Delta\omega t + \phi^{+1})}{dt} = \frac{1}{2\pi}d\left(\arctan(\frac{i^*_{q\_\mathrm{DC}}^{+1}}{i^*_{d\_\mathrm{DC}}^{+1}})\right)/dt \tag{14}$$

The deviation of the given frequency between the actual power grid frequency can be determined by Equation (14), and so the adaptive correction of frequency can be realized.

It is worth noting that in the case of no feedback control on the given frequency, directly adding the correction amount calculated each time to the initial given frequency will lead to a large amplitude oscillation in the set frequency. Therefore, during frequency correction, the correction amount should only be superimposed on the initial given frequency the first time and then on the latest corrected frequency. The correction process is an iterative process, and the expression is shown as follows:

$$\begin{cases} f_{\mathrm{set}}^{(0)} = f^{(0)} \\ f_{\mathrm{set}}^{(1)} = f_{\mathrm{set}}^{(0)} + \Delta f^{(0)} \\ f_{\mathrm{set}}^{(2)} = f_{\mathrm{set}}^{(1)} + \Delta f^{(1)} \\ \vdots \\ f_{\mathrm{set}}^{(n)} = f_{\mathrm{set}}^{(n-1)} + \Delta f^{(n-1)} \end{cases} \tag{15}$$

where $f_{\mathrm{set}}^{(n)}$ is the set frequency after n times of correction; $f^{(0)}$ is the initial given frequency, which is usually taken as the power frequency of the power grid; and $\Delta f^{(n)}$ is the amount of deviation correction between $f_{\mathrm{set}}^{(n)}$ and the actual grid frequency.

Based on the above analysis, combined with the inverter dual-loop control strategy [24], the sequence decomposition and compensation control strategy of the CHB-FFCL without PLL is obtained, as shown in Figure 7.

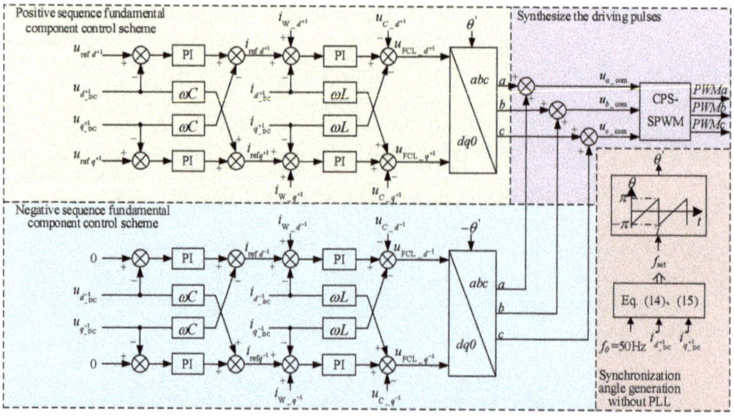

Figure 7. The overall control block diagram of the CHB-FFCL designed to achieve FRT.

Combining the control block diagram shown in Figure 5, ignoring the inductor's resistance, the open-loop transfer function of the control system can be obtained, as shown in Equation (16).

$$G_{\text{open}}(s) = \frac{G_v(s)G_i(s)K_{\text{pwm}}}{LCs^2 + G_i(s)K_{\text{pwm}}Cs + 1} \tag{16}$$

where

$$K_{\text{pwm}} = \frac{U_{\text{dc}}}{U_c} \tag{17}$$

$$G_v(s) = K_{vp} + \frac{2K_{vi}\omega_c s}{s^2 + 2\omega_c s + \omega_0^2} \tag{18}$$

$$G_i(s) = K_{ip} \tag{19}$$

and U_c is the amplitude of the triangular carrier wave, the value here being 1; ω_c is the bandwidth of the controller; and ω_0 is the resonant frequency.

Through this control strategy, the flexible control of the output voltage of the CHB-FFCL is realized, thereby improving the FRT capability of the microgrid in the case of asymmetric faults.

5. Simulation and Discussion

5.1. Effectiveness Analysis of the Proposed Non-PLL Strategy

The three-phase asymmetric fault was set in the simulated power grid, and the resultant current waveform is shown in Figure 8. In the figure, when there was no fault, the three-phase current amplitude was 10 A, and the initial phase angle was 0°. During 0.3–0.5 s, the positive sequence fundamental current and negative sequence fundamental current were injected into the power grid. The amplitude and initial phase angle of the former were 20 A and 45°, respectively. The amplitude and initial phase angle of the latter were 10 A and −15°, respectively. The three-phase asymmetric fault in the power grid could be simulated. DDSRF-PLL and the proposed non-PLL strategy were respectively used to extract the dq-axis components of the positive and negative sequence fundamental currents of the power grid.

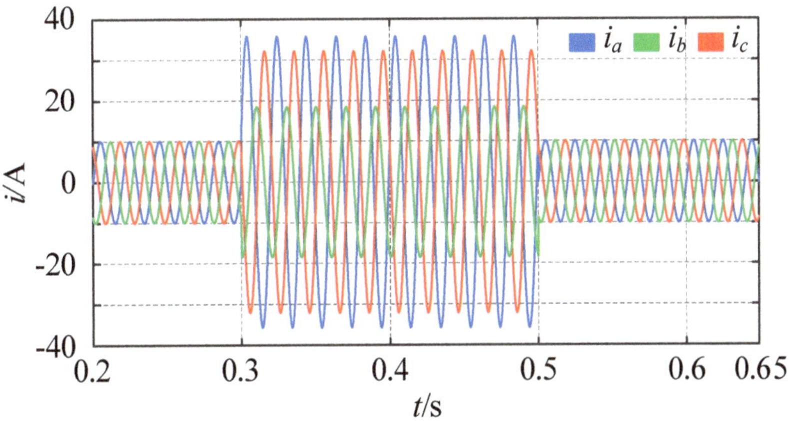

Figure 8. Simulated three−phase current waveform in the case of an asymmetric fault in the power grid.

Since the proposed non-PLL scheme only needs to ensure that the angular frequency of the rotating coordinate system is equal to the angular frequency of the positive sequence fundamental electric quantity, it does not need to ensure that the initial phase angle is equal to it, which is different from the DDSRF-PLL method. Therefore, for the same asymmetric current vector, when the output values of the DDSRF-PLL and the proposed non-PLL scheme reach a steady state, their phase angles for the generated dual synchronous reference coordinate axis relative to the stationary coordinate axis of phase a are different, so the DC electrical quantity components extracted by the two methods are also different. However, for the same asymmetric electric quantity, the amplitudes of the positive sequence and negative sequence fundamental electric quantities obtained from Equation (8) are equal. Therefore, this study first compared the extracted amplitudes to verify the effectiveness of the proposed non-PLL scheme and set the DDSRF-PLL bandwidth to 38.4 Hz. The simulation results are shown in Figure 9.

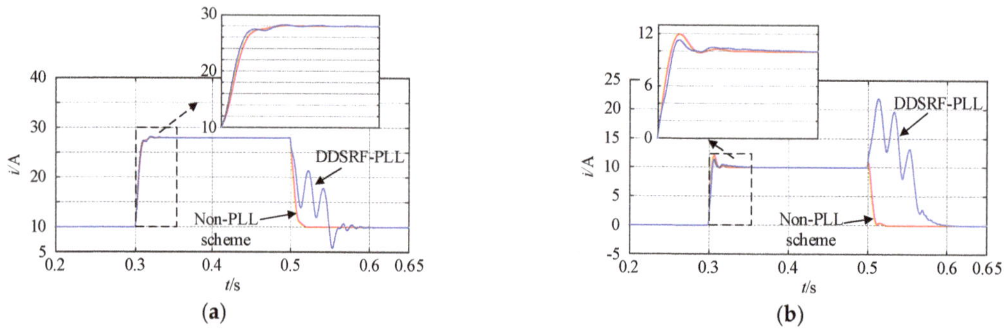

Figure 9. Amplitude separation results for positive and negative sequence fundamental components of current. (**a**) Positive sequence fundamental component. (**b**) Negative sequence fundamental component.

It can be seen that the proposed non-PLL scheme and DDSRF-PLL could quickly extract the amplitudes of positive and negative sequence fundamental currents when a fault occurred (0.3 s), and the extracted results conformed to the parameters of the set fault (positive sequence fundamental current vector: $i^{+1} = 10\angle 0° + 20\angle 45° = 28\angle 30.36°$ A; negative sequence fundamental current vector: $i^{-1} = 10\angle -15°$). It can be seen from the detailed diagram that the proposed non-PLL scheme could reach the new stable state more quickly and smoothly than DDSRF-PLL. At the end of the fault, DDSRF-PLL was prone to oscillate for a long time, and the proposed non-PLL scheme still had good stability performance.

Then, the performances of these two methods were compared to extract each dq-axis DC component. The dq-axis DC components of each sequence current mapped in the double synchronous reference coordinate system were extracted, as shown in Figure 10. In the figure, Δt_1 and Δt_2 are, respectively, the transition times required for the non-PLL scheme and DDSRF-PLL to reach a new steady state when a fault occurs. Comparing Δt_1 and Δt_2, the dynamic characteristics of the proposed scheme can be measured.

It can be seen from Figure 10a–d that the tracking results of other axes all satisfied the relationship $\Delta t_1 < \Delta t_2$, except that the q^+-axis used for DDSRF-PLL feedback control satisfied the relationship $\Delta t_1 > \Delta t_2$ and the $|\Delta t_1 - \Delta t_2|$ value on the other three axes was obviously greater than that on the q^+-axis. This shows that in case of a fault, given the premise that the control parameters of the DDSRF-PLL are appropriate (shown as the short transition time of the q^+-axis), the DDSRF-PLL has a speed advantage only when detecting the DC components on the controlled axis and that the advantage is relatively insignificant, while the speed of reaching the steady state value in the other three axes is significantly slower than with the proposed non-PLL scheme. In addition, at the end of the fault, the DDSRF-PLL had large oscillations in the transition phase of the four axes, and the transition time was long. As a comparison, the proposed non-PLL scheme could still maintain a smooth and rapid transition with excellent dynamic characteristics.

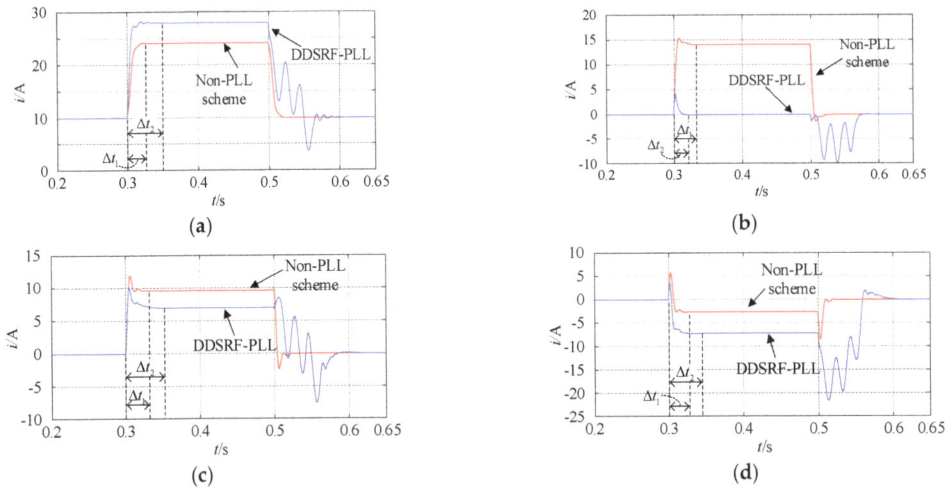

Figure 10. Separation results for the positive and negative sequence components of current in the double synchronous reference coordinate system. (**a**) $d^{+1}-$axis. (**b**) $q^{+1}-$axis. (**c**) $d^{-1}-$axis. (**d**) $q^{-1}-$axis.

When phase jumps occur in electrical quantities, PLL based on feedback control is prone to instability, while the non-PLL scheme can better deal with phase jumps due to the fact that a feedback control loop is not required. In this study, the phase jumps of current in the three-phase unbalanced system were simulated and analyzed. The amplitudes of positive sequence and negative sequence fundamental currents were extracted through the non-PLL scheme and DDSRF-PLL, respectively, and the results are shown in Figure 11. The amplitude of the positive sequence fundamental current was 10 A, and the initial phase angle was 0°. Setting the negative sequence unbalance factor to 0%, 30%, 60% and 90%, respectively, we obtained simulation results with the corresponding labels "0", "1", "2" and "3". At 0.3 s, the phase of the positive sequence fundamental current jumped −60°; the jump process can be clearly seen in Figure 11a. In the figure, trajectories 1 and 2 are current vector trajectories before and after phase jumps respectively. It can be seen from the extraction results for positive sequence and negative sequence fundamental amplitudes in Figure 11b,c that the impact of phase jumps on DDSRF-PLL was significantly greater than for the non-PLL scheme. With the increase in grid unbalance, the impact on DDSRF-PLL also increases. When the negative sequence unbalance factor is too large (such as 90%), DDSRF-PLL cannot recover to a stable state. As a comparison, the non-PLL scheme was less affected and could quickly self-recover to the stable operation state. With the increase in grid unbalance, the impact on the extraction of positive sequence components did not increase significantly. This scheme has good stability.

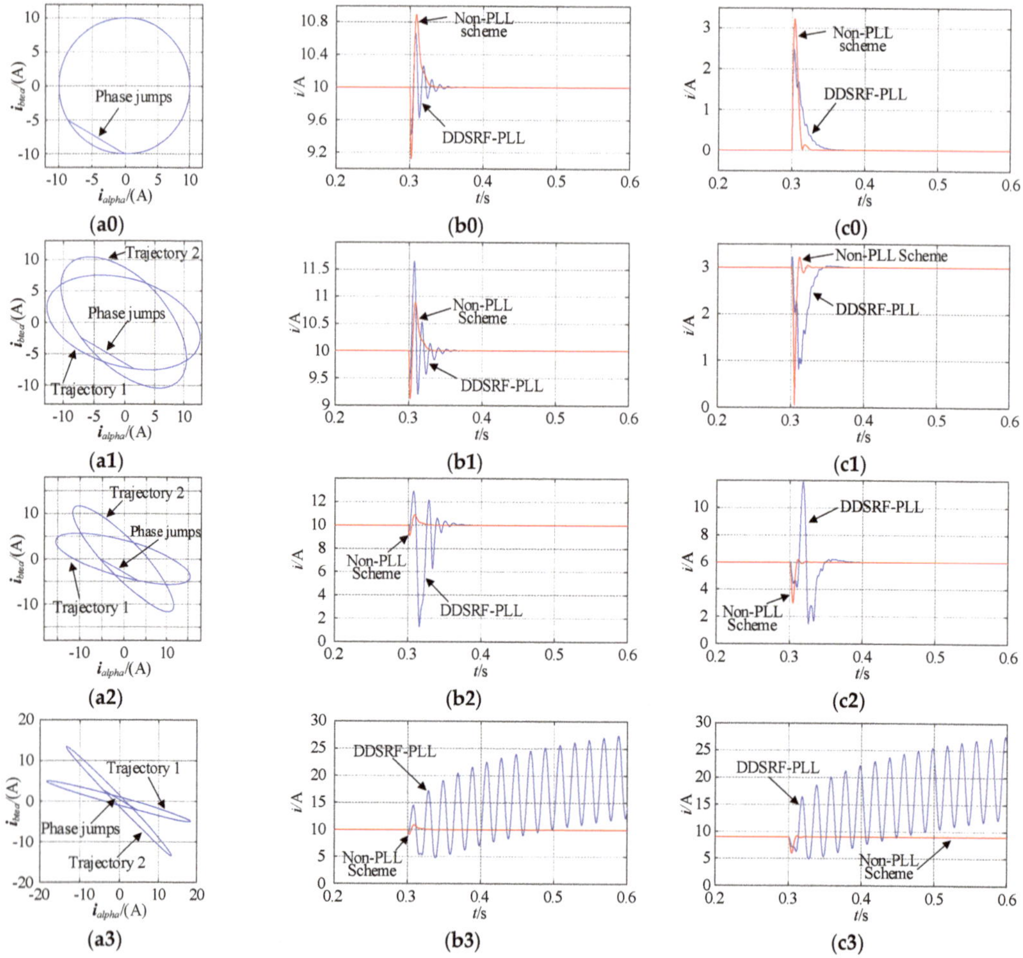

Figure 11. Comparison of simulation results for phase jump faults occurring in the unbalanced power grid. (**a0**–**a3**) Trajectories of the current vector. (**b0**–**b3**) Comparison of amplitude extraction results for the positive sequence fundamental current. (**c0**–**c3**) Comparison of amplitude extraction results for the negative sequence fundamental current.

5.2. Effectiveness Verification of the Proposed Frequency Adaptive Correction Method

To verify the effectiveness of the proposed self-adaptable frequency rectifying method, it was assumed that a complex three-phase fault occurred in the power grid; its voltage waveform is shown in Figure 12. Within 0–2 s, the power grid operated normally without fault. Starting from 2 s, a negative sequence fundamental voltage component with an amplitude of 0.6 p.u. and an initial phase angle of $-30°$ was superimposed on the normal voltage. The fundamental frequency of the power grid decreased to 49.5 Hz in 4 s and increased to 50.5 Hz in 8 s. The *dq*-axis components of each sequence voltage of the three-phase voltage signal were extracted and the simulation results are shown in Figure 13.

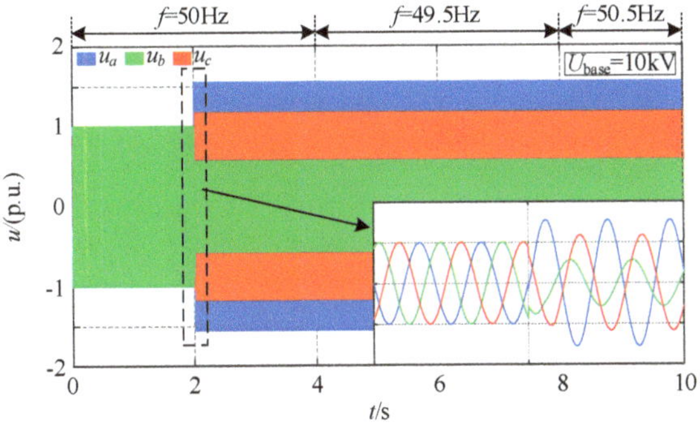

Figure 12. Voltage waveforms when a complex three−phase fault occurred.

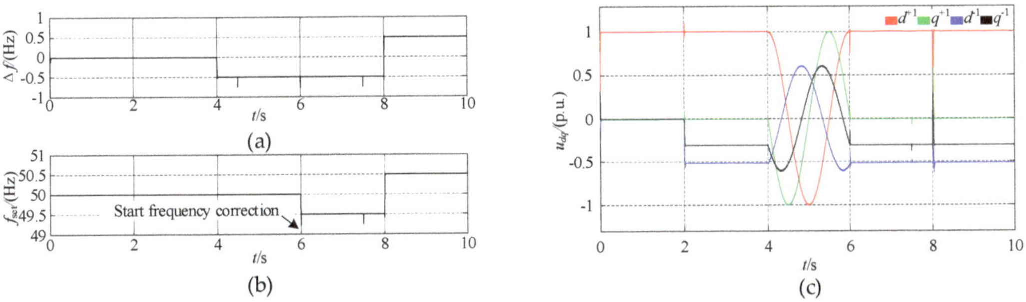

Figure 13. Simulation result diagram of the effect of frequency correction. (**a**) Frequency correction value (compared to the initial set frequency). (**b**) The set frequency of the positive sequence fundamental wave rotating coordinate system. (**c**) Extraction results for the *dq*-axis components of each sequence voltage.

The deviation between the initially given frequency of the fundamental wave rotating coordinate system and the actual frequency of the power grid calculated by applying Equation (14) is shown in Figure 13a. It can be seen that the calculated correction value can accurately reflect the deviation. Using Equation (15), the calculated deviation value Δf was used to iteratively correct the set frequency, the correction of which started at 6 s. The set frequency obtained is shown in Figure 13b. It can be seen that the set frequency could quickly track the actual grid frequency after the correction. Based on the set frequency, the *dq*-axis components of each sequence voltage extracted by the non-PLL method are shown in Figure 13c. It is easy to see that when the initial set frequency is consistent with the grid frequency (0–4 s), the *dq*-axis components of each sequence voltage can be accurately and quickly extracted through the proposed non-PLL method. However, when the two frequencies were inconsistent (4–6 s), the *dq*-axis components extracted were AC electrical quantities, and it can be seen from the figure that the waveforms of these AC components conformed to Equation (13), which verifies the inference made in Section 4 that the decoupling network is still effective in the case of frequency deviation. Obviously, the *dq*-axis components extracted after frequency correction, which started at 6 s, are DC electrical quantities equal to the value at synchronization, which verifies the correctness of the frequency correction strategy. At the moment of 8 s, the actual frequency of the power grid suddenly changed from 49.5 Hz to 50.5 Hz (which is still within the change range of the Chinese national standard), and the set frequency was also rapidly adjusted to 50.5 Hz, which verifies the adaptive ability of the proposed frequency correction scheme.

5.3. Verification of the CHB-FFCL's Ability to Improve FRT in Microgrids

In order to verify the ability of the CHB-FFCL to improve FRT in the microgrid grid-connected system, a power system, shown in Figure 1, was built for simulation analysis. The system used three-phase, three-wire connection for power supply. In order to simplify the research, the microgrid only used energy storage for power supply. The main parameters of the simulation model are shown in Table 1. In addition, the sampling frequency of the simulation system was set to 100 kHz. The asymmetric short-circuit fault was set on the distribution network side for analysis.

Table 1. Main parameters of the system.

Equipment	Parameter	Value
CHB-FFCL	Filter capacitor/μF	0.5
	Filter inductor/mH	8
	Number of cascades	5
	Number of multiples	3
	DC voltage of H-bridge/V	750
	Modulator frequency/kHz	1.8
	Output power/MVA	0.125
	Voltage control loop *Kvp*, *Kvi*	1.5, 10
	Current control loop *Kip*	10
Distribution network system	Frequency/Hz	50
	Line voltage/kV	10.5
	Capacity of main transformer/MVA	10
Microgrid system	Frequency/Hz	50
	Line voltage/kV	10.5
	Capacity of main transformer/MVA	0.8

Based on the above parameters, the stability of the control system was analyzed first. According to the system's open-loop transfer function (Equation (16)), the Nyquist diagram shown in Figure 14 was obtained. Substituting the parameters, it could be deduced that the open-loop transfer function has no pole in the right half s plane, that is, the pole number $p = 0$. It can be seen from the enlarged view of Figure 14b that the trajectory does not surround the point $(-1, j0)$, that is, $N = 0$. According to the Nyquist stability criterion, $Z=P-N=0$, the closed-loop control system is stable and the set parameters were reasonable.

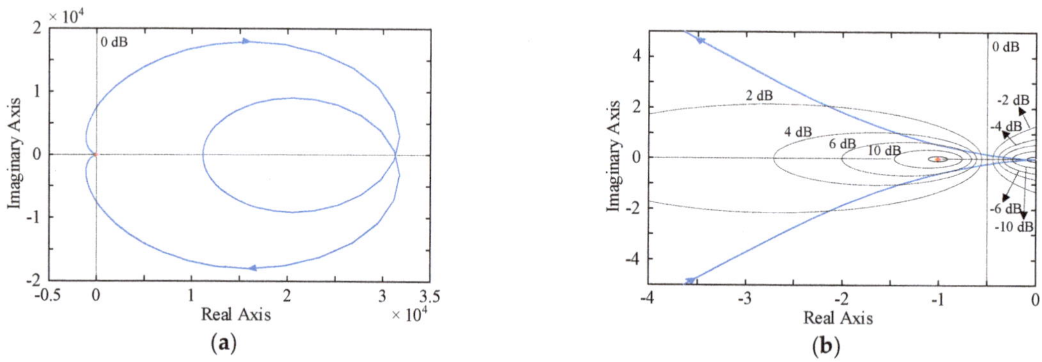

Figure 14. Nyquist diagram of the control system. (**a**) Full view. (**b**) Zoomed in (−4, 0).

In this paper, a *bc* two-phase short-circuit was selected for the analysis of an asymmetric fault. The simulation time was set to 0.6 s. The fault occurred at 0.3 s and disappeared at 0.5 s. The simulation waveform when the CHB-FFCL was not put into operation is shown in Figure 15. It can be seen that the fault voltage and current had obvious three-phase unbalance. As transformer T2 in Figure 1 is connected by Y-D form, the *b*-phase voltage of the PCC rises, making the voltage and current waveform different from the normal *bc*-phase short-circuit waveform. The output power of the microgrid connected to the grid increased, and it was difficult for the output power to maintain the PQ control target value. To protect the local load and ensure the safety of the grid-connected inverter, the microgrid system may be switched to island operation.

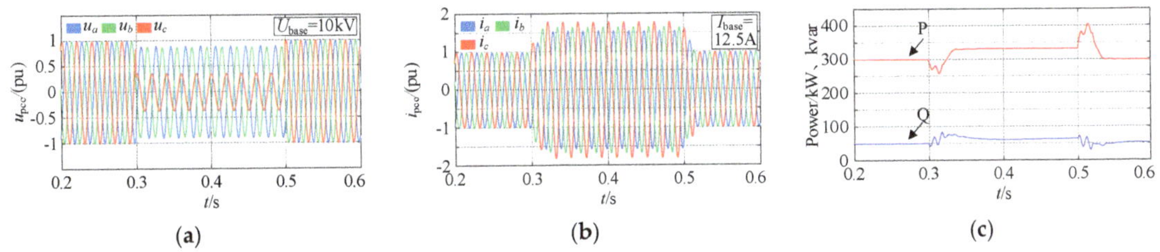

Figure 15. Simulation waveforms for when a two−phase short−circuit fault occurs on the distribtion network side. (**a**) The voltage of the PCC. (**b**) The current injected into the PCC. (**c**) Grid−connected output power of the microgrid.

The application of the CHB-FFCL for voltage regulation and fault current limiting can improve the FRT capability of the microgrid in the case of an external fault. In order to compare the DDSRF-PLL with the proposed non-PLL scheme, this simulation first applied the DDSRF-PLL to provide DC component signals for the sequence decomposition and compensation control of the CHB-FFCL. The CHB-FFCL was switched on at the moment when a fault occurred (0.3 s) and switched off at the end of the fault (0.5 s). The simulation waveform on the microgrid side is shown in Figure 16.

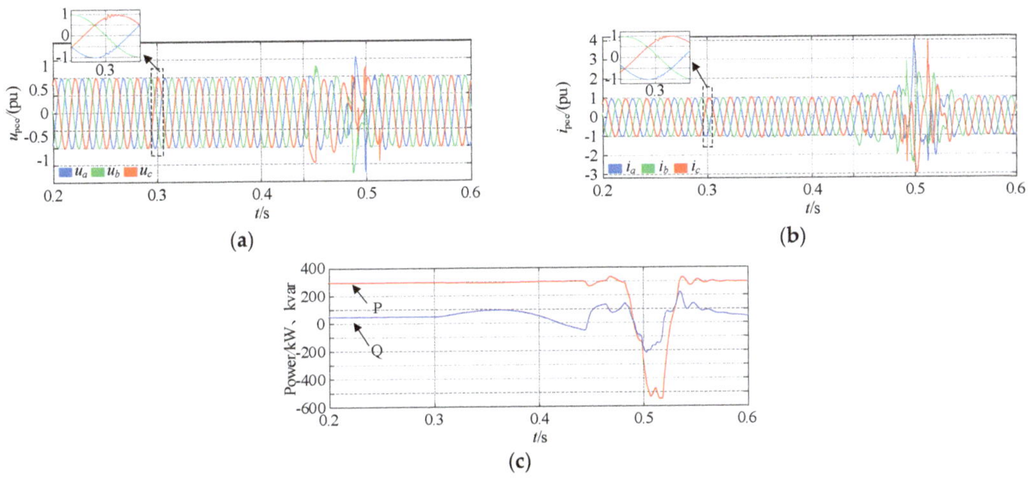

Figure 16. Simulation waveform using DDSRF−PLL. (**a**) The voltage of the PCC. (**b**) The current injected into the PCC. (**c**) Grid−connected output power of the microgrid.

It can be seen from the figure that when the CHB-FFCL was put into operation the PCC voltage and current were soon controlled. However, when the system ran to 0.44 s, it became unstable, and the oscillation could not be eliminated until the fault ended, which indicated that the CHB-FFCL controlled by applying DDSRF-PLL can have a good control effect at the beginning of a fault but may not remain stable for a long time. The control of the grid-connected output power also oscillated greatly with the system instability. In such a case, it is difficult for the grid-connected inverter to continue to operate in PQ control mode, and the microgrid may go off-grid and fail to achieve stable and long-term FRT.

As a comparison, the proposed non-PLL was used to provide control signals for the CHB-FFCL; the simulation waveform is shown in Figure 17. It can be seen from the figure that in this case, the CHB-FFCL could also quickly control the voltage and current of the PCC to ensure that the phase angle did not jump. Moreover, the effect of the CHB-FFCL could remain stable for a long time with good operation stability. As the non-PLL scheme does not have the problem of frequency oscillation, when the power grid transits from the end of the fault to the normal and stable state, the output compensation voltage of the CHB-FFCL is still accurate, which can assist the power grid in transitioning to the stable operation state smoothly. When the power grid is stable, the CHB-FFCL is switched off (the CHB-FFCL was switched off at 0.52 s here) and its switch-off action will not impact the system. In addition, the negative sequence unbalance factors of the compensated PCC voltage and current were both within 2%, so the waveform quality meets the requirements of the Chinese national standard. The grid-connected output power of the microgrid is maintained at a constant value, meeting the requirements of microgrid PQ control, and the FRT of the microgrid is realized.

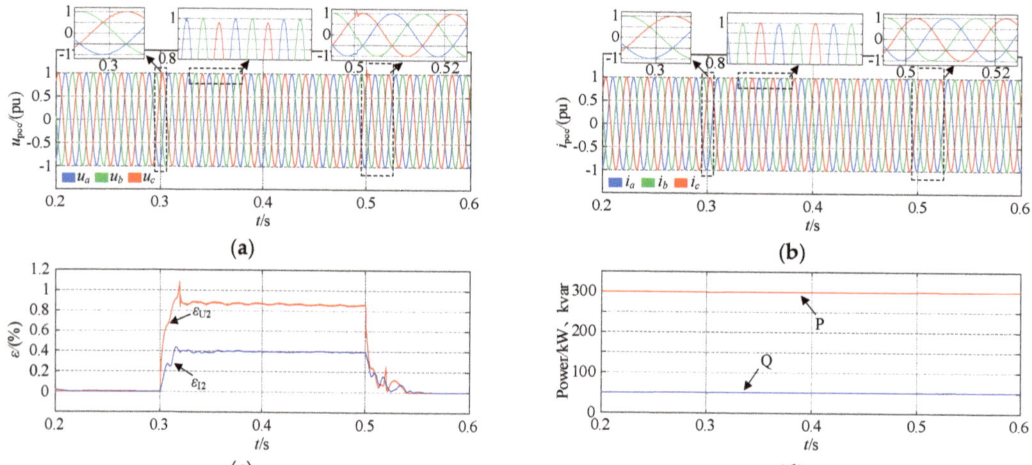

Figure 17. Simulation waveform using the proposed non−PLL scheme. (**a**) The voltage of the PCC. (**b**) The current injected into the PCC. (**c**) Negative sequence unbalance factor of the PCC voltage and current. (**d**) Grid−connected output power of the microgrid.

6. Conclusions

This paper has proposed a scheme to improve the FRT capability of microgrids by using the CHB-FFCL to reduce the impact of faults in external power grids on microgrids. The main conclusions are as follows:

(1) Based on a cascaded inverter structure, the CHB-FFCL has good withstand voltage level and dynamic output characteristics. It can realize voltage regulation and current limiting functions, thus assisting the microgrid to realize FRT.

(2) By setting the angular frequency of the positive sequence fundamental rotating coordinate system so that it is consistent with the power frequency of the power grid, the *dq*-axis DC components of each of the sequence electrical quantities can be extracted without PLL. When there is a deviation between the set frequency and the actual frequency of the power grid, the decoupling network is still effective. Therefore, the correction value of the frequency can be calculated from the AC components of the *dq*-axis electrical quantities so

as to realize the iterative correction of the frequency. The frequency adaptive correction ability of the proposed non-PLL scheme is guaranteed.

(3) Compared with DDSRF-PLL, the proposed non-PLL scheme does not have the defect of frequency oscillation, and because there is no need to design a controller it avoids the problem of complex parameter setting when designing the controller. At the same time, the dynamic response speed of the FCL control system was improved. The simulation results showed that the proposed non-PLL scheme ensured the accuracy of the control and had excellent tracking speed and stability. The CHB-FFCL, which uses non-PLL technology, can eliminate the impact of external asymmetry short-circuit faults on microgrid systems and improve the FRT capability of microgrid systems.

Author Contributions: Conceptualization, F.Z. and G.W.; methodology, G.W.; software, G.W.; validation, B.Q., X.X. and G.W.; formal analysis, F.Z.; investigation, G.W.; resources, F.Z.; data curation, G.W.; writing—original draft preparation, G.W.; writing—review and editing, F.Z.; visualization, G.W.; supervision, S.Z.; project administration, F.Z. All authors have read and agreed to the published version of the manuscript.

Funding: This research is supported by the science and technology project of the State Grid Hubei Corporation of China (Research on key technologies of distribution network considering dynamic planning of microgrids with mobile energy storage, 5215C0220001).

Institutional Review Board Statement: Not applicable.

Informed Consent Statement: Not applicable.

Data Availability Statement: Not applicable.

Conflicts of Interest: The authors declare no conflict of interest.

References

1. Yang, Y.; Blaabjerg, F.; Zou, Z. Benchmarking of Grid Fault Modes in Single-Phase Grid-Connected Photovoltaic Systems. *IEEE Trans. Ind. Appl.* **2013**, *49*, 2167–2176. [CrossRef]
2. Firouzi, M.; Gharehpetian, G.B. Improving Fault Ride-Through Capability of Fixed-Speed Wind Turbine by Using Bridge-Type Fault Current Limiter. *IEEE Trans. Energy Convers.* **2013**, *28*, 361–369. [CrossRef]
3. Rashid, G.; Ali, M.H. A modified bridge-type fault current limiter for fault ride-through capacity enhancement of fixed speed wind generator. *IEEE Trans. Energy Convers.* **2014**, *29*, 527–534.
4. Alam, M.S.; AL-Ismail, F.S.; Abido, M.A.; Hossain, M.A. Fault current limiter of VSC-HVDC systems using variable resistive bridge controller. *Ain Shams Eng. J.* **2021**, *3*, 2643–2654. [CrossRef]
5. Zhang, Y.; Dougal, R.A. State of the art of fault current limiters and their applications in smart grid. In Proceedings of the 2012 IEEE Power and Energy Society General Meeting, San Diego, CA, USA, 22–26 July 2012; pp. 1–6.
6. Safaei, A.; Zolfaghari, M.; Gilvanejad, M.; Ghare-hpetian, G.B. A survey on fault current limiters: Development and technical aspects. *Int. J. Electr. Power Energy Syst.* **2020**, *118*, 105729. [CrossRef]
7. Yang, M.; Wang, X.; Sima, W.; Yuan, T.; Sun, P.; Liu, H. Air-Core-Transformer-Based Solid-State Fault-Current Limiter for Bidirectional HVdc Systems. *IEEE Trans. Ind. Electron.* **2022**, *69*, 4914–4925. [CrossRef]
8. Shuai, Z.; Yao, P.; Shen, Z.J.; Tu, C.; Jiang, F.; Cheng, Y. Design Considerations of a Fault Current Limiting Dynamic Voltage Restorer (FCL-DVR). *IEEE Trans. Smart Grid* **2014**, *6*, 14–25. [CrossRef]
9. Shu, Z.; Chen, Y.; Deng, C.; Zheng, F.; Zhong, H. Pareto Optimal Allocation of Flexible Fault Current Limiter Based on Multi-Objective Improved Bat Algorithm. *IEEE Access* **2021**, *9*, 12762–12778. [CrossRef]
10. Keller, J.; Kroposki, B. *Understanding Fault Characteristics of Inverter-Based Distributed Energy Resources*; NREL/TP-550-46698; National Renewable Energy Laboratory: Golden, CO, USA, 2010.
11. Mohamed YA, R.I.; El-Saadany, E.F. A control method of grid-connected PWM voltage source inverters to mitigate fast voltage disturbances. *IEEE Trans. Power Syst.* **2009**, *24*, 489–491. [CrossRef]
12. Guo, X.; Liu, W.; Zhang, X.; Sun, X.; Lu, Z.; Guerrero, J.M. Flexible Control Strategy for Grid-Connected Inverter Under Unbalanced Grid Faults Without PLL. *IEEE Trans. Power Electron.* **2014**, *30*, 1773–1778. [CrossRef]
13. Qiu, L.; Tang, J.; Cao, Y.; Luo, R.; Wanliang, X.; Lin, L. An Improved Phase-locked Loop Based on Second-order Generalized Integrator. *High Power Convert. Technol.* **2017**, *02*, 2095–3631.
14. Panda, R.K.; Mohapatra, A.; Srivastava, S.C. Enhancing Inertia of Solar Photovoltaic based Microgrid through Notch Filter based PLL in SRF Control. *IET Gener. Transm. Distrib.* **2019**, *14*, 379–388. [CrossRef]
15. Xu, H.; Hu, J.; He, Y. Operation of Wind-Turbine-Driven DFIG Systems Under Distorted Grid Voltage Conditions: Analysis and Experimental Validations. *IEEE Trans. Power Electron.* **2012**, *27*, 2354–2366. [CrossRef]

16. Yang, H.; Nian, H. Stability analysis of grid-connected converter based on interconnected system impedance modeling under unbalanced grid conditions. In Proceedings of the International Conference on Electrical Machines & Systems, Hangzhou, China, 22–25 October 2014.
17. Cespedes, M.; Jian, S. Modeling and mitigation of harmonic resonance between wind turbines and the grid. In Proceedings of the Energy Conversion Congress & Exposition, Phoenix, AZ, USA, 17–22 September 2011.
18. Nian, H.; Cheng, P.; Zhu, Z.Q. Coordinated direct power control of DFIG system without phase-locked loop under unbalanced grid voltage conditions. *IEEE Trans. Power Electron.* **2016**, *31*, 2905–2918. [CrossRef]
19. Cheng, P.; Nian, H. Direct power control of voltage source inverter in a virtual synchronous reference frame dur-ing frequency variation and network unbalance. *IET Power Electron.* **2016**, *9*, 502–511. [CrossRef]
20. Grainger, J.J.; Stevenson, W.D. *Elements of Power System Analysis*; McGraw-Hill: New York, NY, USA, 1994.
21. Sun, Y.; Ma, M.; Ling, F.; Long, H.; Li, F.; Zhang, X. Fault Chain Diagnostic Method Based on Virtual Characteristic Quantity of dq-axis Currents for PV Grid-Connected In-verters. In Proceedings of the 2020 IEEE 9th International Power Electronics and Motion Control Conference (IPEMC2020-ECCE Asia), Nanjing, China, 29 November 2020–2 December 2020; pp. 688–692. [CrossRef]
22. Zhichao, H.; Shengzhen, Y.; Hongyan, Y.; Xingming, F. A control method for SVPWM-based STATCOM using active-reactive current decouple under dq coordinate system. In Proceedings of the 2012 International Conference and Exposition on Electrical and Power Engineering, Iasi, Romania, 25–27 October 2012; pp. 191–196. [CrossRef]
23. Rodríguez, P.; Pou, J.; Bergas, J.; Candela, J.I.; Burgos, R.P.; Boroyevich, D. Decoupled Double Syn-chronous Reference Frame PLL for Power Converters Control. *IEEE Trans. Power Electron.* **2007**, *22*, 584–592. [CrossRef]
24. Shi, J.; Li, S.; Zhao, N.; Niu, Q.; Xu, M. Research on Improved Double Loop Control of Three-phase Inverters with Output LC Filter. In Proceedings of the 2022 IEEE 5th International Electrical and Energy Conference (CIEEC), Nangjing, China, 27–29 May 2022; pp. 2244–2249. [CrossRef]

symmetry

MDPI

Article

A Residual Voltage Data-Driven Prediction Method for Voltage Sag Based on Data Fusion

Chen Zheng [1], Shuangyin Dai [1], Bo Zhang [1], Qionglin Li [1], Shuming Liu [1], Yuzheng Tang [1], Yi Wang [1], Yifan Wu [2] and Yi Zhang [2,*]

[1] Electric Power Research Institute of State Grid Henan Electric Power Company, Zhengzhou 450052, China; zhengchen725@163.com (C.Z.); daishuangyin@163.com (S.D.); 13592602687@126.com (B.Z.); 13525571757@139.com (Q.L.); 15286807325@126.com (S.L.); tangyuzheng1989@163.com (Y.T.); 19937672868@163.com (Y.W.)

[2] College of Electrical Engineering and Automation, Fuzhou University, Fuzhou 350100, China; 200127146@fzu.edu.cn

* Correspondence: zhangyi@fzu.edu.cn

Abstract: Voltage sag is the most serious power quality problem in the three-phase symmetrical power system. The influence of multiple factors on the voltage sag level and low computational efficiency also pose challenges to the prediction of residual voltage amplitude of voltage sag. This paper proposes a voltage sag amplitude prediction method based on data fusion. First, the multi-dimensional factors that influence voltage sag residual voltage are analyzed. Second, these factors are used as input, and a model for predicting voltage sag residual voltage based on data fusion is constructed. Last, the model is trained and debugged to enable it to predict the voltage sag residual voltage. The accuracy and feasibility of the method are verified by using the actual power grid data from East China.

Keywords: voltage sag; residual voltage prediction; data fusion

1. Introduction

Voltage sag is an inevitable short-term disturbance phenomenon that appears during power system operation, and it has become the most serious power quality problem. Voltage sag affects voltage quality and causes considerable economic losses [1,2]. The influencing degree of voltage sag mainly depends on the residual voltage amplitude. Accurate prediction of the residual voltage of the voltage sag can help in clearly understanding the impact of voltage dip on equipment and users. Furthermore, it can provide the basis for users to prevent or control voltage sag, which is very important for reducing the impact of voltage sag [3].

In refs. [4,5], the authors used the Monte Carlo method to express the uncertainty of fault occurrence, including fault type, fault location, etc., and subsequently predicted the probability distribution of voltage sag residual voltage amplitude. The Monte Carlo method cannot consider the influence of the actual environment on the prediction.The authors in [6] pointed out that the influences of generator scheduling and time-varying failure rate on the random prediction of voltage sag should be considered during the voltage sag prediction process. In ref. [7], the authors proposed an online algorithm for residual voltage amplitude prediction based on the harmonic footprint and constructed a general prediction function suitable for all voltage disturbances.

A few methods use monitoring data to predict the voltage sag index. In ref. [8], the authors used the monitoring data to analyze and discuss the randomness of voltage sag. Furthermore, the authors described the voltage sag as a random process and predicted a sag event. In ref. [9], the authors used measured data to predict the voltage sag frequency index of non-monitoring points. The authors in [10] used homologous aggregation and fuzzy c-means theory to reduce the redundancy of the measured data and predict the residual voltage. The aforementioned prediction methods based on monitoring data need long monitoring times and low amounts of monitoring data, which lead to poor prediction accuracies.

This paper proposes a data-driven method for voltage sag residual voltage prediction based on data fusion. The original contributions of this study are summarized as follows:

(1) This paper comprehensively considers the factors influencing the grid and user sides in order to predict voltage sag residual voltage. The corresponding input parameters are selected from different kinds of data and the factors that influence the residual voltage are considered more comprehensively than the Monte Carlo-based method.
(2) This paper builds a prediction method based on data fusion to predict the residual voltage. The amount of data is increased through data fusion, and the problems of low prediction accuracy and low amount of available monitoring data are improved.
(3) This paper presents a residual voltage prediction method for voltage sag based on data fusion, which can be integrated into power-quality-monitoring systems and used for the prevention, evaluation, and treatment of voltage sag. This method can provide users with residual voltage information and help them avoid voltage sag and formulate reasonable voltage sag prevention or treatment measures. The accuracy and efficiency of the method are verified using actual data, which gives it a strong engineering application value.

The remainder of this paper is organized as follows. The data sources of voltage sag residual voltage are described in Section 1. The residual voltage prediction method based on the improved gradient descent method is presented in Section 2. Case studies and analyses based on real power system data are shown in Section 3. The paper is concluded in Section 4.

2. Factors and Data Sources of Voltage Sag Residual Voltage

2.1. Influencing Factors of Voltage Sag Residual Voltage

Motor startup and other factors can cause voltage sag; however, these types of voltage sag are often not serious and existing measures can effectively eliminate their impact [11]. The main cause of voltage sag is a power grid fault [12]. Therefore, this paper mainly considers the impact of power grid faults during the voltage sag residual voltage prediction process.

After a grid fault occurs, the residual voltages at different nodes have varying amplitudes under different grid operation modes, and the impacts on different nodes are also different. Therefore, the influence of grid operation mode should also be considered in the prediction of the residual voltage of voltage sag. In addition, voltage sags are closely related to users. Different users are affected differently depending on the types of sags transmitted to them. Therefore, the influence of the user side should be considered in the prediction of voltage sag residual voltage amplitude. In summary, voltage sag residual voltage is related to fault conditions, grid operation mode, and the influence of the user side. The above factors are considered as the factors influencing the voltage sag residual voltage and used for selecting relevant data for voltage sag residual voltage prediction.

2.2. Data Sources

Voltage sag data can be obtained by the Monte Carlo random fault simulation method as follows: First, the fault parameters, such as fault type, fault distance, fault line, etc., are used as random variables and the simulation calculation is carried out by the Monte Carlo method. Second, the sag level of each bus in the area is obtained. Last, the simulated data are obtained. The simulated data can reflect the influence of system operation mode and fault on residual voltage, but cannot directly reflect the influence of external factors such as weather.

When a voltage sag occurs in the power grid, the power quality monitoring system, energy management system, user power consumption information collection system, and other systems collect a large number of records. There are many kinds of data related to voltage sag, and the relationship between them is complicated. In this paper, the relevant factors affecting the voltage sag residual voltage are obtained from the above multi-source system. The monitoring data not only reflect the influence of system operation mode and power grid faults, but also reflect the influence of external environmental factors and user-side influence. However, the data require long-term monitoring and consist of a small number of samples.

In summary, for the voltage sag residual voltage prediction, the data attributes reflecting the factors that influence the voltage sag residual voltage are selected from the simulated and measured data. These factors are shown in Table A1 of Appendix A.

3. Voltage Sag Residual Voltage Prediction Method

3.1. Multiple Regression Model and Gradient Descent Method

Voltage sag residual voltage is affected by many factors. Therefore, this paper uses the multiple regression model to characterize the effects of various factors on voltage sag residual voltage. This model can be expressed as follows:

$$y = f(x_1, x_2, x_3 \ldots, \theta_1, \theta_2, \theta_3 \ldots) \quad (1)$$

where y is the dependent variable, x_1, x_2, x_3 ... denote the independent variables, and θ_1, θ_2, θ_3 ... represent the regression coefficients corresponding to x_1, x_2, x_3 ..., respectively.

The gradient descent method is a type of line search framework algorithm with negative gradient as the search direction [13]. In order to minimize the loss function, the optimization model parameters are obtained iteratively until convergence is attained, and the best-matching target task is obtained.

The basic calculation process is as follows: Solving the model parameters for the loss function $H(x_1, x_2 \ldots)$ can transform it into an optimization problem of min $H(x_1, x_2 \ldots)$. The calculation of the gradient $g(x_1, x_2 \ldots)$ is shown in (2). It is straightforward to conclude that the negative gradient direction $-\nabla H$ is the fastest direction of decrease of the loss function. The iterative format of the gradient descent method is given by (3). Equations (2) and (3) shown below are iteratively used to reduce the loss function and approach the minimum point, and the parameters of the optimization model can be obtained after convergence.

$$g(x_1, x_2, x_3 \cdots) = \nabla H \quad (2)$$

$$x_{k+1} = x_k - s_k g_k \quad (3)$$

In (3), k is the number of iterations, s_k is the kth iteration step size, and g_k is the kth iteration gradient size.

3.2. Multiple Regression Model Based on Improved Gradient Descent Method

3.2.1. Model Parameters

In order to realize the voltage sag residual voltage prediction using data fusion, this paper constructs a multiple regression model based on the improved gradient descent method. This improvement requires the construction of multiple regression models for the two types of data. The simulated data model is represented by $D_s = f(x)$, and the input parameter is the attribute name corresponding to the simulated data in Table A1, including the common part of the two types of data. The model of the measured data is represented by $D_m = g(y)$, and the input parameter is the attribute name corresponding to the measured data in Table A1, including the common part of the two types of data. The output results of D_s and D_m are the residual voltage amplitudes, expressed by V_s and V_m, respectively.

As the duration of sag mainly depends on the setting value of the protection device [10], it is not predicted in this paper. It can be observed from Table A1 that there is a common input data attribute with a consistent description in the simulated and measured data. This part is represented by the set A_{shared} in this paper, and the corresponding regression coefficient is represented by θ_{shared}. The non-common part attributes of the two types of data are represented by A_s and A_m, and the corresponding regression coefficients are represented by θ_s and θ_m, respectively. The subscripts s and m represent the simulated and measured data models, respectively. During the process of searching for the optimal model parameters using the improved gradient descent method, the loss functions $H_s(x)$ and $H_m(z)$ of the two models are given as

$$H_s(x) = \frac{1}{2n}\sum_{i=1}^{n}(f(x_i) - V_s(i))^2 \tag{4}$$

$$H_m(z) = \frac{1}{2h}\sum_{j=1}^{h}(g(z_j) - V_m(j))^2 \tag{5}$$

where n and h are the simulated and measured data sample capacities, respectively; x_i and $V_s(i)$ correspond to the input and output of the ith simulation data sample, respectively; and z_j and $V_m(j)$ correspond to the input and output of the jth measured data sample, respectively.

3.2.2. Model Update Strategy

This paper presents an improved gradient descent method. When the output of the two models is the residual voltage amplitude, the functions of A_{shared} in both D_s and D_m are identical. Therefore, considering the simulated and measured data as the source and target domains, respectively, the prior knowledge obtained by simulation in D_s is migrated to D_m for updating θ_{shared}. After completing the migration, the updated θ_{shared} is substituted back into the original model and the updated θ_s and θ_m are recalculated to complete the single-learning process. At this time, D_m not only reflects the role of the factors influencing the voltage sag residual voltage from the information point of view, but also reveals the impacts of relevant factors on the voltage sag residual voltage from the physical point of view. The model realizes physical–information integration through knowledge migration and improves its information space and learning performance.

The traditional gradient descent method uses a constant, s_k, whose value is obtained through trial and error. This method has the following problems: (1) The learning degree of the model is different in different stages of training; and (2) It takes a certain amount of time to obtain the ideal value of s_k by multiple attempts.

In view of the above problems, this paper considers improving the step update strategy of the model. The Armijo–Goldstein criterion is introduced into the step size update process. There are two formulas for the Armijo–Goldstein criterion, which can be expressed by (6) and (7). The value of s_k satisfying (6) and (7) is called an acceptable step size factor. After the introduction of the Armijo–Goldstein criterion, s_k can be automatically updated to an acceptable step size factor and the algorithm will have superlinear convergence.

$$f(x_k + s_k d_k) \le f(x_k) + s_k \rho {g_k}^T d_k \tag{6}$$

$$f(x_k + s_k d_k) \ge f(x_k) + s_k (1-\rho) {g_k}^T d_k \tag{7}$$

In (6) and (7), $f(x)$ is the objective function, d_k is the search direction, g_k is the gradient size, and $\rho \in (0, 0.5)$ to ensure the superlinear convergence of the algorithm.

3.3. Overall Process

To summarize, the process of voltage sag residual voltage prediction based on data fusion is as follows:

(1) Acquire the measured data that can reflect the factors influencing the residual voltage from the multi-source system. The simulated data are obtained by a random sag simulation calculation based on the Monte Carlo method.
(2) Carry out data preprocessing on the simulated and measured data to adapt them to the model, and use the above data as the model input.

(3) Build a multiple regression model based on the improved gradient descent method. During the iterative process of the gradient descent method, the model updates the model parameters and adaptively adjusts the step size based on knowledge transfer and the Armijo–Goldstein criterion until convergence is reached.
(4) After the training is completed, the model learns the knowledge from the physical and information aspects and can predict the residual voltage amplitude.

Figure 1 shows the overall flow chart of the voltage sag residual voltage prediction method proposed in this paper.

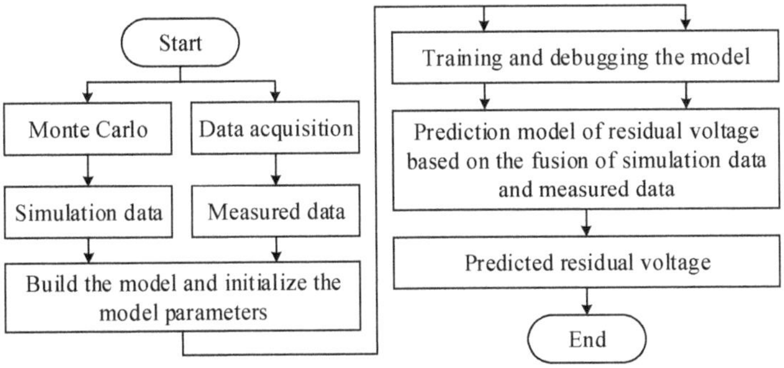

Figure 1. Flow chart of voltage sag residual voltage prediction.

4. Case Study

4.1. Data Source and Input

This paper selects 57 voltage sag measured data samples from January 2021 to December 2021 in a city in eastern China. The simulated data are obtained using Monte Carlo random simulations. The simulation calculation process is mainly realized by Bonnevillr Power Administration (BPA) [14], which is power system simulation software commonly used by power companies. The region selected for study has long been affected by voltage sags, and there are many industrial users that have a large proportion of sensitive equipment in the region, which is representative for this study.

4.2. Residual Voltage Prediction

4.2.1. Prediction Results

The simulated and measured data are obtained according to Table 1 and are subsequently preprocessed. In the measured data, 44 samples are randomly selected as the training set, and the remaining 13 samples are used as the test set. The simulation dataset for knowledge transfer consists of 5280 data points obtained from the BPA-based Monte Carlo random simulation calculation.

The model uses root-mean-square error (RMSE) and mean absolute error (MAE) as the evaluation metrics. A total of 44 measured training data points and 5280 simulated training data points are input into the model for training. After training, the RMSE of the model on the training set is 0.1678, the MAE is 0.1377, and the model is able to predict the residual voltage amplitude.

$$RMSE = \sqrt{\frac{1}{n}\sum_{i=1}^{n}(x_i - x_i')^2} \tag{8}$$

$$MAE = \frac{1}{n}\sum_{i=1}^{n}|x_i - x_i'| \tag{9}$$

In (8) and (9), i is the number of samples, x_i is the actual value, and $x_i^{'}$ is the predicted value.

Table 1. Evaluation metric for each method on the test set.

	Proposed Method	M1	M2
RMSE	0.1475	0.4228	0.2526
MAE	0.1379	0.3646	0.1801

After the training, the method is tested on the test set data to verify its generalization ability. In order to evaluate the model performance, it is compared with the following methods: 1. The traditional gradient descent method [13], denoted as M1 here; and 2. A support vector machine [15] (SVM, rbf-kernel, $\varepsilon = 0.0198$), denoted as M2 here. Figure 2 shows the prediction results of each method, Table 1 shows the evaluation indices of each method and Table 2 shows the relative error between the predicted and actual values of each method.

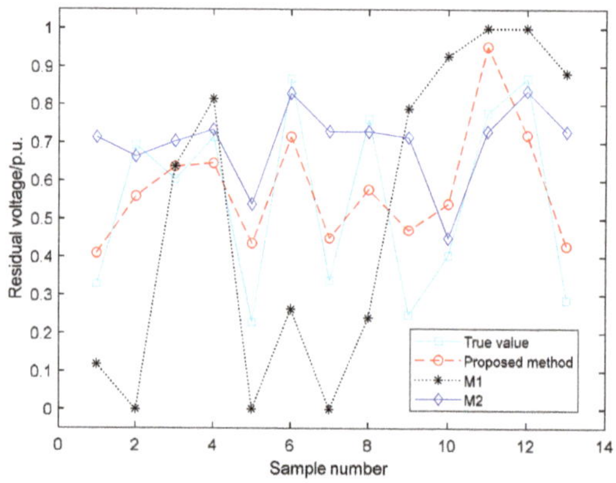

Figure 2. Test set prediction results.

Table 2. Relative error between the predicted and real values of each method.

Sample \ Method	Proposed Method (%)	M1 (%)	M2 (%)
1	24.43	−63.86	117.10
2	−19.35	−100.00	−4.25
3	5.34	5.53	16.42
4	−9.67	13.90	2.72
5	90.67	−100.00	136.10
6	−17.48	−69.65	−4.33
7	33.31	−100.00	116.63
8	−24.24	−68.46	−4.31
9	89.75	218.50	188.14
10	33.82	130.08	11.42
11	22.46	28.67	−6.15
12	−17.15	15.24	−3.86
13	49.25	207.83	154.51

The above results show that the RMSE and MAE of the proposed method are better than those of the other two methods. In addition, it can be observed from Table 2 that the relative prediction error of the method proposed in this paper is smaller than those of the other two methods. The method M1 has a large overall error and a large deviation from the error. At the same time, many output results of the method exceed the upper and lower limits of voltage amplitude and are limited by the threshold, indicating its poor convergence. Compared with M1, the output results obtained by the proposed method do not exceed the voltage amplitude limit. The convergence is improved and the error is considerably reduced.

The model structure of M2 is more complex than that of the proposed method. However, its evaluation metric is still slightly inferior compared to that of the proposed method. Although the complex model structure adopted in M2 can improve the error to a certain extent, the problem of small sample space still significantly impacts the model performance. The method proposed in this paper uses data fusion to improve the updating strategy of the model in M1 and enhance the data mining performance. Therefore, the proposed method achieves better results compared to M2.

4.2.2. Number of Iterations

Under the same convergence conditions, the numbers of iterations between the proposed method and M1 are compared in Table 3. The number of iterations of M2 is not compared here because of the different training methods.

Table 3. Duration and the number of iterations of single training.

Method	Iterations
Proposed method	16,101
M1	310,780

Compared with M1, the number of iterations of the proposed method is significantly reduced. This is because the proposed method modifies the update mode of the model parameters and step size through data fusion. This modification improves the learning performance of the training process and accelerates the learning process. Thus, the number of iterations is significantly reduced.

In summary, the accuracy of this method is better than that of the other two methods, and the convergence performance is also significantly improved compared with M1. Therefore, in the actual power system, this method can predict the residual voltage of the possible voltage sag at fixed intervals by collecting relevant data. This prediction can help users to prevent and control the voltage sag and reduce economic losses.

5. Conclusions

In this paper, a voltage sag residual voltage data-driven prediction method based on data fusion was proposed and verified. The following conclusions were obtained:

(1) This method analyzed the relevant factors affecting the residual voltage of voltage sag from multiple dimensions. The relevant influencing factors were selected as input in order to consider the factors influencing the residual voltage of voltage sag more comprehensively.

(2) This method considered different data characteristics and realized the prediction of voltage sag residual voltage through data fusion. Consequently, the prediction accuracy and convergence rate were improved.

(3) The method was convenient and practical, which was verified by examples. In the future, it can be used to predict the residual voltage of voltage sag and assist the analysis of voltage sag level and consequences.

Author Contributions: Conceptualization, Y.W. (Yifan Wu) and Y.Z.; methodology, Y.W. (Yifan Wu); software, Y.W. (Yifan Wu); validation, Y.W. (Yifan Wu), C.Z., and S.D.; formal analysis, C.Z. and B.Z.; investigation, C.Z. and S.L.; resources, Q.L. and Y.T.; writing—original draft preparation, Y.W. (Yifan Wu) and C.Z.; writing—review and editing, Y.W. (Yifan Wu) and Y.Z.; supervision, Y.W. (Yi Wang); project administration, C.Z. All authors have read and agreed to the published version of the manuscript.

Funding: This research is supported by the science and technology project of the headquarters of the State Grid Corporation of China (Key technologies and applications of digital mirroring for analysis and treatment of voltage sags in distribution networks, 5400-202124153A-0-0-00).

Conflicts of Interest: The authors declare no conflict of interest.

Abbreviations

BPA	Bonnevillr Power Administration
RMSE	Root-mean-square error
MAE	Mean absolute error
SVM	Support vector machine

Nomenclature

The subscript 's' represents the simulation data model and the subscript 'm' represents the measured data model.

D_s	Simulated data model
D_m	Measured data model
A_s	Unique parameters of simulated data model
A_m	Unique parameters of measured data model
A_{shared}	Common parameters of simulated and measured data model
θ_s	Unique regression coefficient of simulated data model
θ_m	Unique regression coefficient of measured data model
θ_{shared}	Common regression coefficient of simulated and measured data model
$H_s(x)$	Loss function of simulated data model
$H_m(z)$	Loss function of measured data model
g_k	Gradient of the kth iteration
s_k	Step size of the kth iteration

Appendix A

Table A1. Influencing factors of voltage sag residual voltage.

Data Sources	Attribute Name
Simulated data	Total load Fault impedance
Measured data	Weather Season Time Power user type Proportion of sensitive load Line status Fault cause
Common part of simulated and measured data	Monitoring bus Monitoring bus voltage level Duration of voltage sag Fault type Fault phase Fault location Distance-to-fault Residual Voltage

References

1. Zhang, Y.; Li, W.; Lin, F.; Zhang, Y.; Huang, Y.; Yang, C. Voltage Sag Mitigation Strategy for Industrial Users Based on Process Electrical Characteristics-physical Attribute. *Proc. CSEE* **2021**, *41*, 632–642.
2. Liu, X.; Xiao, X.; Wang, Y. Voltage Sag Severity and Its Measure and Uncertainty Evaluation. *Proc. CSEE* **2014**, *34*, 644–658.
3. Goswami, A.K. Voltage sag assessment in a large chemical industry. *IEEE Trans. Ind. Appl.* **2012**, *48*, 1739–1746. [CrossRef]
4. Naidu, S.R.; de Andrade, G.V.; da Costa, E.G. Voltage Sag Performance of a Distribution System and Its Improvement. *IEEE Trans. Ind. Appl.* **2012**, *48*, 218–224. [CrossRef]
5. dos Santos, A.; Correia de Barros, M.T. Predicting Equipment Outages Due to Voltage Sags. *IEEE Trans. Power Deliv.* **2016**, *31*, 1683–1691. [CrossRef]
6. Park, C.H.; Jang, G.; Thomas, R.J. The Influence of Generator Scheduling and Time-Varying Fault Rates on Voltage Sag Prediction. *IEEE Trans. Power Deliv.* **2008**, *23*, 1243–1250. [CrossRef]
7. Stanisavljević, A.M.; Katić, V.A. Magnitude of Voltage Sags Prediction Based on the Harmonic Footprint for Application in DG Control System. *IEEE Trans. Ind. Electron.* **2019**, *66*, 8902–8912. [CrossRef]
8. dos Santos, A.; Rosa, T.; de Barros, M.T.C. Stochastic Characterization of Voltage Sag Occurrence Based on Field Data. *IEEE Trans. Power Deliv.* **2019**, *34*, 496–504. [CrossRef]
9. Zambrano, X.; Hernandez, A.; Izzeddine, M.; de Castro, R.M. Estimation of Voltage Sags From a Limited Set of Monitors in Power Systems. *IEEE Trans. Power Deliv.* **2017**, *32*, 656–665. [CrossRef]
10. Wang, Y.; Yang, M.H.; Zhang, H.Y.; Wu, X.; Hu, W.X. Data-driven prediction method for characteristics of voltage sag based on fuzzy time series. *Int. J. Electr. Power Energy Syst.* **2022**, *134*, 107394. [CrossRef]
11. Cheng, H.; Ai, Q.; Zhang, Z.; Zhu, Z. *Power Quality*; Tsinghua University Press: Beijing, China, 2006.
12. Si, X.; Li, Q.; Yang, J.; Xu, Y.H.; Zhang, B. Analysis of voltage sag characteristics based on measured data. *Electr. Power Autom. Equip.* **2017**, *37*, 144–149.
13. Sun, Y. *Application of Gradient Descent Method in Machine Learning*; Southwest Jiaotong University: Chengdu, China, 2018.
14. Wang, J.; Zhang, Y.; Chen, J.; Wu, M. Evaluation of voltage sag in provincial power grid and optimization of potential power supply points for industrial users. *Electr. Power Autom. Equip.* **2021**, *41*, 201−207+224.
15. Eskandarpour, R.; Khodaei, A. Leveraging. Accuracy-Uncertainty Tradeoff in SVM to Achieve Highly Accurate Outage Predictions. *IEEE Trans. Power Syst.* **2018**, *33*, 1139–1141. [CrossRef]

symmetry

MDPI

Article

Cascaded Multilevel Inverter-Based Asymmetric Static Synchronous Compensator of Reactive Power

Martynas Šapurov [1,2,*], Algirdas Baskys [1,2], Raimondas Pomarnacki [1], Artūras Serackis [1], Mindaugas Jankauskas [1], Van Khang Huynh [3], Vytautas Bleizgys [1,2], Aldas Dervinis [2], Edvardas Bielskis [4], Sarunas Paulikas [1], Nerijus Paulauskas [1] and Darius Guršnys [1]

[1] Faculty of Electronics, Vilnius Gediminas Technical University, Naugarduko St. 41, LT-03227 Vilnius, Lithuania; algirdas.baskys@vilniustech.lt (A.B.); raimondas.pomarnacki@vilniustech.lt (R.P.); arturas.serackis@vilniustech.lt (A.S.); m.jankauskas@vilniustech.lt (M.J.); vytautas.bleizgys@vilniustech.lt (V.B.); sarunas.paulikas@vilniustech.lt (S.P.); nerijus.paulauskas@vilniustech.lt (N.P.); darius.gursnys@vilniustech.lt (D.G.)

[2] State Research Institute Center for Physical Sciences and Technology, Sauletekio Av. 3, LT-10257 Vilnius, Lithuania; aldas.dervinis@ftmc.lt

[3] Department of Engineering Sciences, University of Agder, Postboks 422, 4604 Kristiansand, Norway; huynh.khang@uia.no

[4] Institute of Regional Development, Vilnius University Šiauliai Academy, P. Visinskio Str. 38, LT-76352 Siauliai, Lithuania; edvardas.bielskis@su.lt

* Correspondence: martynas.sapurov@ftmc.lt; Tel.: +370-684-70305

Abstract: The topology of the static synchronous compensator of reactive power for a low-voltage three-phase utility grid capable of asymmetric reactive power compensation in grid phases has been proposed and analysed. It is implemented using separate, independent cascaded H-bridge multilevel inverters for each phase. Every inverter includes two H-bridge cascades. The first cascade operating at grid frequency is implemented using thyristors, and the second one—operating at high frequency is based on the high-speed MOSFET transistors. The investigation shows that the proposed compensator is able to compensate the reactive power in a low-voltage three-phase grid when phases are loaded by highly asymmetrical reactive loads and provides up to three times lower power losses in the compensator as compared with the situation when the compensator is based on the conventional three-level inverters implemented using IGBT transistors.

Keywords: reactive power; compensator; renewable energy; cascaded inverter; low-voltage grid; asymmetric compensation; smooth compensation

1. Introduction

The amount of electricity generated using decentralized renewable energy sources is constantly growing. Such energy generation is called Distributed Generation. Power electronics-based electronic converters are widely used in distributed energy grids to convert the non-standard electricity produced by the renewable energy sources to the standard one and to improve the quality of delivered energy. One of the problems encountered in distributed energy grids related to energy quality is reactive power compensation [1–3]. Electro-mechanically commutated Capacitor Banks or Static Var Compensator (SVC) based on the thyristor-switched capacitors and thyristor-switched reactors can be used to solve this problem [4–6]. However, the working conditions of renewable energy sources (photovoltaic and wind power plants) are constantly changing due to changes in weather conditions [7]. For this reason, the energy supply situation in such grids is constantly changing, so reactive energy compensators have to be characterized by a short response time. Therefore, it is appropriate to use a more advanced voltage source inverter-based

Static Synchronous Compensator (STATCOM) [8–10] that is characterized by a short response time in the distributed energy grids.

Inverter-based reactive power compensators are implemented using an inverter that acts as a voltage source and generates AC voltage using the PWM technique. The amplitude of the inverter voltage is controlled by the changing of the modulation index (by changing the duration of voltage pulses generated by the inverter). According to energy transfer theory, from one voltage source to another through the filter with the impedance X, the quantities of active and reactive power can be obtained by solving equations [11]

$$P = \frac{U_I \cdot U_G \cdot \sin\delta}{X}, \tag{1}$$

$$Q = -\frac{U_G^2 - U_I \cdot U_G \cdot \cos\delta}{X}, \tag{2}$$

where P is active power, Q is reactive power, U_I is amplitudes of inverter voltage, U_G is amplitudes of grid voltage, and δ is the phase angle between voltages U_I and U_G. According to Equation (1), in order to keep the active power $P = 0$ and to generate only reactive power, the phase angle between the voltage sources has to be $\delta = 0°$, i.e., the voltage generated by the reactive power compensator inverter has to be synchronized with the grid voltage. It is seen from Formula (2) that at $\delta = 0°$, the capacitive reactive power (Q has to be positive in (2)) is supplied to the grid if the voltage amplitude of the inverter is higher than the voltage amplitude of the utility grid $(U_I > U_G)$. The higher the amplitude of the inverter voltage, the more capacitive reactive power is being supplied to the utility grid. In the case when the voltage amplitude of the inverter is lower in comparison to the voltage amplitude of the utility grid $(U_I < U_G)$, the inductive reactive power is consumed (Q has to be negative in (2)). The lower the inverter voltage, the more inductive reactive power is being consumed. If the amplitude of the inverter voltage is equal to the utility grid voltage, the reactive power is not produced or consumed.

The reactive power to be compensated in each phase of the three-phase grid is often different, i.e., asymmetric, so the compensator has to be adapted to asymmetric reactive power compensation [12,13]. In such a case, the three-phase inverter must be able to supply to each phase of the grid the voltage with independently controlled amplitude. Despite that there are publications dedicated to the development of a three-phase inverter control method to compensate for the reactive power asymmetrically [14], the works [15–18] state that the STATCOM-type compensator, based on a three-phase inverter, which includes three arms, is incapable of asymmetric reactive power compensation. This statement is supported by the fact that there are no STATCOM-type compensators on the market to control reactive power in a three-phase grid asymmetrically. Because of this, the hybrid STATCOM, which includes conventional STATCOM and the thyristor-switched reactor (TSR), or thyristor-switched capacitor (TSC) compensators, can be used for asymmetric compensation. STATCOM compensates for a symmetrical component of reactive power, while the asymmetric component is compensated using TSR or TSC. However, TSR and TSC compensators are slow in comparison to STATCOM compensators; because of this, during the asymmetric compensation the response of the hybrid STATCOM is slow as well.

Based on the above, it can be concluded that accurate asymmetric short response compensation of the reactive power in a three-phase grid can be performed using just the compensator with the separate single-phase inverters for each phase.

The key contribution of this work is that the topology of the static synchronous compensator capable of asymmetric reactive power compensation in a low-voltage three-phase utility grid has been proposed. The compensator is based on the employment of separate, independent cascaded H-bridge multilevel inverters for each phase.

2. Topology of Static Synchronous Reactive Power Compensator Based on Independent Cascaded H-Bridge Multilevel Inverters

A cascaded inverter is a type of multilevel inverter topology where switching devices are connected in series. Cascaded inverters are widely used in renewable energy plants, variable-speed motor drives and reactive power compensators for electric grids [19–21]. Cascaded inverter topology in comparison to a conventional three-level inverter allows a decrease in current and voltage harmonics and reduced switching losses of the inverter by distributing power over two inverters. In the proposed reactive power compensator, the hybrid modulation technique [22–24] is used for the forming of the compensator output voltage. In order to reduce switching losses, the majority of inverter power is produced using the first cascade operating at low (grid) frequency. It generates the square waveform voltage with the amplitude close to the utility grid voltage amplitude and with a frequency equal to grid voltage frequency. This square waveform voltage includes not just the first harmonic, the frequency of which is equal to grid frequency, but the higher odd harmonics as well. The second cascade of the cascaded inverter performs the elimination of higher harmonics produced by the first cascade using the PWM technique operating at high frequency and low power. The spectrum of ideal square pulse consists of the infinite sum of sine waves of odd harmonic order. The amplitude of each harmonic can be calculated using the following equation [25]:

$$A_{fj} = \frac{2 \cdot A}{j \cdot \pi}\left(\sin\left(\frac{j \cdot m \cdot \pi}{2}\right) + \sin\left(j \cdot \left(\pi - \frac{m \cdot \pi}{2}\right)\right)\right), \tag{3}$$

where j—harmonic number, A_{fN}—amplitude of N-th order harmonic, A—amplitude of square pulse voltage, and m—modulation index. The lowest order harmonic in the second cascade voltage spectrum has to be the third harmonic, and as it is seen from (3), it has the highest amplitude as compared with other higher-order harmonics. If the amplitude of the square wave voltage generated by the first cascade is 320 V, the amplitude of the third harmonic calculated using (3) is A_{f3} = 120 V. Since the second cascade has to generate the pulsed voltage for the cancellation of the higher-order harmonics created by the first cascade, the supply voltage of the second cascade has to be $U_{DC2} \geq 120$ V.

Usually, the switches of reactive power compensator inverters are implemented using IGBT transistors that have good enough dynamic characteristics but are characterized by a relatively high voltage drop in state ON. In the proposed topology of a static synchronous reactive power compensator based on three independent cascaded H-bridge multilevel inverters, the first cascade operates at low frequency; therefore, the impact of the dynamic characteristics of inverter switches on power losses is not sufficient. Because of this, the thyristors that have slow dynamics but are characterized by low voltage drop in the state ON, i.e., guarantee low conducting losses, could be used instead of IGBT transistors to minimize the power losses in the first cascade. Since the amplitude of the PWM voltage produced by the second cascade is relatively low, the switches of this cascade can be realized using low-voltage MOSFET transistors that are characterized by very high speed and low voltage drop in state ON, and allow achieving lower power losses as compared with the IGBT transistors as well.

The topology of the proposed three-phase asymmetric static synchronous compensator based on independent cascaded H-bridge multilevel inverters for a four-wire grid is presented in Figure 1. Every cascaded inverter provides seven levels of output voltage. The first cascade of every cascaded inverter is implemented using thyristors (T1.1–T1.4), the second—using MOSFET transistors (Q1.1–Q1.4). Each cascade is supplied by a separate isolated voltage source (DC1.1 and DC1.2), i.e., six galvanically isolated DC voltage sources were used, which could be implemented using distributed renewable energy. The cascaded inverter of every phase includes an LCL output filter for connection to the utility grid and for the filtering of current harmonics as well. The parameters of the filter components were as follows: $L1.1 = 1.5 \times 10^{-8}$ H; $L1.2 = 1.5 \times 10^{-3}$ H; $C = 6.2 \times 10^{-9}$ F.

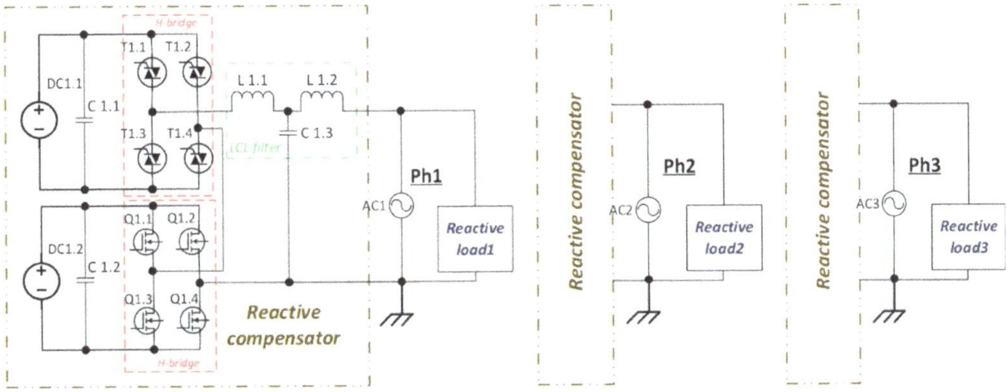

Figure 1. The topology of three-phase static synchronous compensator based on independent cascaded H-bridge multilevel inverters.

It is necessary to stress that the proposed topology is more complex than the classic one based on the three-arm three-phase inverter. However, it has an important advantage—it allows asymmetric compensation of the reactive power in the three-phase grid to be performed.

3. Investigation of Reactive Power Compensator Output Voltage and Current Spectrums

Matlab/Simulink model of static synchronous reactive power compensator based on three independent cascaded H-bridge multilevel inverters was built. The model of cascaded inverter for one of the phases is presented in Figure 2. The signals for the control of thyristors of the first cascade that produces the square waveform voltage are generated using the block *PWM 1*, which creates the signals by comparing the grid frequency ($f_{sq} = f_{gr}$) square signal with low-frequency triangle signal ($f_{tri} = 2f_{gr}$). The amplitude of the square signal, and as a consequence, the amplitude of the inverter output voltage, is set up by the variation in modulation index *m*.

Since the purpose of the second cascade of the cascaded inverter is the elimination of the higher harmonics produced by the first cascade, the spectrum of the waveform generated by the second cascade has to include the same higher harmonics with the same amplitudes but with the opposite phases as compared with the spectrum of the square waveform. This waveform is generated using the PWM technique. The control signals for the MOSFET transistors of the second cascade are formed using the same principle as it is used in the sinusoidal PWM method [26]. The only difference is that instead of the comparison of the sinus signal with the high-frequency triangle signal, the comparison of the waveform that has to generate the second cascade has to be provided. The spectrum of this waveform has to include the third and higher odd harmonics. The sinus waveforms with the frequencies that correspond to odd harmonic frequencies up to the 21st harmonic with the amplitudes calculated using (3) are generated employing the blocks *Harmonic j*, where j = 1, 3, 5, . . . , 21. These blocks generate odd harmonic sinus waves from the 3rd to 21st harmonic and allow control of the modulation index *m* and delay *T*. All harmonics are summed using block *Summ*. The resulting waveform is presented in Figure 3. Control signals for the second cascade MOSFET transistors are obtained in the second cascade PWM generator PWM 2 using a voltage comparator by applying a high-frequency (equal to PWM carrier frequency) triangular signal to one input and a waveform with the sum of harmonics to another input of the comparator. The direct and inverted PWM signals obtained at the outputs of the comparator are fed to an inverter for the control of MOSFET transistors.

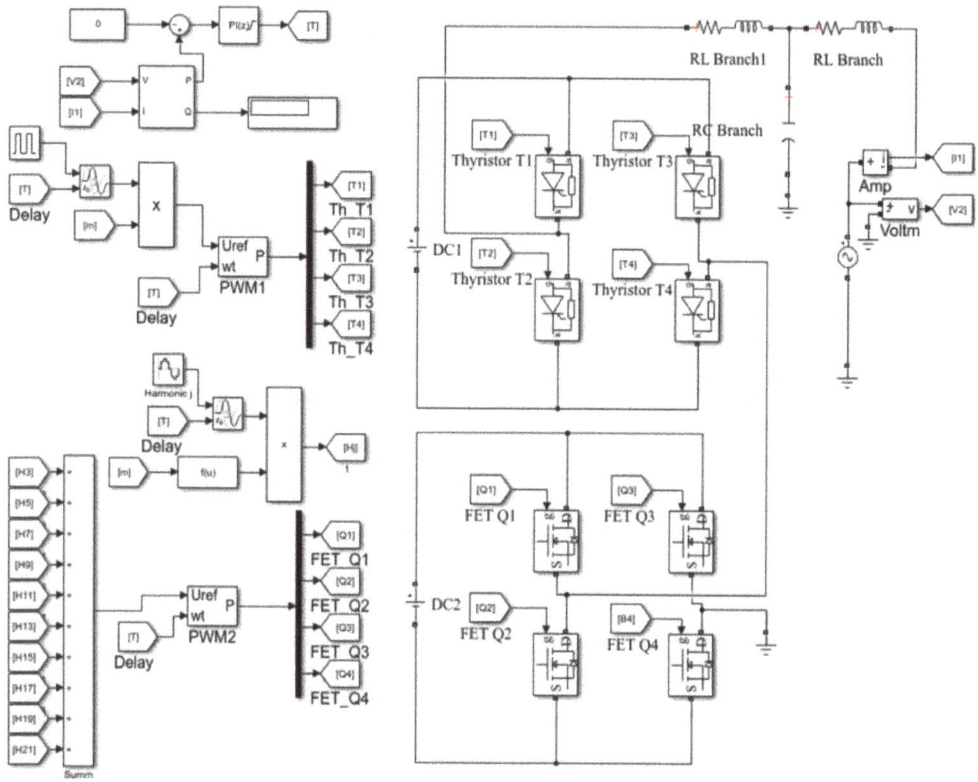

Figure 2. *Matlab/Simulink* model of H-bridge cascaded inverter for the one phase.

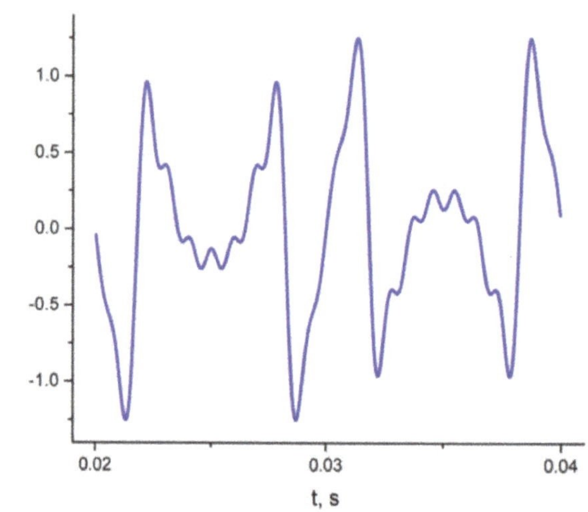

Figure 3. The waveform that includes all odd harmonics ranging from 3rd to 21st harmonic for the forming of signals for the control of MOSFET transistors of the second cascade of the cascaded inverter.

The output voltage of the cascaded inverter has to be in phase with the grid voltage for control of the reactive power. However, the voltage phase is shifted by the output filter of the inverter, and this shift depends on the inverter load, i.e., it is not constant. Therefore, it is necessary to control the phase of the cascaded inverter output voltage. The PI controller is used for this purpose. Since the direct measurement of the grid phase is complicated, the zero value of active power provided to the grid, which corresponds to the zero-phase difference (see Equation (1)), is used as the set point for the PI controller. The active power measurement for the feedback of the phase control system is performed using current and voltage measurement blocks *Amp* and *Voltm* (Figure 2). The PI controller controls the shift of all generated signals introducing the shift in time *T* for the control of the inverter output voltage phase.

The analysis of the reactive power compensator based on three independent cascaded H-bridge multilevel inverters was performed for the reactive load that varies in every phase of the compensator in a range from -4000 to $+4000$ VAr.

Waveforms and spectrums of the output voltage and current of the first cascade of the cascaded inverter are presented in Figure 4. The second cascade of the cascaded inverter has to generate the waveforms with the spectrum, which would allow cancelling the harmonics produced by the first cascade in order to provide quality reactive power. The second cascaded inverter in the proposed compensator provides harmonic cancellation up to the 21st harmonic, i.e., up to frequency $f_{21} = 1050$ Hz. Waveforms and spectrums of cascaded inverter output voltage and current with harmonic cancellation are presented in Figure 5. The cancellation of harmonics up to the 21st harmonic using the second cascade of the cascaded inverter allows reaching the 1.74% THD of output current.

The waveforms of the current of DC1 and DC2 voltage sources of the cascaded inverter are presented in Figure 6. They are obtained for a +4000 VAr reactive load. As it is seen, the current pulsates in both directions between the load and the DC sources of the compensator because of the reactive nature of the load. Active power is used only to compensate for the losses in the circuits between the DC sources and the reactive load.

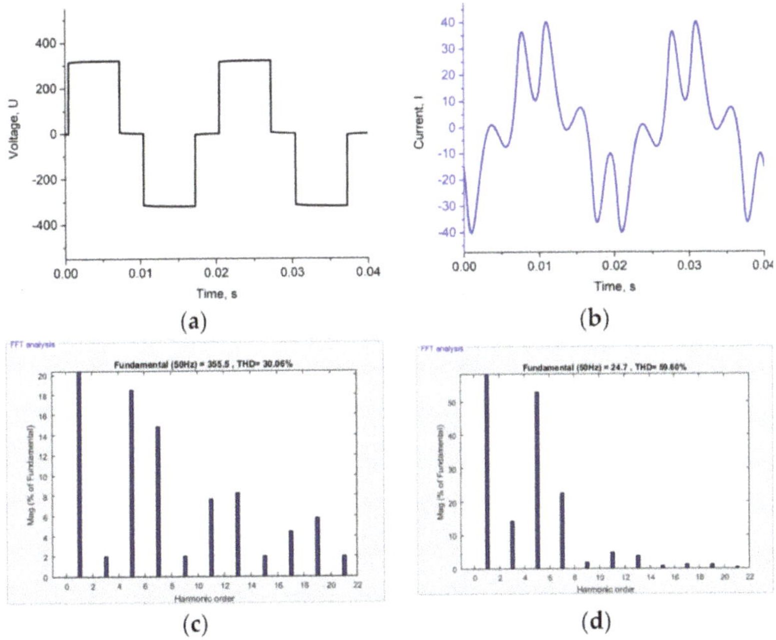

Figure 4. Waveforms and spectrums of output voltage and current of the first cascade of the cascaded inverter: (**a**) waveform of output voltage; (**b**) waveform of output current; (**c**) spectrum of output voltage; (**d**) spectrum of output current (the amplitude of 1st harmonic in voltage and current spectrums is 100%).

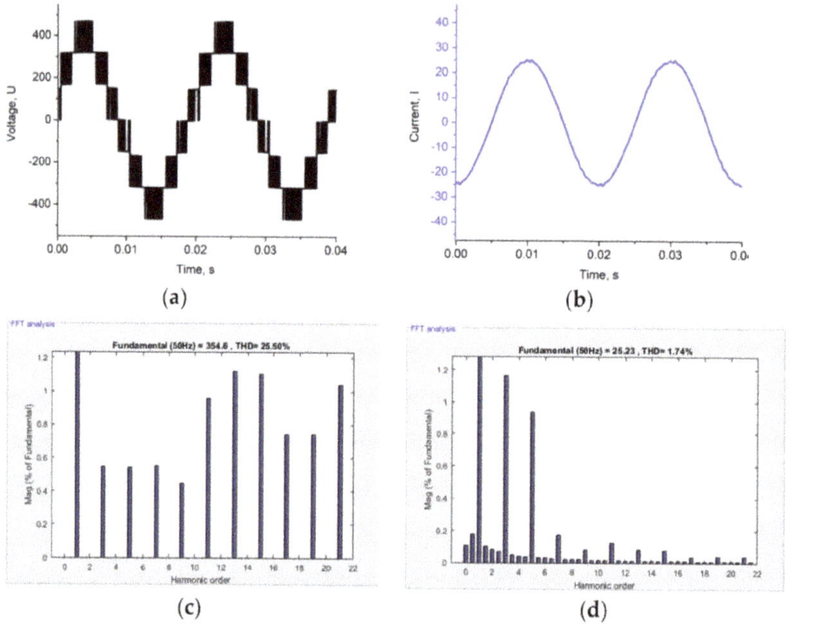

Figure 5. Waveforms and spectrums of cascaded inverter output voltage and current with the cancelation of harmonics: (**a**) output voltage waveform; (**b**) output current waveform; (**c**) spectrum of output voltage; (**d**) spectrum of output current (the amplitude of 1st harmonic in voltage and current spectrums is 100%).

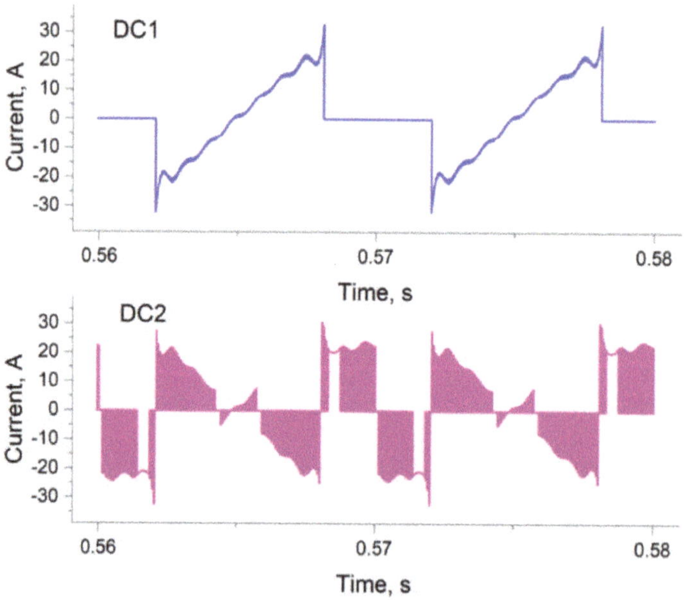

Figure 6. Waveforms of current of DC1 and DC2 voltage sources of cascaded inverter at 4000 VAr load.

4. Investigation of Asymmetric Reactive Power Compensation Capability Using Compensator Based on Independent Cascaded H-Bridge Multilevel Inverters

The investigation of the operation of the proposed reactive power compensator was performed using *Matlab/Simulink* software. The analysis was carried out according to the following scheme: first, there was now reactive power in the utility grid. Later, different reactive loads were connected to each phase: at time moment 0.025 s—minus 4.4 kVAr load to the first phase; at time moment 0.075 s—2.0 kVAr load to the second phase; and at time moment 0.125 s—3.0 kVAr to the third phase. The asymmetric loads were implemented by connecting inductances and capacitors with different values of parameters to the four-wire utility grid. Each phase load was connected between the appropriate phase and neutral. In order to properly distinguish transients, the compensator was set to start the reactive power compensation in every phase 0.04 s after the reactive load to this phase was connected. The same experiment was carried out for the case when a conventional three-phase STATCOM type compensator was used. The topology of conventional STATCOM is presented in Figure 7 [27]. It is based on the conventional three-phase inverter containing three arms implemented using IGBT transistor switches Q1–Q6. The same arm at the same time is used to form voltages of two phases in the conventional three-phase inverter. Therefore, it is not possible to form the phase voltages independently; because of this, the conventional device does not allow the asymmetric compensation of reactive power. The obtained reactive power transients for each phase in the three-phase grid are presented in Figure 8.

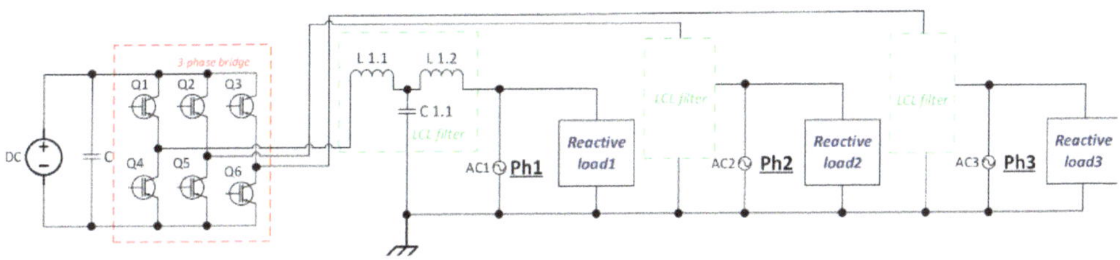

Figure 7. The topology of the conventional three-phase STATCOM compensator.

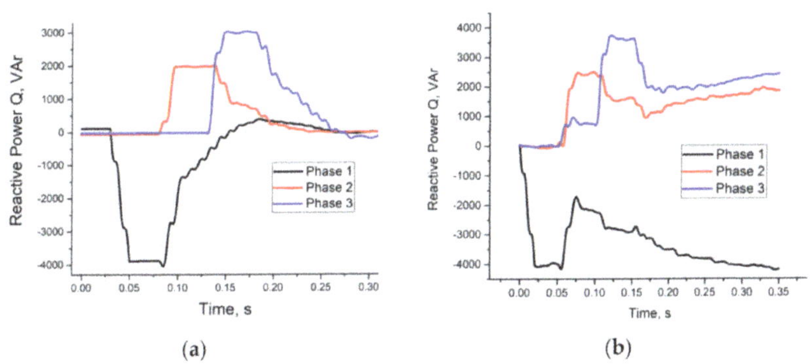

Figure 8. Transients of reactive power in three-phase grid with the asymmetric reactive loads (1st phase load—minus 4.4 kVAr, 2nd—2.0 kVAr, 3rd—3.0 kVAr) when compensation is provided using a compensator based on three independent cascaded H-bridge multilevel inverters (**a**) and conventional STATCOM compensator (**b**).

The obtained results show that using the proposed compensator based on independent cascaded H-bridge multilevel inverters, the reactive power was compensated approximately during 0.1 s after the compensation was started, i.e., the proposed compensator is able to compensate the reactive power in the three-phase four-wire grid even when phases are loaded highly asymmetrically: one phase by inductive and the other two phases by different capacitive loads. This concludes that such an approach is capable of quality asymmetric reactive power compensation in the three-phase grid. However, the transients presented in Figure 8b show that the conventional three-phase STATCOM compensator is not capable of compensating for the asymmetric reactive power loads. It is seen that after the transition, the reactive power in the phases is not compensated. It happens because it is impossible to control the voltages of the individual phases independently using the classical three-phase three-arm inverter. This can be observed in Figure 9, where the simulated response of output voltages of conventional STATCOM to the change in voltage in one of the phases is presented. It is seen that the change in the voltage in one phase causes voltage changes in other phases.

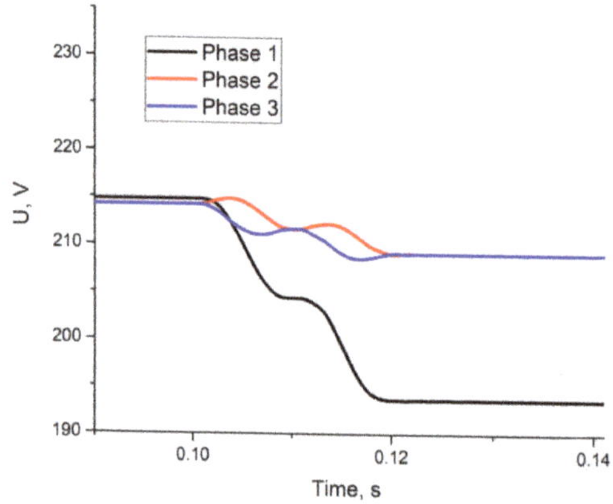

Figure 9. Conventional STATCOM three-phase output voltage response to change in voltage in the first phase by reducing PWM control signal pulse durations of switches connected to the first phase by 10%, at moment 0.1 s. Initially, control signals for all phases were fixed for nominal voltage.

The dependences of reactive power and the THD of output current on output voltage amplitude for the proposed compensator were obtained. The results for one of the phases are presented in Figure 10. It is seen that the THD of the compensator current is highly dependent on produced reactive power and reaches the highest values at low reactive load. This is because at low reactive load the amplitude of the fundamental harmonic decreases more strongly than the amplitudes of the higher harmonics. Meanwhile, at higher loads, the THD value is only a few percent.

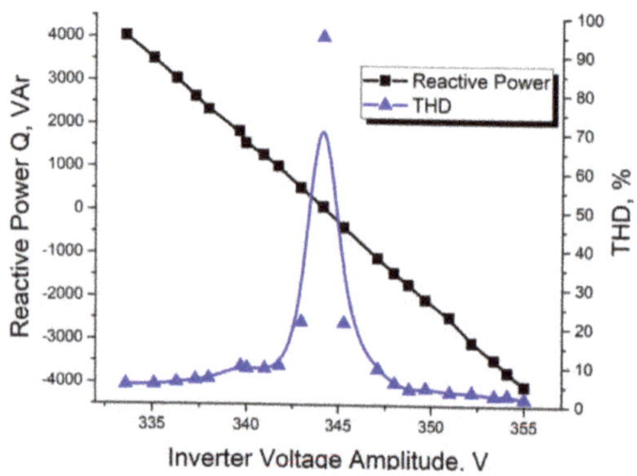

Figure 10. The dependences of reactive power produced by the compensator and THD of compensator current on compensator output voltage amplitude.

5. Investigation of Efficiency of Reactive Power Compensator Based on Independent Cascaded H-Bridge Multilevel Inverters

The efficiency of the reactive power compensator is important because higher efficiency allows not just a reduction in the energy losses but the simplification of the means for the cooling of inverter switches, and a reduction in the size of the compensator as well. Since the transistors of the inverter switches have to be shunted by the reverse diodes, the losses in the inverter depend not only on the characteristics of transistors but on parallel diodes as well. The power losses in parallel diodes are highly dependent on the mode of inverter operation, i.e., on the direction of current through the inverter switches.

The power losses in the inverters depend not only on the power of the load and the parameters of the switches, but on the carrier frequency as well. Therefore, the efficiency of cascaded multilevel inverters of the proposed reactive power compensator was investigated at various carrier frequencies and produced reactive power. In order to prove the advantage of the proposed topology based on cascaded H-bridge multilevel inverters, the obtained results were compared with the case when the compensator is based on the three independent conventional single-phase three-level inverters implemented using IGBT transistors. The *Matlab/Simulink* model of this compensator was created for this purpose. The model of the compensator for one of the phases is presented in Figure 11. The block *PWM* is used for the forming of the PWM control signals for the IGBT transistors of the inverter. The signals are formed using the sinusoidal PWM method based on the comparing of the sinus waveform with the high frequency (equal to PWM carrier frequency) triangle signal. The amplitude of the inverter output voltage is controlled by setting the value of modulation index m, which determines the duty cycle of the PWM control signal of inverter transistors. The PI controller is used to control the phase of inverter output voltage. It has to be synchronous with the grid voltage to keep the active power close to zero. The voltage phase is controlled by introducing the shift in time T for PWM control signals of inverter IGBT transistors.

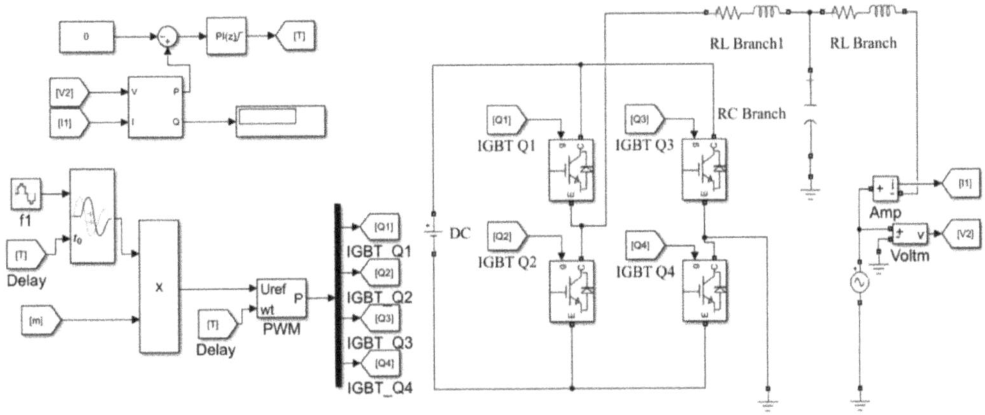

Figure 11. *Matlab/Simulink* model of the reactive power compensator based on the three conventional single-phase inverters for one phase.

The obtained dependences of power losses in reactive power compensator inverters on the carrier frequency at various values of generated reactive power are presented in Figure 12a,b. The dependences of difference between power losses in the cascaded H-bridge multilevel inverter-based compensator and in the compensator implemented using conventional three-level inverters are given in Figure 13. The types and main parameters of transistors and thyristors, for which the simulation of compensators was performed, are presented in Tables 1–3.

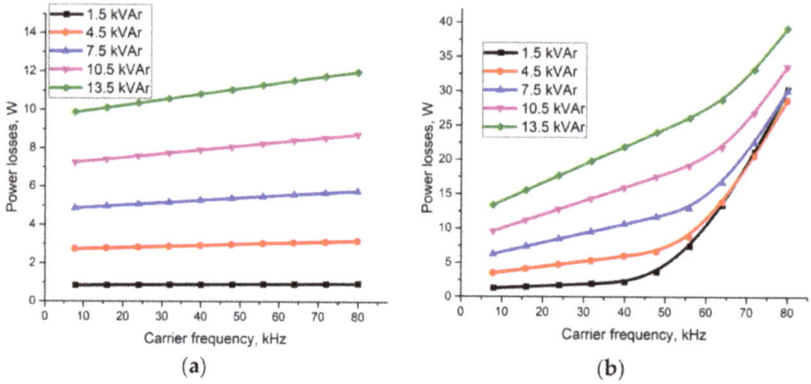

Figure 12. Dependencies of power losses in the reactive power compensator inverters on the carrier frequency at various reactive power loads: (**a**) in cascaded multilevel inverters; (**b**) in conventional three-level inverters.

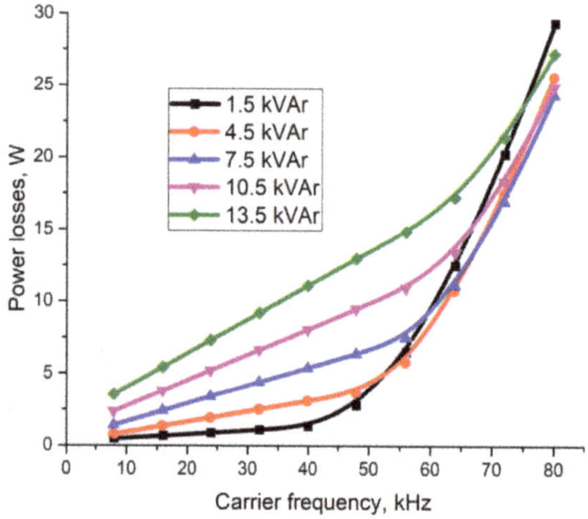

Figure 13. The dependencies of the difference in power losses between the conventional three-level and cascaded multilevel inverters of reactive power compensator on the carrier frequency at various reactive power loads.

Table 1. MOSFET transistors of cascaded multilevel inverters: IXFP36N20X3M (200 V, 18 A).

R_{DS} (on) (Resistance Drain-Source) at I_D = 18 A	Input Capacitance	Output Capacitance	Reverse Transfer Capacitance
38 mΩ	1425 pF	280 pF	1.2 pF

Table 2. Thyristors of cascaded multilevel inverters: MGTO1200 (1200 V, 20 A).

V_T (Voltage Drop) at I_T = 20 A	V_{T0} (Threshold Voltage)	R_T (Slope Resistance)
1.27 V	0.86 V	13.2 mΩ

Table 3. IGBT transistors of conventional three-level inverters: FGA20N120FTD (1200 V, 20 A).

V_{CE} (Sat) at I_C = 20 A	Input Capacitance	Output Capacitance	Reverse Transfer Capacitance
1.6 V	3080 pF	95 pF	60 pF

The obtained results (Figure 12) show that the compensator based on the cascaded multilevel inverters provides lower power losses as compared with the situation when the compensator is based on the conventional three-level inverters. The power losses consist of switching and conducting losses. Switching losses depend on the dynamic characteristics of the inverter transistors and their parallel diodes. At low carrier frequency, the difference in power losses is not significant. As an example, at 4500 VAr reactive power load of every phase and 7.5 kHz carrier frequency, the total power losses in all three conventional inverters of compensator are about 12.6 W and 9.6 W when instead of conventional inverters, the cascaded inverters are employed, i.e., the difference is 24%. However, this difference rises strongly as the frequency increases. The power losses at the same load and 80 kHz carrier frequency are 37 W and 12 W, accordingly, i.e., the difference increases up to 68%. This can be explained by the fact that the low-voltage MOSFET transistors characterized by high speed can be employed in the second cascade of the proposed reactive power compensator for the creating of the high-frequency components of the generated three-phase voltages. However, the compensator based on the conventional inverters has to be implemented using high-voltage IGBT transistors that are relatively slow and, because of this, cause high switching losses. The lower power losses at low switching frequency in the proposed compensator are caused by lower conducting losses of thyristors and low-voltage MOSFET transistors as compared with the conducting losses of high-voltage IGBT transistors.

6. Conclusions

1. The accurate asymmetric compensation of the reactive power with the short response time in the three-phase low-voltage grid can be performed using just a static synchronous compensator with separate single-phase voltage source inverters for each phase.
2. To minimize the power losses in the reactive power compensator implemented using cascaded H-bridge multilevel inverters, the first cascade of every inverter that operates at grid frequency has to be implemented using thyristors that are slow but are characterized by the low voltage drop in the state ON, and the second cascade has to be realized using low-voltage high-speed MOSFET transistors that provide low switching losses and low voltage drop in the state ON.
3. Using the proposed asymmetric compensator, the reactive power can be compensated for fully during 0.1 s after the compensation was started, when the phases are loaded asymmetrically by the following reactive loads: first phase—by minus 4.4 kVAr, second—by 2.0 kVAr and third—by 3.0 kVAr.
4. The power losses in the proposed asymmetric compensator based on the cascaded multilevel inverters, at 4500 VAr reactive power load of every phase and 7.5 kHz carrier frequency are about 24%, and at 80 kHz—68% lower as compared with the situation when the reactive power compensator is implemented using conventional three-level inverters based on IGBT transistors.

Author Contributions: Conceptualization, M.Š., A.B., A.S.; methodology, V.B., A.D.; validation, M.Š.; investigation, R.P., E.B., S.P.; writing—original draft preparation, A.B., N.P.; writing—review and editing, D.G., V.K.H., M.J.; visualization, M.Š.; funding acquisition, V.K.H. All authors have read and agreed to the published version of the manuscript.

Funding: The project "Industrial Internet methods for electrical energy conversion systems monitoring and diagnostics" benefits from an EUR 993,000 grant from Iceland, Liechtenstein and Norway through the EEA Grants. The aim of the project is to provide research in the field of energy conversion systems and to develop artificial intelligence and virtual emulator-based prognostic and diagnostic methodologies for these systems. Project contract with the Research Council of Lithuania (LMTLT) no. is S-BMT-21-5 (LT08-2-LMT-K-01-040).

Institutional Review Board Statement: Not applicable.

Informed Consent Statement: Not applicable.

Data Availability Statement: Not applicable.

Conflicts of Interest: The authors declare no conflict of interest.

References

1. Charalambous, A.; Hadjidemetriou, L.; Zacharia, L.; Bintoudi, A.D.; Tsolakis, A.C.; Tzovaras, D.; Kyriakides, E. Phase Balancing and Reactive Power Support Services for Microgrids. *Appl. Sci.* **2019**, *9*, 5067. [CrossRef]
2. Christian, L.E.; Putranto, L.M.; Hadi, S.P. Design of Microgrid with Distribution Static Synchronous Compensator (D-STATCOM) for Regulating the Voltage Fluctuation. In Proceedings of the 2019 IEEE 7th International Conference on Smart Energy Grid Engineering (SEGE), Oshawa, ON, Canada, 12–14 August 2019; pp. 48–52.
3. He, Z.; Ma, F.; Xu, Q.; Chen, Y.; Li, C.; Li, M.; Guerrero, J.M.; Luo, A. Reactive Power Strategy of Cascaded Delta-Connected STATCOM Under Asymmetrical Voltage Conditions. *IEEE J. Emerg. Sel. Top. Power Electron.* **2017**, *5*, 784–795. [CrossRef]
4. Saxena, N.K.; Kumar, A. Reactive power control in decentralized hybrid power system with STATCOM using GA, ANN and ANFIS methods. *Int. J. Electr. Power Energy Syst.* **2016**, *83*, 175–187. [CrossRef]
5. Edrah, M.; Lo, K.L.; Anaya-Lara, O. Reactive power control of DFIG wind turbines for power oscillation damping under a wide range of operating conditions. *IET Gener. Transm. Distrib.* **2016**, *10*, 3777–3785. [CrossRef]
6. Sufyan, M.; Rahim, N.A.; Eid, B.; Raihan, S.R.S. A comprehensive review of reactive power control strategies for three phase grid connected photovoltaic systems with low voltage ride through capability. *J. Renew. Sustain. Energy* **2019**, *11*, 042701. [CrossRef]
7. Bielskis, E.; Baskys, A.; Sapurov, M. Single Stage Microinverter Based on Two-Switch DC-DC Flyback Converter. *Elektronika Elektrotechnika* **2017**, *23*, 29–32. [CrossRef]
8. Sarkar, M.N.I.; Meegahapola, L.G.; Datta, M. Reactive Power Management in Renewable Rich Power Grids: A Review of Grid-Codes, Renewable Generators, Support Devices, Control Strategies and Optimization Algorithms. *IEEE Access* **2018**, *6*, 41458–41489. [CrossRef]
9. Ardeshna, N.K.; Chowdhury, B.H. Optimizing micro-grid operations in the presence of wind generation. In Proceedings of the 2008 40th North American Power Symposium, Calgary, AB, Canada, 28–30 September 2008; pp. 1–7.
10. Barrios-Martínez, E.; Ángeles-Camacho, C. Technical comparison of FACTS controllers in parallel connection. *J. Appl. Res. Technol.* **2017**, *15*, 36–44. [CrossRef]
11. Erickson, R.W.; Maksimovic, D. *Fundamentals of Power Electronics*, 2nd ed.; Springer: Norwell, MA, USA, 2001; ISBN 978-0-7923-7270-7.
12. Šapurov, M.; Bleizgys, V.; Baskys, A.; Dervinis, A.; Bielskis, E.; Paulikas, S.; Paulauskas, N.; Macaitis, V. Asymmetric Compensation of Reactive Power Using Thyristor-Controlled Reactors. *Symmetry* **2020**, *12*, 880. [CrossRef]
13. Lee, T.-L.; Hu, S.-H.; Chan, Y.-H. Design of D-STATCOM for voltage regulation in Microgrids. In Proceedings of the 2010 IEEE Energy Conversion Congress and Exposition, Atlanta, GA, USA, 12–16 September 2010; pp. 3456–3463.
14. Chang, W.-N.; Liao, C.-H. Design and Implementation of a STATCOM Based on a Multilevel FHB Converter with Delta-Connected Configuration for Unbalanced Load Compensation. *Energies* **2017**, *10*, 921. [CrossRef]
15. Chaudhari, P.; Rane, P.; Bawankar, A.; Shete, P.; Kalange, K.; Moghe, A.; Panda, J.; Kadrolkar, A.; Gaikwad, K.; Bhor, N.; et al. Design and implementation of STATCOM for reactive power compensation and voltage fluctuation mitigation in microgrid. In Proceedings of the 2015 IEEE International Conference on Signal Processing, Informatics, Communication and Energy Systems (SPICES), Kozhikode, India, 19–21 February 2015; pp. 1–5.
16. Wang, Q.; Wang, B.; Xu, W.; Xu, J. Research on STATCOM for reactive power flow control and voltage stability in microgrid. In Proceedings of the 2018 13th IEEE Conference on Industrial Electronics and Applications (ICIEA), Wuhan, China, 31 May–2 June 2018; pp. 2474–2479.
17. Pană, A.; Băloi, A.; Molnar-Matei, F. From the Balancing Reactive Compensator to the Balancing Capacitive Compensator. *Energies* **2018**, *11*, 1979. [CrossRef]
18. Ayala-Chauvin, M.; Kavrakov, B.S.; Buele, J.; Varela-Aldás, J. Static Reactive Power Compensator Design, Based on Three-Phase Voltage Converter. *Energies* **2021**, *14*, 2198. [CrossRef]
19. Choudhury, S.; Bajaj, M.; Dash, T.; Kamel, S.; Jurado, F. Multilevel Inverter: A Survey on Classical and Advanced Topologies, Control Schemes, Applications to Power System and Future Prospects. *Energies* **2021**, *14*, 5773. [CrossRef]
20. Maheswari, K.T.; Bharanikumar, R.; Arjun, V.; Amrish, R.; Bhuvanesh, M. A comprehensive review on cascaded H-bridge multilevel inverter for medium voltage high power applications. *Mater. Today Proc.* **2021**, *45*, 2666–2670. [CrossRef]

21. Shuvo, S.; Hossain, E.; Islam, T.; Akib, A.; Padmanaban, S.; Khan, M.Z.R. Design and Hardware Implementation Considerations of Modified Multilevel Cascaded H-Bridge Inverter for Photovoltaic System. *IEEE Access* **2019**, *7*, 16504–16524. [CrossRef]
22. Chandwani, H.B.; Matnani, M.K. A review of modulation techniques for hybrid multilevel inverter. In Proceedings of the 2012 1st International Conference on Emerging Technology Trends in Electronics, Communication & Networking, Surat, Gujarat, India, 19–21 December 2012; pp. 1–7.
23. Hu, W.; Guo, J.; Liu, J.; Yuan, Y. Hybrid phase-shifted modulation method for hybrid cascaded H-bridge inverters. *J. Power Electron.* **2021**, *21*, 1556–1566. [CrossRef]
24. Jana, P.; Chattopadhyay, S.; Maiti, S.; Bajpai, P.; Chakraborty, C. Hybrid modulation technique for binary asymmetrical cascaded multilevel inverter for PV application. In Proceedings of the 2016 IEEE International Conference on Power Electronics, Drives and Energy Systems (PEDES), Trivandrum, India, 14–17 December 2016; pp. 1–6.
25. Haykin, S.; Van Veen, B. *Signals and Systems*; John Wiley & Sons: Hoboken, NJ, USA, 2003; ISBN 0-471-16474-7.
26. Rashid, M.H. *Power Electronics Handbook*, 4th ed.; Elsevier Inc.: Amsterdam, The Netherlands, 2018; ISBN 978-0-12-811407-0.
27. PQC-STATCON. Available online: https://search.abb.com/library/Download.aspx?DocumentID=9AKK105713A9163&LanguageCode=en&DocumentPartId=&Action=Launch (accessed on 6 June 2021).

MDPI AG
Grosspeteranlage 5
4052 Basel
Switzerland
Tel.: +41 61 683 77 34

Symmetry Editorial Office
E-mail: symmetry@mdpi.com
www.mdpi.com/journal/symmetry

www.ingramcontent.com/pod-product-compliance
Lightning Source LLC
Chambersburg PA
CBHW061453210926
51850CB00008B/1697